Missing Data in Longitudinal Studies

Strategies for Bayesian Modeling and Sensitivity Analysis

MONOGRAPHS ON STATISTICS AND APPLIED PROBABILITY

General Editors

J. Fan, V. Isham, N. Keiding, T. Louis, R. L. Smith, and H. Tong

Monographs on Statistics and Applied Probability 109

Missing Data in Longitudinal Studies
Strategies for Bayesian Modeling and Sensitivity Analysis

Michael J. Daniels
University of Florida
Gainesville, U.S.A.

Joseph W. Hogan
Brown University
Providence, Rhode Island, U.S.A.

Chapman & Hall/CRC
Taylor & Francis Group
Boca Raton   London   New York

Chapman & Hall/CRC is an imprint of the
Taylor & Francis Group, an informa business

Chapman & Hall/CRC
Taylor & Francis Group
6000 Broken Sound Parkway NW, Suite 300
Boca Raton, FL 33487-2742

© 2008 by Taylor & Francis Group, LLC
Chapman & Hall/CRC is an imprint of Taylor & Francis Group, an Informa business

No claim to original U.S. Government works
Printed in the United States of America on acid-free paper
10 9 8 7 6 5 4 3 2 1

International Standard Book Number-13: 978-1-58488-609-9 (Hardcover)

Library of Congress Cataloging-in-Publication Data

Daniels, M. J.
    Missing data in longitudinal studies : strategies for Bayesian modeling and sensitivity analysis / Michael J. Daniels and Joseph W. Hogan.
        p. cm. -- (Monographs on statistics and applied probability ; 109)
    Includes bibliographical references and index.
    ISBN 978-1-58488-609-9 (alk. paper)
    1. Missing observations (Statistics) 2. Longitudinal method. 3. Sensitivity theory (Mathematics) 4. Bayesian statistical decision theory. I. Hogan, Joseph W. II. Title. III. Series.

QA276.D3146 2007
519.5--dc22   2007040408

Visit the Taylor & Francis Web site at
http://www.taylorandfrancis.com

and the CRC Press Web site at
http://www.crcpress.com

To Marie, Mary and Mia (M.J.D.)

To Dawn, Jack, Luke and Patrick (J.W.H.)

Contents

Preface

Motivations

A considerable amount of research in public health, medicine, and the life and social sciences is driven by longitudinal studies that collect repeated observations over time. An almost inevitable complication in drawing inferences is that data will be missing. For studies with long follow up periods, the proportion of individuals with missing data can be substantial.

Despite a vast and growing literature, and recent advances in statistical software, the appropriate handling of missing data in *longitudinal studies* remains a vexing statistical problem, and has enough complexity to make it a worthy topic of study in its own right; more than just the standard set of longitudinal modeling and missing data tools are needed.

With repeated measures, the best information about missing responses frequently comes from observed responses on the same variable; hence methods for covariance modeling take on heightened importance. Missingness in longitudinal studies is frequently a result of dropout, mortality and other processes that are well characterized as event times; hence methods for joint modeling of repeated measures and event times play a central role.

And as with any analysis of incomplete data, parameters or models of interest cannot be inferred without resorting to assumptions that employ various degrees of subjective judgment. All too often, the assumptions used for data analysis are motivated more by convenience than by substance; this seems especially true for longitudinal data, where the models or data structures are complex enough on their own without the added complication of missing data. Hence a principled approach is needed for formalizing subjective assumptions in contextually interpretable ways.

The Bayesian approach

When data are incomplete, inferences about parameters of interest cannot be carried out without the benefit of subjective assumptions about the distribution of missing responses. They are subjective because data cannot be used to critique them. Some of these assumptions are used with such regularity that we forget they are being made; for example, when commercial software such as SAS or Stata is used to analyze incomplete longitudinal data using a random effects model, the *missing at random (MAR)* assumption is being used; when the Kaplan-Meier estimator is used to summarize a survival curve from censored event times, *non-informative censoring* is being assumed. Neither assumption can be formally checked, so the validity of inferences relies on subjective judgment.

For this reason, it is our view that analysis of incomplete data cries out for the Bayesian approach in order to formalize the subjective component of inference. Specifically, we argue that subjective assumptions should be formulated in terms of prior distributions for parameters that index the conditional distribution of missing data given observed data. In this framework, default assumptions such as missing at random can be encoded using point mass priors, or assumptions can be elaborated by centering priors away from MAR, introducing uncertainty about the assumptions, or both.

Ideally, the role of the statistician is to write down a model that enables critical thinking about the missing data assumptions, based on questions like these: Can missing data be predicted from observables using an extrapolation? Are dropouts systematically different from completers? If so, how? And what is the degree of uncertainty about the assumptions?

This is not a new idea (see Rubin, 1977), but in our reading of the literature it is surprisingly underutilized. One possible reason is the difficulty associated with parameterizing complex models so that missing data assumptions are coherently represented with priors. Another potential reason is lack of knowledge about how to use appropriate software. With this in mind, we have included 11 worked examples and case studies. These illustrate concepts ranging from basic modeling of repeated measures to structured models for covariance to use of informative priors under nonignorable missingness. Several of these examples have been carried out using WinBUGS software, which is available at http://www.mrc-bsu.cam.ac.uk/bugs/. Code used for analyses in the book, and datasets where available, are posted at http://www.stat.brown.edu/~jhogan.

Audience

The book is intended for statisticians, data analysts and other scientists involved in the collection and analysis of longitudinal data. Our emphasis is on clinical and public health research, but many of the ideas will be applicable to other fields. We assume familiarity with statistical inference at the level of Casella and Berger (2001), and with regression at the level of Kutner et al. (2003). Familiarity with regression for longitudinal data and principles of Bayesian inference is helpful, but these ideas are reviewed in Chapters 2 and 3, where numerous references to books and key papers are given.

Readers of this book are likely to have at least a passing familiarity with basic methods for handling missing data, but the literature is vast and there are distinct points of view on how to approach inference. Several outstanding texts can be consulted to gain an appreciation, including Little and Rubin (2002); Schafer (1997); van der Laan and Robins (2003); Tsiatis (2006); and Molenberghs and Kenward (2007).

Although this is primarily a research monograph, we expect that it will be a suitable supplemental text for graduate courses on longitudinal data, missing data, or Bayesian inference. We have used material from the book for our own graduate courses at University of Florida and Brown University.

Content

The book is composed of ten chapters, roughly divided into three main parts. Chapters 1 through 4 make up the first part, which covers needed background material. Chapter 1 provides detailed descriptions of motivating examples that are used throughout the book; Chapters 2 and 3 provide background on longitudinal data models and Bayesian inference, respectively; and Chapter 4 illustrates key concepts with worked data analysis examples.

The second part includes Chapters 5 through 7; it introduces missing data mechanisms for longitudinal data settings, and gives in-depth treatment of inference under the ignorability assumption. Chapter 5 gives detailed coverage of Rubin's missing data taxonomy (Rubin, 1976) applied to longitudinal data, including MAR, ignorability, and implications for posterior inference. The notion of a *full-data model* is introduced, and the chapter primarily describes assumptions that are needed to infer features or parameters of the full-data model from incomplete data. Chapter 6 discusses inference under the ignorability assumption, with emphasis on the importance of covariance modeling. In Chapter 7, several case studies are used to illustrate.

Chapters 8 through 10 make up the third part, where the focus shifts to nonignorable missingness. In Chapters 8 and 9, we reiterate the central idea that inference about a full-data distribution from incomplete data requires untestable assumptions. We divide the full-data distribution into an observed-data model and an extrapolation model, and introduce *sensitivity parameters* that govern the untestable assumptions about the extrapolation. Chapter 8 reviews common approaches such as selection models, mixture models, and shared parameter models, and focuses on parameterization of missing data assumptions in each model.

In Chapter 8, the primary focus is on the likelihood (data model), whereas Chapter 9 is concerned with formulation of priors that reflect assumptions about missing data. In Chapter 9, we show how to construct priors that reflect specific missing data mechanisms (MAR, MNAR), how to center the models at MAR, and give suggestions about calibrating the priors. Both mixture and selection modeling approaches are addressed. Finally, Chapter 10 provides three detailed case studies that illustrate analyses under MNAR.

Acknowledgments

We began this project in 2005; prior to that and since then, we have benefited from friendship, cooperation and support of family, friends and colleagues.

Our research in this area has been encouraged and positively influenced by interactions with many colleagues in statistics and biostatistics; Alicia Carriquiry, Peter Diggle, Constantine Gatsonis, Rob Kass, Nan Laird, Tony

Lancaster, Xihong Lin, Mohsen Pourahmadi, Jason Roy, Dan Scharfstein and Scott Zeger deserve special mention for mentorship, support and/or collaboration. Ian Abramson and the Department of Mathematics at UC San Diego provided generous hospitality during a sabbatical for J.H. The National Institutes of Health have funded our joint and individual research on missing data and Bayesian inference over the past several years.

Our subject-matter collaborators have generously allowed us to use their data for this book. Thanks are due to Charles Carpenter and Ken Mayer at Brown University and Lytt Gardner at the CDC (HERS data); Dave MacLean at Pfizer (Growth Hormone Study); Bess Marcus at Brown University (both CTQ studies); and David Abrams at NIH and Brian Hitsman at Brown University (OASIS data).

Graduate students Li Su and Joo Yeon Lee at Brown University, and Keunbaik Lee and Xuefeng Liu at University of Florida provided expert programming and assistance with preparation of data examples. Li Su served as our expert consultant on WinBUGS and developed code for many of the examples.

Several colleagues took time to read the text and offer invaluable suggestions. Rob Henderson carefully read the entire book and provided insightful comments that led to major improvements in the final version. Brad Carlin, Allison DeLong, and Tao Liu each provided detailed and timely feedback on one or more chapters. And through what turned out to be a much longer process than we ever imagined, Rob Calver at Chapman & Hall never failed to keep the wind in our sails with his encouragement, patience and good humor.

And finally, it is difficult to overstate the support and patience provided by our friends and – most of all – our families, at just the right times during this project. Without Marie, Mary and Mia; and Dawn, Jack, Luke and Patrick, it simply could not have been completed.

Gainesville   M. J. Daniels
Providence   J. W. Hogan
December, 2007

CHAPTER 1

Description of Motivating Examples

1.1 Overview

This chapter describes, in detail, several datasets that will be used throughout the book to motivate the material and to illustrate data analysis methods. For each dataset, we describe the associated study, the primary research goals to be met by the analyses, and the key complications presented by missing data. Empirical summaries are given here; for detailed analyses we provide references to subsequent chapters.

These datasets derive primarily from our own collaborations, introducing the potential for selection bias in the coverage of topics; however, we have selected these because they cover a range of study designs and missing data issues. Although the data are primarily from clinical trials, we also have data from a large observational study. The datasets have both continuous and discrete longitudinal endpoints. Both continuous-time and discrete-time dropout processes are covered here. In one of the studies, an intermediate goal involves capturing the joint distribution of two processes evolving simultaneously.

It is our hope that readers can find within these examples connections to their own work or data-analytic problem of interest. The literature on missing data continues to grow, and it is impossible to address every situation with representative examples, so we have included at the end of each chapter a list of supplementary reading to assist readers in their navigation of current research.

In the descriptions of studies and datasets, we use the terms *dropout, missing at random*, and *missing not at random* under the assumption that the reader has at least a casual acquaintance with their meaning. Precise definitions and implications for analyses are treated in considerable detail in upcoming chapters; readers also may want to consult Little and Rubin (2002), Tsiatis (2006), and Molenberghs and Kenward (2007). In short, missing data are said to be missing at random (MAR) if the probability of missingness depends only on observable data; when probability of missingness depends on the missing observations themselves, even after conditioning on observed information, missing data are said to be missing not at random (MNAR).

1.2 Dose-finding trial of an experimental treatment for schizophrenia

1.2.1 Study and data

These data were collected as part of a randomized, double-blind clinical trial of a new pharmacologic treatment of schizophrenia (Lapierre et al., 1990). Other published analyses of these data can be found in Hogan and Laird (1997a) and Cnaan et al. (1997).

The trial compares three doses of the new treatment (low, medium, high) to the standard dose of halperidol, an effective antipsychotic that has known side effects. At the time of the trial, the experimental therapy was thought to have similar antipsychotic effectiveness with fewer side effects; the trial was designed to find the appropriate dosing level. The study enrolled 245 patients at 13 different centers, and randomized them to one of the four treatment arms. The intended length of follow-up was 6 weeks, with measures taken weekly except for week 5. Schizophrenia severity was assessed using the Brief Psychiatric Rating Scale, or BPRS, a sum of scores of 18 items that reflect behaviors, mood, and feelings (Overall and Gorham, 1988). The scores ranged from 0 to 108 with higher scores indicating higher severity. A minimum score of 20 was required for entry into the study.

1.2.2 Questions of interest

The primary objective is to compare mean change from baseline to week 6 between the four treatment groups.

1.2.3 Missing data

Dropout in this study was substantial; only 139 of the 245 participants had a measurement at week 6. The mean BPRS on each treatment arm showed differences between dropouts and completers (see Figure 1.1). Reasons for dropout included adverse events (e.g., side effects), lack of treatment effect, and withdrawal for unspecified reasons; a summary appears in Table 1.1. Reasons such as lack of observed treatment effect are clearly related to the primary efficacy outcome, but others such as adverse events and participant withdrawal may not be.

1.2.4 Data analyses

In Section 4.3 we use data on those with complete follow-up to illustrate posterior inference for a standard random effects model. In Section 7.3 the model is re-fitted using all the data under an MAR assumption.

For some individuals, dropout occurs for reasons that clearly are related

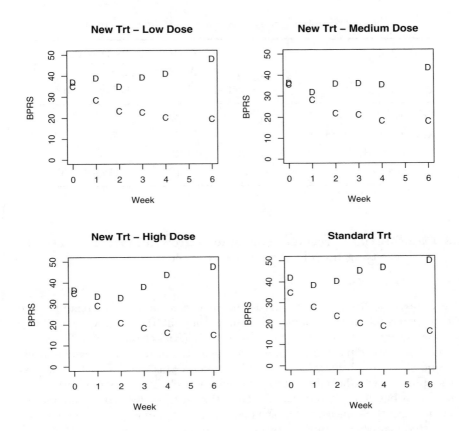

Figure 1.1 *Schizophrenia trial: mean BPRS scores, stratified on dropout for lack of treatment effect (plotting symbol D for dropouts and C for non-dropouts). Among those experiencing lack of effect, 10 had BPRS measured at week 6; hence BPRS scores appear at week 6 among 'dropouts'. Measurements were not taken at week 5.*

to outcome, and for others the connection between dropout and outcome is less clear (e.g., dropout due to adverse side effects); hence MAR may not be entirely appropriate. Using this trial as a motivating example, model formulation for settings with both MAR and MNAR missingness is discussed in Section 8.4.5. We use a pattern mixture approach, treating MAR dropouts as censoring times for the MNAR dropout process.

Table 1.1 *Schizophrenia trial: dropout by reason, stratified by treatment arm.*

| Treatment Arm | Completed | Reason for Dropout | | |
		Adverse Event	Lack of Effect	Other
New Trt – Low Dose	27	2	21	11
New Trt – Medium Dose	40	1	7	13
New Trt – High Dose	33	2	13	12
Standard Trt	34	12	11	6

1.3 Clinical trial of recombinant human growth hormone (rhGH) for increasing muscle strength in the elderly

1.3.1 Study and data

The data come from a randomized clinical trial conducted to examine the effects of recombinant human growth hormone (rhGH) therapy for building and maintaining muscle strength in the elderly (Kiel et al., 1998). The study enrolled 161 participants and randomized them to one of four treatment arms: placebo (P), growth hormone only (G), exercise plus placebo (EP), and exercise plus growth hormone (EG). Various muscle strength measures were recorded at baseline, 6 months, and 12 months. Here, we focus on mean quadriceps strength (QS), measured as the maximum foot-pounds of torque that can be exerted against resistance provided by a mechanical device.

1.3.2 Questions of interest

The primary objective of our analyses is to compare mean QS at month 12 in the four treatment arms among all those randomized (i.e., draw inference about the intention to treat effect).

1.3.3 Missing data

Roughly 75% of randomized individuals completed all 12 months of follow-up, and most of the dropout was thought to be related to the unobserved responses at the dropout times. Table 1.2 summarizes mean and standard deviation of QS for available follow-up data, both aggregated and stratified by dropout time.

Table 1.2 *Growth hormone trial: sample means (standard deviations) stratified by treatment group and dropout pattern s. Patterns defined by number of observations and n_s is the number in pattern s.*

Treatment	s	n_s	Month		
			0	6	12
EG	1	12	58 (26)		
	2	4	57 (15)	68 (26)	
	3	22	78 (24)	90 (32)	88 (32)
	All	38	69 (25)	87 (32)	88 (32)
G	1	6	68 (17)		
	2	5	77 (33)	81 (42)	
	3	30	67 (22)	64 (21)	63 (20)
	All	41	69 (23)	66 (25)	63 (20)
EP	1	7	65 (32)		
	2	2	87 (52)	86 (51)	
	3	31	65 (24)	81 (25)	73 (21)
	All	40	66 (26)	82 (26)	73 (21)
P	1	8	66 (29)		
	2	5	53 (19)	62 (31)	
	3	28	67 (23)	62 (20)	63 (19)
	All	41	65 (24)	62 (22)	63 (19)

1.3.4 Data analyses

These data are used several times throughout the book to illustrate various models. In Section 4.2 we analyze data from individuals with complete follow-up to illustrate multivariate normal regression and model selection procedures for choosing an appropriate variance-covariance structure; Example 7.2 illustrates the use of multivariate normal models under an MAR constraint, with emphasis on how the choice of variance-covariance structure affects inferences about the mean. In Section 10.2 we use more general pattern mixture models that permit MNAR to illustrate strategies for sensitivity analysis, incorporation of informative priors, and model fit.

1.4 Clinical trials of exercise as an aid to smoking cessation in women: the Commit to Quit studies

1.4.1 Studies and data

The Commit to Quit studies were randomized clinical trials to examine the impact of exercise on the ability to quit smoking among women. The women in each study were aged 18 to 65, smoked five or more cigarettes per day for at least 1 year, and participated in moderate or vigorous intensity activity for less than 90 minutes per week.

The first trial (hereafter CTQ I; Marcus et al., 1999) enrolled 281 women and tested the effect on smoking cessation of supervised vigorous exercise vs. equivalent staff contact time to discuss health care (henceforth called the 'wellness' arm); the second trial (hereafter CTQ II; Marcus et al., 2005) enrolled 217 female smokers and was designed to examine the effect of moderate partially supervised exercise. Other analyses of these data can be found in Hogan et al. (2004b), who illustrate weighted regression using inverse propensity scores; Roy and Hogan (2007), who use principal stratification methods to infer the causal effect of compliance with vigorous exercise; and Liu and Daniels (2007), who formulate and apply new models for the joint distribution of smoking cessation and weight change.

In each study, smoking cessation was assessed weekly using self-report, with confirmation via laboratory testing of saliva and exhaled carbon monoxide. As is typical in short-term intervention trials for smoking cessation, the target date for quitting smoking followed an initial 'run-in' period during which the intervention was administered but the participants were not asked to quit smoking. In CTQ I, measurements on smoking status were taken weekly for 12 weeks; women were asked to quit at week 5. In CTQ II, total follow-up lasted 8 weeks, and women were asked to quit in week 3.

1.4.2 Questions of interest

In each study, the question of interest was whether the intervention under study reduced the rate of smoking. This can be answered in terms of the effect of *randomization* to treatment vs. control, or in terms of the effect of *complying* with treatment vs. not complying (or vs. complying with control). The former can be answered using an intention to treat analysis, contrasting outcomes based on treatment arm assignment. The latter poses additional challenges but can be addressed using methods for inferring causal effects; we refer the reader to Roy and Hogan (2007) for details.

In our analyses, we frame the treatment effect in terms of either (a) time-averaged weekly cessation rate following the target quit date or (b) cessation rate at the final week of follow-up. More details about analyses are given below.

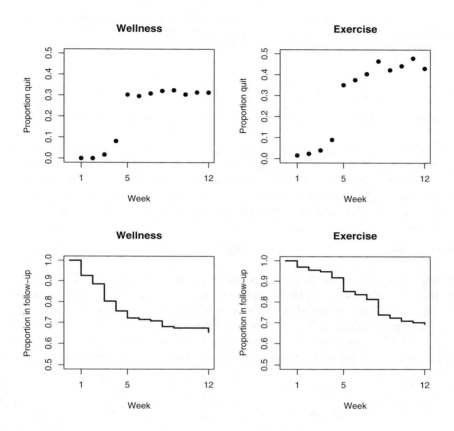

Figure 1.2 *CTQ-I: graph of proportion quit by week (top panels) and proportion remaining in the study (bottom panels), stratified by treatment group. Proportion quit at each week is calculated among those remaining in follow-up. Week 5 is the target 'quit week'.*

1.4.3 Missing data

Each of the studies had substantial dropout: in CTQ I, 31% dropped out on the exercise arm and 35% on the control arm; for CTQ II, 46% dropped out on the exercise arm and 41% on the control arm.

Figure 1.2 shows, for CTQ I, weekly cessation rates based on available data at each time point, coupled with proportion remaining in the study; Figure 1.3 shows the same data stratified by dropout status (yes/no), making clear that dropout is related at least to observed smoking status during the study. There also exists some empirical support for the notion that dropout is related to

missing outcomes in smoking cessation studies, in the sense that dropouts
are more likely to be smoking once they have dropped out (Lichtenstein and
Glasgow, 1992).

1.4.4 Data analyses

Analyses of CTQ I using standard regression models under MAR

In Section 4.4 we illustrate and compare random effects logistic regression and
marginalized transition models for estimating conditional (subject-specific)
and marginal (population-averaged) treatment effects using data from individuals with complete follow-up. Model selection procedures also are illustrated.
In Example 7.4 the data are reanalyzed under MAR using all available data.

Analysis of CTQ II using auxiliary variables

Weight change is generally associated with smoking cessation. In Section 7.5
we illustrate the use of auxiliary information on longitudinal weight changes
to inform the distribution of smoking cessation outcomes in making treatment
comparisons in CTQ II. The weight change data are incorporated through a
joint model for longitudinal smoking cessation and weight, and the marginal
distribution of smoking cessation is used for treatment comparisons.

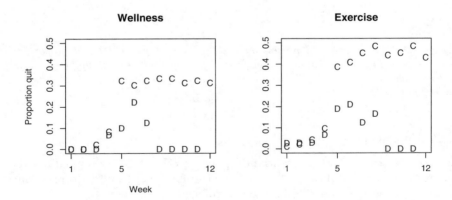

Figure 1.3 *CTQ-I: proportion quit by week, separately by treatment group, stratified
by dropout status (C = completer, D = dropout).*

1.5 Natural history of HIV infection in women: HIV Epidemiology Research Study (HERS) cohort

1.5.1 Study and data

The HIV Epidemiology Research Study (HERS) was a longitudinal cohort study of the natural history of HIV in women (Smith et al., 1997). Between 1993 and 1996, the HERS enrolled 1310 women who were either HIV-positive or at high risk for infection; 871 were HIV-positive at study entry. Every 6 months for up to 5 years, several outcomes were recorded for each participant, including standard measures of immunologic function and viral burden, plus a comprehensive set of measures characterizing health status and behavioral patterns (e.g., body mass index, depression status, drug use behavior). Our analyses of HERS data will focus on modeling CD4 progression relative to timing of highly active antiviral therapy (HAART) initiation, using data where some individuals have incomplete follow-up.

1.5.2 Questions of interest

The HERS is a multi-site study with several published investigations on different aspects of HIV epidemiology and associated statistical methodology. Relative to the full scope of HERS, our objectives for illustrating data analyses are necessarily simplified. Our interest is in characterizing the trajectory of CD4 relative to the initiation of HAART. Figure 1.4 shows an aggregated scatterplot of 4327 CD4 counts vs. time to HAART initiation from a sample of 393 individuals who were observed to initiate HAART during the study period. A small number of individual trajectories are highlighted. The data exhibit considerable within- and between-individual variability, and nonlinear trajectories relative to initiation of treatment. The goal of our analysis is to infer the population CD4 curve relative to initiation of HAART.

1.5.3 Missing data

All individuals enrolled in HERS were scheduled for 5 year follow-up (12 visits); in our subsample of 393 women, only 195 completed follow-up. The remainder had incomplete data due to missed visits, dropout, and death (36). Missingness due to death necessitates careful definition of the target quantities for inference. Our treatment of the HERS data does not explicitly address this issue; readers are referred to Zheng and Heagerty (2005) and Rubin (2006) for recent treatment.

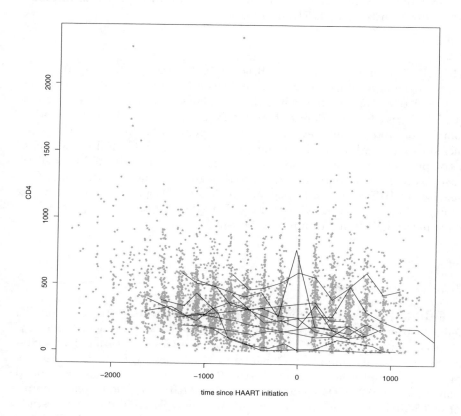

Figure 1.4 *HER Study: Plot of CD4 vs. time for 393 HIV infected women who received HAART during their enrollment in HERS. Horizontal axis is centered at time of the first visit where receipt of HAART is reported.*

1.5.4 Data analyses

An analysis of these data is given in Chapter 6. The nonlinearity of the time trend motivates a flexible approach to modeling; we use penalized splines under MAR to characterize the population CD4 curve, with emphasis on the importance of choosing an appropriate variance-covariance model and its effect on the fitted curve.

1.6 Clinical trial of smoking cessation among substance abusers: OASIS study

1.6.1 Study and data

The OASIS Trial is an NIH-funded study designed to compare standard (ST) vs. enhanced (ET) counseling interventions for smoking cessation among alcoholics. The trial enrolled 298 individuals, randomized to standard vs. more intensive counseling. Assessment of smoking status was made at 1, 3, 6, and 12 months following randomization. Previous analyses of these data appear in Lee et al. (2007).

1.6.2 Questions of interest

The primary goal of our analysis is comparison of smoking cessation rates at 12 months by treatment randomization (i.e., intention to treat effect).

Table 1.3 *OASIS Trial: number (proportion) quit, smoking and missing at each month. $S/(Q+S)$ denotes empirical smoking rate among those still in follow-up; $(S+M)/(Q+S+M)$ denotes empirical smoking rate after counting missing values as smokers. ET = enhanced intervention, ST = standard intervention.*

Treatment		Month			
		1	3	6	12
ET	Quit	26 (.18)	13 (.09)	17 (.11)	16 (.11)
($n = 149$)	Smoking	123 (.83)	70 (.47)	63 (.42)	51 (.34)
	Missing	—	66 (.44)	69 (.46)	82 (.55)
	$\dfrac{S}{Q+S}$	.83	.84	.79	.76
	$\dfrac{S+M}{Q+S+M}$	.83	.91	.89	.89
ST	Quit	23 (.15)	14 (.09)	15 (.10)	11 (.07)
($n = 149$)	Smoking	126 (.85)	80 (.54)	78 (.52)	78 (.52)
	Missing	—	55 (.37)	56 (.38)	60 (.40)
	$\dfrac{S}{Q+S}$	.85	.85	.84	.88
	$\dfrac{S+M}{Q+S+M}$	.85	.91	.90	.93

1.6.3 Missing data

Dropout rate was relatively high in the OASIS study (40% on ST, 55% on ET). Table 1.3 summarizes proportion quit, smoking, and missing at each month, stratified by treatment arm, and includes two commonly used estimates of overall smoking rate. The first is derived using only those still in follow-up, and the second assumes those with missing observations are smokers.* In Table 1.4 the association between status at times t_{j-1} and t_j (for $j = 2, 3, 4$) indicates that smoking status at t_{j-1} is predictive of dropout at t_j, and motivates the use of models that assume MAR at a minimum.

Table 1.4 *OASIS Trial: number and proportion of transitions from status at measurement times t_{j-1} to t_j (with t_j corresponding to months 1, 3, 6, and 12 for $j = 1, 2, 3, 4$), stratified by treatment. Figures in parentheses are row proportions. ET = enhanced intervention, ST = standard intervention.*

Treatment	Status at t_{j-1}	Status at t_j ($j = 2, 3, 4$) Quit	Smoking	Missing	Total
ET	Quit	26 (.46)	21 (.38)	9 (.16)	56
	Smoking	18 (.07)	152 (.59)	86 (.34)	256
	Missing	2 (.01)	11 (.08)	122 (.90)	135
ST	Quit	24 (.46)	17 (.33)	11 (.21)	52
	Smoking	12 (.04)	204 (.72)	68 (.24)	284
	Missing	4 (.04)	15 (.14)	92 (.83)	111

1.6.4 Data analyses

These data are analyzed in detail in Section 10.3, using both MAR and MNAR models. The first analysis uses a pattern mixture model where, conditional on dropout time, the longitudinal smoking outcomes follow a Markov transition model. The model is fit under MAR assumptions, then elaborated to allow for MNAR mechanisms. Sensitivity analyses and the use of informative priors elicited from experts are illustrated.

We also use a selection model approach, allowing for MNAR dropout. The two models are compared in terms of inference about treatment effect and assumptions about missing data.

* In our analyses of these data, we make a simplifying assumption that those with missing outcome at *month 1* are smokers.

1.7 Equivalence trial of competing doses of AZT in HIV-infected children: Protocol 128 of the AIDS Clinical Trials Group

1.7.1 Study and data

ACTG 128 is a randomized equivalence trial of high vs. low dose of AZT for the treatment of HIV in children (Brady et al., 1998). The trial enrolled 426 children and randomized them to a regimen of either 180mg or 90mg AZT six times daily. Because AZT is associated with potentially harmful side effects and drug toxicity, the trial was designed to provide information on whether the lower dose provided efficacy comparable to the high dose while reducing the rate of side effects. The key clinical outcomes are CD4 cell count, scheduled for measurement every 3 months, and neurocognitive test results, scheduled for measurement every 6 months. Our focus here is on CD4 cell counts; both the CD4 and neurocognitive outcomes have been been analyzed elsewhere, with attention to handling dropout and noncompliance; see Hogan and Laird (1996); Hogan and Daniels (2002), and Hogan et al. (2004a).

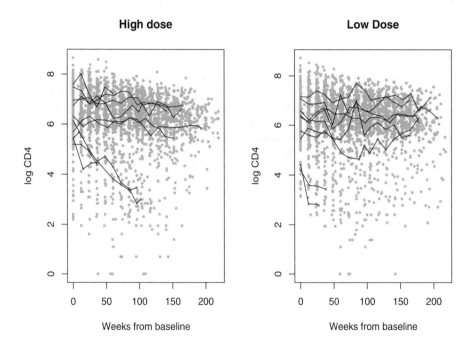

Figure 1.5 *Pediatric AIDS Trial: CD4 count vs. time for all individuals, stratified by treatment, with a subsample of individual trajectories highlighted.*

1.7.2 Questions of interest

The primary objective of our analyses is comparison of change in CD4 count from baseline to the end of the study, using the intention to treat principle; that is, we seek to compare change in CD4 among all participants randomized to high or low dose AZT.

1.7.3 Missing data

Dropout occurs for various reasons, including lack of treatment effect and drug toxicity. A key feature of this study is that dropout time is measured on a continuum, which requires some added flexibility in the modeling process. For purposes of keeping the example straightforward, we do not use information about reason for dropout in our analyses, but see Hogan and Laird (1996) and Hogan and Daniels (2002) for analyses that do. Figure 1.5 shows CD4 counts aggregated over all individuals, with a subsample of individual trajectories highlighted. The plot suggests that dropouts (those with shorter trajectories) tend to have a lower CD4 count at baseline and a more pronounced negative slope over time.

1.7.4 Data analyses

These data are analyzed in Section 10.4 using a mixture of varying coefficient models (Hogan et al., 2004a). The modeling approach allows regression coefficients such as slope over time to vary as smooth functions of dropout times. The conditional distribution of CD4 given dropout is then averaged over the dropout distribution, which can be modeled nonparametrically. The results of our analysis are compared to the standard random effects approach.

CHAPTER 2

Regression Models
for Longitudinal Data

2.1 Overview

This chapter reviews some modern approaches to formulation and interpretation of regression models for longitudinal data. Section 2.2 outlines notation for longitudinal data and describes basic regression approaches. In Section 2.3 we describe the generalized linear model (GLM) for univariate data, which forms the basis of many regression models for longitudinal data. Sections 2.4 and 2.5 describe different approaches to regression modeling based on whether the mean is specified directly, in terms of marginal means; or conditionally, in terms of latent variables, random effects, or response history. Many conditionally specified models are multilevel models that partition the variance-covariance structure in a natural way, leading to low dimensional parameterizations.

In Section 2.6 we focus on semiparametric models highlighting those that permit flexibility in modeling time trends and allow covariate effects to vary smoothly with time (e.g., varying coefficient models). Finally, Section 2.7 reviews key issues in the interpretation of longitudinal regression models, highlighting longitudinal vs. cross-sectional effects and key assumptions about time-varying covariates.

The literature on regression models for longitudinal data is vast, and we make no attempt to be comprehensive here. Our review is designed to highlight predominant approaches to regression modeling, emphasizing those models used in later chapters. For recent accounts, readers are referred to Davidian and Giltinan (1998); Verbeke and Molenberghs (2000); Diggle et al. (2002); Fitzmaurice et al. (2004); Laird (2004); Weiss (2005); Hedeker and Gibbons (2006) and Molenberghs and Verbeke (2006).

2.2 Preliminaries

2.2.1 Longitudinal data

Appealing to first principles, one can think of longitudinal data as arising from the joint evolution of response and covariates,

$$\{Y_i(t), \boldsymbol{x}_i(t) : t \geq 0\}.$$

If the process is observed at a discrete set of time points $\mathscr{T} = \{t_1, \ldots, t_J\}$ that is common to all individuals, the resulting response data can be written as a $J \times 1$ vector

$$
\begin{aligned}
\boldsymbol{Y}_i &= \{Y_i(t) : t \in \mathscr{T}\} \\
&= (Y_{i1}, \ldots, Y_{iJ})^{\mathrm{T}}.
\end{aligned}
$$

The covariate process $\{\boldsymbol{x}_i(t) : t \geq 0\}$ is $1 \times p$. At time t_j, the observed covariates are collected in the $1 \times p$ vector

$$
\begin{aligned}
\boldsymbol{x}_{ij} &= \boldsymbol{x}_i(t_j) \\
&= (x_{i1}(t_j), \ldots, x_{ip}(t_j)) \\
&= (x_{ij1}, \ldots, x_{ijp}).
\end{aligned}
$$

Hence the full collection of observed covariates is contained in the $J \times p$ matrix

$$
\boldsymbol{X}_i = \begin{bmatrix} \boldsymbol{x}_{i1} \\ \boldsymbol{x}_{i2} \\ \vdots \\ \boldsymbol{x}_{iJ} \end{bmatrix}.
$$

When the set of observation times is common to all individuals, we say the responses are *balanced* or *temporally aligned*. It is sometimes the case that observation times are *unbalanced*, or *temporally misaligned*, in that they vary by subject. In this case the set $\mathscr{T}$ is indexed by i, such that

$$
\mathscr{T}_i = \{t_{i1}, \ldots, t_{iJ_i}\},
$$

and the dimensions of $\boldsymbol{Y}_i$ and $\boldsymbol{X}_i$ are $J_i \times 1$ and $J_i \times p$, respectively.

In regression, we are interested in characterizing the effect of covariates $\boldsymbol{X}$ on a longitudinal dependent variable $\boldsymbol{Y}$. Formally, we wish to draw inference about the joint distribution of the vector $\boldsymbol{Y}_i$ conditionally on $\boldsymbol{X}_i$. Likelihood-based regression models for longitudinal data require a specification of this joint distribution using a *model* $p(\boldsymbol{y} \mid \boldsymbol{x}, \boldsymbol{\theta})$. The parameter $\boldsymbol{\theta}$ typically is a finite-dimensional vector of parameters indexing the model; it might include regression coefficients, variance components, and parameters indexing serial correlation.

The joint distribution of responses can specified directly or conditionally. In this chapter we differentiate models on how the mean is specified. A model with a *directly* specified mean characterizes $E(\boldsymbol{Y} \mid \boldsymbol{x})$ without resorting to latent structures such as random effects. Models with *conditionally* specified means include Markov models and various types of multilevel models, where the mean is given conditionally on previous responses or on random effects that reflect aspects of the joint response distribution. Mostly, we deal with conditionally specified models that use a multilevel format, for example involving subject-specific random effects or latent variables $\boldsymbol{b}$ to partition within-

and between-subject variation. The usual strategy is to specify a model for the joint distribution of responses and random effects, factored as

$$p(\boldsymbol{y}, \boldsymbol{b} \mid \boldsymbol{x}) \quad = \quad p(\boldsymbol{y} \mid \boldsymbol{b}, \boldsymbol{x}) \, p(\boldsymbol{b} \mid \boldsymbol{x}).$$

The distribution of interest, $p(\boldsymbol{y} \mid \boldsymbol{x})$, is obtained by integrating over $\boldsymbol{b}$. For example, if the conditional mean is given by a function like

$$E(\boldsymbol{Y} \mid \boldsymbol{x}, \boldsymbol{b}) = g(\boldsymbol{x}\boldsymbol{\beta} + \boldsymbol{b}),$$

then

$$E(\boldsymbol{Y} \mid \boldsymbol{x}) = \int g(\boldsymbol{x}\boldsymbol{\beta} + \boldsymbol{b}) \, p(\boldsymbol{b} \mid \boldsymbol{x}) \, d\boldsymbol{b},$$

which may not always take a closed form.

2.2.2 Regression models

As we review several different regression models, the intent is to give the reader a sense of the rich variety of models that can be used to characterize longitudinal data, and to demonstrate how these fit coherently into a single framework. As a result, missing data strategies described in later chapters can be applied very generally. Methods described here will be familiar to those with experience analyzing longitudinal data (e.g., multivariate normal regression model, random effects models), but others represent fairly new developments. Examples include marginalized transition models (Heagerty, 2002), varying coefficient models (Zhang, 2004), and regression splines (Eilers and Marx, 1996; Lin and Zhang, 1999; Ruppert et al., 2003). Here we focus on *specification* and *interpretation*; Chapter 3 covers various aspects of inference.

Because many regression models for longitudinal data have their foundation in the generalized linear model (GLM) for cross-sectional data (McCullagh and Nelder, 1989), our review begins with a concise description of GLMs. Coverage of models for *longitudinal* data begins with random effects models; these build directly on the GLM structure by introducing individual-level random effects to capture between-subject variation. Conditional on the random effects, within-level variation can be described by a simpler model, such as a GLM. Random effects models are very attractive in that they naturally partition variation in the dependent variable into its between- and within-subject components, and they can be used to model both balanced and unbalanced data. At the same time, there is sometimes the disadvantage that the implied marginal distributions of responses can be opaque.

Directly specified models have a natural construction when the error distribution is multivariate normal; for binary, count, and other discrete data, the choice of an appropriate joint distribution is less obvious. Our review touches on some recent developments for discrete longitudinal responses, such as the marginalized transition model (Heagerty, 2002). For a detailed review

of likelihood-based models of multivariate discrete response, see Chapter 11 of
Diggle et al. (2002), Chapter 7 of Laird (2004), and Chapter 11 of Fitzmaurice
et al. (2004).

For all models covered in the first part of this chapter, $E(Y \mid x)$ takes a
known functional form, usually linear in some transformed scale. Section 2.6
describes models in which the regression function is unknown but can be at
least partially specified in terms of unspecified smooth functions. The latter
type of model is typically called *semiparametric*, because one or more components of the regression function are left unspecified, while distributional
assumptions are made about the error structure. Nonlinear and semiparametric models have a close connection to the GLM structure; we emphasize
that connection and illustrate that regression models as a whole can be very
generally characterized (Hastie and Tibshirani, 1990; Ruppert et al., 2003).

The final element of our review concerns interpretation of covariate effects in longitudinal models. Because the response and covariates change with
time, models of longitudinal data afford the opportunity to infer both within-
and between-subject covariate effects; however the importance of underlying
assumptions to the interpretation of covariate effects should not be underesti-
mated. Section 2.7 discusses three key aspects of interpretation and specification for longitudinal models: cross-sectional vs. longitudinal effects of a time-
varying covariate, marginal (population-averaged) vs. conditional (subject-
specific) covariate effects, and assumptions governing the use of time-varying
covariates.

2.2.3 Full vs. observed data

The distinction between *full* and *observed* data is particularly important when
drawing inference from incomplete longitudinal data. Throughout Chapters 2
and 3, the models refer to a full-data distribution.

We define the full data as those observations intended to be collected on a
pre-specified interval, such as $[0, T]$. For example, if intended collection times
$t_1, \ldots, t_J$ are common to all individuals, then the full response and covariate
data are $(Y_{i1}, X_{i1}), \ldots, (Y_{iJ}, X_{iJ})$, where $Y_{ij} = Y_i(t_j)$ and $X_{ij} = X_i(t_j)$. In
Chapter 5 we expand the definition of *full data* to include random variables
such as dropout time that characterize the missing data process.

In most applications, interest lies in the effect of covariates on the mean
structure. When data are fully observed, the variance and covariance models
can frequently be treated as nuisance parameters. Correct specification of
variance and covariance allows more efficient use of the data, but it is not
always necessary for obtaining proper inferences about mean parameters.

When data are not fully observed, variance-covariance specification takes
on heightened importance because missing data will effectively be imputed or
extrapolated from observed data, based on modeling assumptions. For longi-

tudinal data, unobserved responses will be imputed from observed responses for the same individual; the assumed correlation structure will usually dictate (at least in part) the functional form of the imputation. This theme recurs throughout the book, and therefore our review pays particular attention to aspects of variance-covariance specification.

2.2.4 Additional notation

Random variables and their realizations are denoted by Roman letters (e.g., X, x), and parameters are represented by Greek letters (e.g., α, θ). Vector- and matrix-valued random variables and parameters are represented using boldface (e.g., $\boldsymbol{x}$, $\boldsymbol{Y}$, $\boldsymbol{\beta}$, $\boldsymbol{\Sigma}$). For any matrix or vector $\boldsymbol{A}$, we use $\boldsymbol{A}^{\mathrm{T}}$ to denote transpose. If $\boldsymbol{A}$ is invertible, then $\boldsymbol{A}^{-1}$ is its inverse, $|\boldsymbol{A}|$ is its determinant, and $\boldsymbol{L} = \boldsymbol{A}^{1/2}$ is the lower triangular matrix square root (Cholesky factor) such that $\boldsymbol{L}\boldsymbol{L}^{\mathrm{T}} = \boldsymbol{A}$. A q-dimensional identity matrix is denoted $\boldsymbol{I}_q$ and a diagonal matrix by $\mathrm{diag}(\boldsymbol{a})$, where $\boldsymbol{a}$ is the vector of diagonal elements. The parameterizations of specific probability distributions used in the text can be found in the Appendix.

2.3 Generalized linear models for cross-sectional data

The generalized linear model (GLM) forms the foundation for many approaches to regression with multivariate responses, such as longitudinal or clustered data. Models such as random effects or mixed effects models, latent variable and latent class models, and regression splines, all highly flexible and general, are based on the GLM framework. Moment-based methods such as generalized estimating equations (GEE) also follow directly from the GLM for cross-sectional data (Liang and Zeger, 1986).

The GLM is a regression model for a dependent variable Y arising from the exponential family of distributions

$$p(y \mid \theta, \psi) = \exp\left\{ (y\theta - b(\theta))/a(\psi) + c(y, \psi) \right\},$$

where a, b, and c are known functions, θ is the *canonical parameter*, and ψ is a *scale parameter*. The exponential family includes several commonly used distributions, such as normal, Poisson, binomial, and gamma. It can be readily shown that

$$
\begin{aligned}
E(Y) &= b'(\theta) \\
\mathrm{var}(Y) &= a(\psi)b''(\theta),
\end{aligned}
$$

where $b'(\theta)$ and $b''(\theta)$ are first and second derivatives of $b(\theta)$ with respect to θ (see McCullagh and Nelder, 1989, Section 2.2.2 for details).

The effect of covariates $\boldsymbol{x}_i = (x_{i1}, \ldots, x_{ip})$ can be modeled by introducing the linear predictor

$$\eta_i = \eta(\boldsymbol{x}_i, \boldsymbol{\beta}) = \boldsymbol{x}_i \boldsymbol{\beta},$$

where $\boldsymbol{\beta} = (\beta_1, \ldots, \beta_p)^{\mathrm{T}}$ is a vector of regression coefficients. Define $\mu_i = \mu(\boldsymbol{x}_i, \boldsymbol{\beta}) = E(Y \mid \boldsymbol{x}_i, \boldsymbol{\beta})$. A smooth, monotone function g links the mean μ_i to the linear predictor η_i via

$$g(\mu_i) = \eta_i = \boldsymbol{x}_i \boldsymbol{\beta}. \tag{2.1}$$

In many exponential family distributions, it is possible to identify a link function g such that $\boldsymbol{X}^{\mathrm{T}}\boldsymbol{Y}$ is the sufficient statistic for $\boldsymbol{\beta}$ (here, $\boldsymbol{X}$ is the $n \times p$ design matrix and $\boldsymbol{Y} = (Y_1, \ldots, Y_n)^{\mathrm{T}}$ is the $n \times 1$ vector of responses). In this case, the *canonical parameter* is $\theta = \eta$. Examples are well-known and widespread: for the Poisson distribution, the canonical parameter is $\log(\mu)$; for binomial distribution, it is the log odds (logit), $\log\{\mu/(1-\mu)\}$.

Although canonical links are sometimes convenient, their use is not necessary to form a GLM. In general, only the specification of a mean and variance function, conditionally on covariates, is required. The mean follows (2.1), and the variance is given by

$$v(\mu_i, \phi) = \phi h(\mu_i),$$

where $h(\cdot)$ is some function of the mean and $\phi > 0$ is a scale factor. Certain choices of g and h will yield likelihood score equations for common parametric regression models based on exponential family distributions. For example, setting $g(\mu) = \log\{\mu/(1-\mu)\}$, $h(\mu) = \mu(1-\mu)$ and $\phi = 1$ yields logistic regression under a Bernoulli distribution. Similarly, Poisson regression can be specified by setting $g(\mu) = \log(\mu)$, $h(\mu) = \mu$, and $\phi = 1$.

2.4 Conditionally specified models

This section focuses on models that specify the mean of $\boldsymbol{Y}$ given $\boldsymbol{X}$ conditionally on random effects; these models also are known by a variety of names, including 'mixed effects models', 'multilevel models', 'random effects models', and 'random coefficient models'. Throughout the text, we use the terms 'random effects' and 'multilevel' models. Readers are referred to Breslow and Clayton (1993) and Daniels and Gatsonis (1999) for a more complete accounting and list of references. This class of models also includes regression models with factor-analytic and latent class structures (see Bartholomew and Knott, 1999 for a full account) and Markov models (where the mean is specified conditional on a subset of past responses).

Conditionally specified models with multiple levels, using random effects or latent variables, provide a highly flexible class of models for handling longitudinal data. The models can be applied either to balanced or unbalanced response profiles and can be used to capture key features of both between- and within-subject variation using relatively few parameters.

The most common random effects models for longitudinal data specify the joint distribution $p(\boldsymbol{y}, \boldsymbol{b} \mid \boldsymbol{x}, \boldsymbol{\theta})$ as

$$p(\boldsymbol{y} \mid \boldsymbol{b}, \boldsymbol{x}, \boldsymbol{\theta}_1) \, p(\boldsymbol{b} \mid \boldsymbol{x}, \boldsymbol{\theta}_2).$$

The parameter $\boldsymbol{\theta}_1$ captures the conditional effect of $\boldsymbol{X}$ on $\boldsymbol{Y}$ and the parameter $\boldsymbol{\theta}_2$ captures features of the distribution of random effects $\boldsymbol{b}$. The distribution unconditional on random effects is obtained by integrating $\boldsymbol{b}$ out of the joint distribution

$$p(\boldsymbol{y} \mid \boldsymbol{x}, \boldsymbol{\theta}_1, \boldsymbol{\theta}_2) = \int p(\boldsymbol{y} \mid \boldsymbol{x}, \boldsymbol{b}, \boldsymbol{\theta}_1) \, p(\boldsymbol{b} \mid \boldsymbol{x}, \boldsymbol{\theta}_2) \, d\boldsymbol{b}; \qquad (2.2)$$

notice the unconditional distribution is indexed by the full set of parameters $\boldsymbol{\theta} = (\boldsymbol{\theta}_1, \boldsymbol{\theta}_2)$.

2.4.1 Random effects models based on GLMs

By including random effects, generalized linear models can be used to model longitudinal and clustered data. For common distributions such as Bernoulli and Poisson, the GLM with random effects can be written in terms of the conditional mean and variance. The model for the conditional mean $\mu_{ij}^b = E(Y_{ij} \mid \boldsymbol{x}_{ij}, \boldsymbol{w}_{ij}, b_i)$ takes the form

$$g\{E(Y_{ij} \mid \boldsymbol{x}_{ij}, \boldsymbol{w}_{ij}, \boldsymbol{b}_i)\} = g(\mu_{ij}^b) = \boldsymbol{x}_{ij}\boldsymbol{\beta} + \boldsymbol{w}_{ij}\boldsymbol{b}_i,$$

where $g(\cdot)$ is a link function and $\boldsymbol{w}_{ij}$ is a design matrix for the subject-specific random effects. This representation of the conditional mean motivates the term 'mixed-effects model' because the coefficients quantify both population-level ($\boldsymbol{\beta}$) and individual-level ($\boldsymbol{b}_i$) effects.

The conditional variance is given by

$$V_{ij}^b = \mathrm{var}(Y_{ij} \mid \boldsymbol{x}_{ij}, \boldsymbol{w}_{ij}, \boldsymbol{b}_i) = \phi h(\mu_{ij}^b)$$

for $\phi > 0$ and a suitably chosen $h(\cdot)$. Finally, within-subject correlation is specified through a covariance function

$$C_{ijk}^b(\boldsymbol{\gamma}) = \mathrm{cov}(Y_{ij}, Y_{ik} \mid \boldsymbol{x}_{ij}, \boldsymbol{x}_{ik}, \boldsymbol{b}_i, \boldsymbol{\gamma}).$$

In many cases it is assumed that $C_{ijk}^b = 0$; i.e., that the random effects capture relevant within-subject correlation (after averaging over their distribution), but this assumption may not always be appropriate for longitudinal responses.

At the second level, the random effects $\boldsymbol{b}_i$ follow some distribution such as multivariate normal. The model for the marginal joint distribution of $(Y_{i1}, \ldots, Y_{iJ} \mid \boldsymbol{X}_i)$ is obtained by integrating over $\boldsymbol{b}_i$ as in (2.2).

The relationship between marginal and conditional models is important to understand, particularly as it relates to interpreting covariate effects. In what follows we give several examples to illustrate.

2.4.2 Random effects models for continuous response

A natural choice for modeling continuous or measured responses is the normal distribution. In random effects models, allowing both within- and between-subject variation to follow a normal distribution, or more generally a Gaussian process, affords considerable modeling flexibility while retaining interpretability.

Example 2.1. *Normal random effects model for continuous responses.*
A common model for continuous longitudinal responses is the normal random effects model. This model illustrates well the concept of a conditionally specified joint distribution because the variance-covariance structure in $p(\boldsymbol{y} \mid \boldsymbol{x}, \boldsymbol{\theta})$ is a by-product of the assumed random effects distribution.

Like many random effects models, it is easiest to describe in two stages. At the first stage, the vector of responses $\boldsymbol{Y}_i$, measured at times $\{t_{i1}, \ldots, t_{i,J_i}\}$, are normal conditionally on a $q \times 1$ vector of random effects $\boldsymbol{b}_i$,

$$\boldsymbol{Y}_i \mid \boldsymbol{x}_i, \boldsymbol{b}_i \sim N(\boldsymbol{\mu}_i^b, \boldsymbol{\Sigma}_i^b),$$

where the superscript b denotes that the mean and covariance are conditional on $\boldsymbol{b}_i$. To incorporate covariate effects, let

$$\boldsymbol{\mu}_i^b = \boldsymbol{x}_i \boldsymbol{\beta} + \boldsymbol{w}_i \boldsymbol{b}_i,$$

where $\boldsymbol{w}_i$ is the design matrix for random effects. The variance matrix $\boldsymbol{\Sigma}_i^b = \boldsymbol{\Sigma}_i^b(\boldsymbol{\phi})$ captures within-subject variation and is parameterized by the $r \times 1$ vector $\boldsymbol{\phi}$ of nonredundant parameters. Hence $\boldsymbol{\theta}_1 = (\boldsymbol{\beta}, \boldsymbol{\phi})$.

When $\boldsymbol{w}_i \subseteq \boldsymbol{x}_i$, as is usually the case, the $\boldsymbol{b}_i$ can be thought of as error terms for one or more of the regression coefficients, which gives rise to the term 'random coefficient model'. For example, if $\boldsymbol{x}_i = \boldsymbol{w}_i$, we obtain

$$\boldsymbol{\mu}_i^b = \boldsymbol{x}_i \boldsymbol{\beta}_i = \boldsymbol{x}_i (\boldsymbol{\beta} + \boldsymbol{b}_i), \tag{2.3}$$

where the random effects $\boldsymbol{b}_i$ can be interpreted as individual-specific deviations from $\boldsymbol{\beta}$.

The within-subject variance $\boldsymbol{\Sigma}_i^b(\boldsymbol{\phi})$ usually has a simplified structure, parameterized through a covariance function $C_{ijk}^b(\boldsymbol{\phi})$. For example, an exponential structure takes the form

$$C_{ijk}^b(\boldsymbol{\phi}) \;=\; \sigma^2 \rho^{|t_{ij} - t_{ik}|},$$

where $\boldsymbol{\phi} = (\sigma^2, \rho)$ and $0 \leq \rho \leq 1$.

At the second level, the q-dimensional vector of random effects is assigned a distribution that can depend on covariates. The (multivariate) normal is a common choice,

$$\boldsymbol{b}_i \mid \boldsymbol{x}_i \sim N(\boldsymbol{0}, \boldsymbol{\Omega}),$$

where $\boldsymbol{\Omega} = \boldsymbol{\Omega}(\boldsymbol{\eta})$ is a $q \times q$ covariance matrix indexed by $\boldsymbol{\eta}$ (hence $\boldsymbol{\theta}_2 = \boldsymbol{\eta}$).

It also is possible to allow $\boldsymbol{\eta}$ to depend on individual-level covariates through appropriate specifications (Daniels and Zhao, 2003).

Upon integrating over $\boldsymbol{b}$, the marginal distribution of $\boldsymbol{Y}_i$ follows the multivariate normal distribution

$$\boldsymbol{Y}_i \mid \boldsymbol{x}_i, \boldsymbol{w}_i \sim N(\boldsymbol{x}_i\boldsymbol{\beta}, \; \boldsymbol{w}_i\boldsymbol{\Omega}\boldsymbol{w}_i^{\mathrm{T}} + \boldsymbol{\Sigma}_i^b). \tag{2.4}$$

The marginal variance $\mathrm{var}(\boldsymbol{Y}_i \mid \boldsymbol{x}_i)$ depends on parameters from both $p(\boldsymbol{y} \mid \boldsymbol{x}, \boldsymbol{b})$ and $p(\boldsymbol{b} \mid \boldsymbol{x})$. Moreover, we see by comparing (2.3) and (2.4) that $\boldsymbol{\beta}$ can be interpreted both as a marginal and a conditional effect of $\boldsymbol{X}$ on $\boldsymbol{Y}$. We discuss this further in Section 2.7. A version of this model is used to analyze the schizophrenia trial in Sections 4.3 and 7.3. □

2.4.3 Random effects models for discrete responses

Random effects specifications can be very useful for modeling longitudinal discrete responses, where the joint distribution rarely takes an obvious form and principles from generalized linear models are not as easily applied. In the case of longitudinal binary data, for example, it is straightforward to show that the joint distribution of a J-dimensional response variable can be represented by a multinomial distribution with 2^J categories. When J is appreciably large, however, parameter constraints must be imposed to make modeling practical. See Laird (2004), Chapter 7, for a more detailed discussion.

Compared to direct specification of the joint distribution, random effects models offer the advantage of being parsimonious, providing a natural decomposition of multiple sources of variation, and applying equally well to balanced and unbalanced response profiles. The regression parameters represent covariate effects in the conditional rather than marginal joint distribution of $\boldsymbol{Y}$, however, and because the link functions are nonlinear transformations of the mean (e.g., log, logit), these do not generally coincide. Therefore care must be taken when interpreting regression effects. The logistic regression with normal random effects illustrates several of these points rather well.

Example 2.2. *Logistic regression with random effects.*
As in Example 2.1, a logistic random effects model is specified in terms of the joint distribution

$$p(\boldsymbol{y}, \boldsymbol{b} \mid \boldsymbol{x}, \boldsymbol{\theta}) \;\; = \;\; p(\boldsymbol{y} \mid \boldsymbol{x}, \boldsymbol{b}, \boldsymbol{\theta}_1)\, p(\boldsymbol{b} \mid \boldsymbol{x}, \boldsymbol{\theta}_2),$$

where $\boldsymbol{\theta} = (\boldsymbol{\theta}_1, \boldsymbol{\theta}_2)$. Conditionally on $\boldsymbol{b}_i$, the distribution of each component Y_{ij} of $\boldsymbol{Y}_i$ follows the Bernoulli model

$$Y_{ij} \mid \boldsymbol{x}_{ij}, \boldsymbol{b}_i \sim \mathrm{Ber}(\mu_{ij}^b),$$

where

$$g(\mu_{ij}^b) = \boldsymbol{x}_{ij}\boldsymbol{\beta} + \boldsymbol{w}_{ij}\boldsymbol{b}_i \tag{2.5}$$

(hence $\boldsymbol{\theta}_1 = \boldsymbol{\beta}$). The random effects distribution follows

$$\boldsymbol{b}_i \mid \boldsymbol{x}_i \sim N(\boldsymbol{0}, \boldsymbol{\Omega}),$$

so $\boldsymbol{\theta}_2$ corresponds to the nonredundant components of $\boldsymbol{\Omega}$.

The parameter $\boldsymbol{\beta}$ characterizes the conditional, or *subject-specific* effect of $\boldsymbol{X}$ on $\boldsymbol{Y}$. By contrast, the marginal — or *population-averaged* — distribution $p(\boldsymbol{y} \mid \boldsymbol{x}, \boldsymbol{\theta})$ must be obtained by integrating over $\boldsymbol{b}$. When $\boldsymbol{w}_i$ is a subset of $\boldsymbol{x}_i$, the marginal mean $\mu_{ij}(\boldsymbol{\beta}, \boldsymbol{\Omega}) = E(Y_{ij} \mid \boldsymbol{x}_{ij}, \boldsymbol{\beta}, \boldsymbol{\Omega})$ is

$$\begin{aligned} \mu_{ij}(\boldsymbol{\beta}, \boldsymbol{\Omega}) &= \int \mu_{ij}^b(\boldsymbol{\beta}) \, p(\boldsymbol{b} \mid \boldsymbol{\Omega}) \, d\boldsymbol{b} \\ &= \int \frac{\exp(\boldsymbol{x}_{ij}\boldsymbol{\beta} + \boldsymbol{w}_{ij}\boldsymbol{b})}{1 + \exp(\boldsymbol{x}_{ij}\boldsymbol{\beta} + \boldsymbol{w}_{ij}\boldsymbol{b})} \, p(\boldsymbol{b} \mid \boldsymbol{\Omega}) \, d\boldsymbol{b}. \end{aligned}$$

where $p(\boldsymbol{b} \mid \boldsymbol{\Omega})$ is the multivariate normal density with mean $\boldsymbol{0}$ and variance $\boldsymbol{\Omega}$. The marginal effect of $\boldsymbol{X}$ differs from the conditional effect in that it is a function of both $\boldsymbol{\beta}$ and $\boldsymbol{\Omega}$, and on the logit scale, it is no longer linear.

Zeger and Liang (1992) show that in some cases, the marginal effect in the logit-normal model is approximately linear on the logit scale, and differs from the conditional effect by a scale factor that depends on $\boldsymbol{\Omega}$. To illustrate, consider the simple case of a logistic regression with a single covariate x_i and random intercept; i.e.,

$$\text{logit}(\mu_{ij}^b) = \beta_0 + b_i + \beta_1 x_i,$$

where $b_i \sim N(0, \nu^2)$. Here, β_1 is the conditional effect of x. It can be shown that the marginal effect of x, denoted by β_1^{m}, can be approximately represented by the logistic model

$$\text{logit}(\mu_{ij}) = \beta_0^{\mathrm{m}} + \beta_1^{\mathrm{m}} x_i,$$

where $\beta_1^{\mathrm{m}} = (c^2\nu^2 + 1)^{-1/2}\beta$, and $c \approx .346$. Hence the marginal or population averaged effect of x is attenuated relative to the conditional or subject-specific effect, with degree of attenuation governed by the magnitude of the random effects variance ν^2. Interpreting the marginal and conditional effects is considered further in Section 2.7.

In Section 4.4 we use this model to characterize the effect of a behavioral intervention on weekly smoking cessation status using longitudinal binary data from the Commit to Quit I study. $\square$

Examples 2.1 and 2.2 assume the random effects $\boldsymbol{b}_i$ follow a normal distribution; this is not necessary and in many cases it may be inappropriate or incorrect. Zhang and Davidian (2001) describe models where the random effects distribution belongs to a flexible class of densities that includes the normal as a special case. Verbeke and Lesaffre (1996) describe random effects distributions that follow discrete mixtures of normal distributions. For simple models, it is sometimes possible to use exploratory analysis in order to ascertain whether a normal or other symmetric distribution is suitable for describing the random effects. In other cases, more formal methods of model choice may be needed.

2.5 Directly specified (marginal) models

This section reviews the family of models in which the joint distribution of $\boldsymbol{Y}$ given $\boldsymbol{X}$ is directly specified by a model $p(\boldsymbol{y} \mid \boldsymbol{x}, \boldsymbol{\theta})$. Usually the most challenging aspect of model specification is finding a suitable parameterization for the correlation and/or covariance, particularly when observations are unbalanced in time or when the number of observations per subject is large relative to sample size. In these cases, sensible decisions about dimension reduction must be made.

For continuous data that can be characterized using a normal distribution or Gaussian process, model specification (though not necessarily selection) can be reasonably straightforward, owing to the natural separation of mean and variance parameters in the normal distribution. The analyst can focus efforts separately on models for mean and covariance structure.

Other types of data pose more significant challenges to the process of direct specification due to a lack of obvious choices for joint distribution models. Unlike the normal distribution, which generalizes naturally to the multivariate and even the stochastic process setting, common distributions like binomial and Poisson do not have obvious multivariate analogues. One problem is that the mean and covariance models share the same parameters, even for simple specifications. Another potential problem is that unlike with the normal model, higher-order associations do not necessarily follow from pairwise associations, and need to be specified or explicitly constrained (Fitzmaurice and Laird, 1993). The joint distribution of J binary responses, for example, has $2^J - J$ parameters governing the association structure. With count data, appropriate specification of even a simple correlation structure is not immediately obvious.

This section describes various approaches to direct model specification, illustrated with examples from the normal and binomial distributions. The first examples use the normal distribution. For longitudinal binary data, we describe an extension of the log-linear model that allows transparent interpretation of both the mean and serial correlation. Another useful approach to modeling association in binary data is the multivariate probit model, which exploits properties of the normal distribution by assuming the binary random variables are manifestations of an underlying normally distributed latent process.

2.5.1 Multivariate normal and Gaussian process models

The multivariate normal distribution provides a highly flexible starting point for modeling continuous response data, both temporally aligned and misaligned. It also is useful for handling situations where the number of observation times is large relative to the number of units being followed. The most straightforward situation is where data are temporally aligned and $n \gg J$, allowing both the mean and variance to be unstructured. When responses are temporally misaligned, or when J is large relative to n, structure must be imposed.

A key characteristic of the normal distribution allowing for flexible modeling across a wide variety of settings is that the mean and variance have separate parameters. The next two examples illustrate a variety of model specifications using the normal distribution.

Example 2.3. *Multivariate normal regression for temporally aligned observations.*

Assume that observations on the primary response variable are taken at a fixed schedule of times $t_1, \ldots, t_J$. For a response vector $\boldsymbol{Y}_i = (Y_{i1}, \ldots, Y_{iJ})^{\mathrm{T}}$ with associated $J \times p$ covariate matrix $\boldsymbol{X}_i$, the multivariate normal regression is written as

$$\boldsymbol{Y}_i \mid \boldsymbol{x}_i \sim N(\boldsymbol{\mu}_i, \boldsymbol{\Sigma}_i),$$

where $\boldsymbol{\mu}_i$ is $J \times 1$ and $\boldsymbol{\Sigma}_i$ is $J \times J$. The mean $\mu_i = E(\boldsymbol{Y}_i \mid \boldsymbol{X}_i = \boldsymbol{x}_i)$ follows a regression model

$$\boldsymbol{\mu}_i = \boldsymbol{x}_i \boldsymbol{\beta},$$

where $\boldsymbol{x}_i$ is the observed $J \times p$ covariate matrix and $\boldsymbol{\beta}$ is a $p \times 1$ vector of regression coefficients.

The covariance matrix is parameterized with a vector of non-redundant parameters $\boldsymbol{\phi}$. To emphasize that the covariance matrix may depend on $\boldsymbol{x}_i$ through $\boldsymbol{\phi}$, we sometimes write

$$\boldsymbol{\Sigma}_i(\boldsymbol{\phi}) = \boldsymbol{\Sigma}(\boldsymbol{x}_i, \boldsymbol{\phi})$$

(Daniels and Pourahmadi, 2002). If $\boldsymbol{\Sigma}_i$ is assumed constant across individuals, it has $J(J+1)/2$ unique parameters, but structure can be imposed to reduce this number (Jennrich and Schluchter, 1986). As an alternative to leaving $\boldsymbol{\Sigma}_i$ fully parameterized, common structures for longitudinal data include banded or Toeplitz (with common parameter along each off-diagonal), and autoregressive correlations of pre-specified order (Núñez Antón and Zimmermann, 2000; Pourahmadi, 2000; Pourahmadi and Daniels, 2002).

The matrix $\boldsymbol{X}_i$ can include information about measurement time, baseline covariates, and the like. If we set $\boldsymbol{x}_{ij} = (1, t_j)$ and $\boldsymbol{\beta} = (\beta_0, \beta_1)^{\mathrm{T}}$, then β_1 corresponds to the average slope over time, where the average is taken over the population from which the sample of individuals is drawn. When J is small enough, $\boldsymbol{x}_{ij}$ can include a vector of time indicators, allowing the mean to be unstructured in time. In Sections 4.2 and 7.2, this model is used to analyze data from the Growth Hormone Study. □

In the previous example, it is sometimes possible to allow both the mean and variance to remain unstructured in time when there are relatively few time points and covariate levels. When time points are temporally misaligned, or when the number of observation times is large relative to the sample size, information at the unique measurement times will be sparse and additional structure needs to be imposed. Our focus in the next example is on covariance parameterization in terms of a covariance function. Further details can be found in Diggle et al. (2002), Chapter 4.

Example 2.4. *Multivariate normal regression model for temporally misaligned observations.*
The main difference in model specification when observations are temporally misaligned has mainly to do with the covariance parameterization. As with Example 2.3, a normal distribution may be assumed, but with covariance $\boldsymbol{\Sigma}_i$ whose dimension and structure depend on the number and timing of observations for individual i. The joint distribution follows

$$\boldsymbol{Y}_i \mid \boldsymbol{x}_i \sim N(\boldsymbol{\mu}_i, \boldsymbol{\Sigma}_i).$$

Also as in Example 2.3, $\boldsymbol{\mu}_i = \boldsymbol{x}_i\boldsymbol{\beta}$. The covariance $\boldsymbol{\Sigma}_i(\boldsymbol{\phi})$ has dimension J_i, and structure is imposed by specifying a model $C(t_{ik}, t_{il}, \boldsymbol{\phi})$ for the elements $\sigma_{ikl} = \text{cov}(Y_{ik}, Y_{il})$. For example, the model

$$C(t_{ik}, t_{il}; \phi_1, \phi_2) \quad = \quad \phi_1 \exp(-\phi_2^{|t_{ik}-t_{il}|}) \tag{2.6}$$

requires only ϕ_1 and ϕ_2 to fully parameterize $\boldsymbol{\Sigma}_i(\boldsymbol{\phi})$ (with the constraints $\phi_1 > 0$ and $\phi_2 \geq 0$). This covariance model implies
 (i) $\text{var}(Y_{ij} \mid \boldsymbol{x}_i, \boldsymbol{\phi}) = \phi_1$ (setting $t_{ik} = t_{il}$)
 (ii) For any two observations Y_{ik}, Y_{il} separated by lag $|t_{ik} - t_{il}|$,

$$\text{corr}(Y_{ik}, Y_{il} \mid \boldsymbol{x}_i, \boldsymbol{\phi}) = \exp(-\phi_2^{|t_{ik}-t_{il}|}).$$

The correlation function above is *stationary* because for any given lag, the correlation is a constant function of (t_{ik}, t_{il}).

As with Example 2.3, the variance and correlation parameters may depend on covariates through an appropriately specified model. If we modify (2.6) such that

$$\sigma_{ikl} = \phi_{1i} \exp(-\phi_2^{|t_{ik}-t_{il}|})$$

(i.e., ϕ_1 now depends on i), then we may model ϕ_{1i} as a function of covariates via

$$\log(\phi_{1i}) = \boldsymbol{x}_i\boldsymbol{\alpha}.$$

In Section 7.6, a semiparametric version of this model will be used to characterize the response of longitudinal CD4 counts to initiation of highly active antiretroviral therapy (HAART) in the HER Study.   □

2.5.2 Directly specified models for discrete longitudinal responses

Although the multivariate normal distribution is a natural choice for characterizing the joint distribution of continuous longitudinal responses, approaches to dealing with discrete observations such as binary, categorical, or count data are less obvious. This section describes two models for longitudinal binary data, both of which can be generalized to handle ordinal or multinomial response.

A challenge in formulating models for the joint distribution of discrete responses is having a sensible parameterization of the correlation while maintaining an interpretable regression structure for the mean. This difficulty arises because the correlation is a function of the mean. One approach is to fully parameterize higher-order interactions and then constrain some of them to be zero (Fitzmaurice and Laird, 1993; Fitzmaurice et al., 1994, 1996). Another is to formulate models that deal explicitly with serial correlation. Our examples here include the marginalized transition model (MTM) (Heagerty, 2002), which is a constrained version of a log-linear model, and a multivariate probit model, which uses a latent multivariate normal structure to induce correlation (Chib and Greenberg, 1998). The probit model is used widely in econometric modeling but less so for biostatistical applications. Model coefficients lack the simplicity of odds ratio interpretation, but the assumption of an underlying latent normal distribution provides computational tractability and flexible modeling of serial covariance structures.

Marginalized transition models

We illustrate the formulation of MTMs using an example with first-order dependence.

Example 2.5. *Marginalized transition model for temporally aligned binary data.*
Let $\boldsymbol{Y}_i = (Y_{i1}, \ldots, Y_{iJ})^{\mathrm{T}}$ denote a vector of binary responses, and let $\boldsymbol{X}_i$ denote the design matrix. The marginal mean given covariates $\boldsymbol{x}_{ij}$ is

$$\mu_{ij} = E(Y_{ij} \mid \boldsymbol{x}_{ij}) = \mathrm{P}(Y_{ij} = 1 \mid \boldsymbol{x}_{ij}).$$

To describe the model, some additional notation is needed. For any time-dependent variable Z, let $\overline{\boldsymbol{Z}}_j = \{Z_1, \ldots, Z_j\}$ denote its history up to and including time j. The MTM likelihood is a transition model where the distribution of Y_{ij} given $(\overline{\boldsymbol{Y}}_{i,j-1}, \boldsymbol{x}_{ij})$ follows a Bernoulli distribution, but is constrained to allow $g(\mu_{ij})$ to be linear in covariates.* Hence the marginal mean retains its form as a regression, but the distributional assumptions are made in the serial correlation model.

We use the logistic regression formulation for illustration. The underlying joint distribution of responses is factored as

$$p(y_1, \ldots, y_J \mid \boldsymbol{x}) = p(y_1 \mid \boldsymbol{x}_1) \, p(y_2 \mid y_1, \boldsymbol{x}_2) \cdots p(y_J \mid \overline{\boldsymbol{y}}_{J-1}, \boldsymbol{x}_J).$$

Each component of the joint distribution is assumed to follow a Bernoulli distribution

$$Y_{ij} \mid \overline{\boldsymbol{y}}_{i,j-1}, \boldsymbol{x}_{ij} \sim \text{Ber}(\phi_{ij}),$$

where $\phi_{ij} = E(Y_{ij} \mid \overline{\boldsymbol{y}}_{i,j-1}, \boldsymbol{x}_{ij})$. For the first-order dependence model, denoted MTM(1), we have

$$p(y_j \mid \overline{\boldsymbol{y}}_{j-1}, \boldsymbol{x}_j) = p(y_j \mid y_{j-1}, \boldsymbol{x}_{ij}).$$

The model is specified in terms of two simultaneous equations. The first allows logit of the marginal mean μ_{ij} to depend linearly on covariates $\boldsymbol{x}_{ij}$; the second characterizes the dependence structure described by the conditional mean ϕ_{ij},

$$\begin{aligned} \text{logit}(\mu_{ij}) &= \boldsymbol{x}_{ij}\boldsymbol{\beta} \\ \text{logit}(\phi_{ij}) &= \Delta_{ij} + y_{i,j-1}\gamma_j, \end{aligned} \qquad (2.7)$$

where $\Delta_{ij} = \Delta(\boldsymbol{x}_{ij})$ is determined by $\boldsymbol{\beta}$ and γ_j. If serial correlation γ_j depends on individual-level covariates $\boldsymbol{w}_{ij} \subseteq \boldsymbol{x}_{ij}$, then we replace γ_j by γ_{ij}, where

$$\gamma_{ij} = \boldsymbol{w}_{ij}\boldsymbol{\alpha}.$$

These constraints imply that for a given value of $\boldsymbol{x}_{ij}$, Δ_{ij} is a deterministic function of γ_j, $\boldsymbol{\beta}$, and y (Heagerty, 2002). To see this clearly, first write

$$\phi_{ij} = \phi(\Delta_{ij}, \gamma_j, y) = E(Y_{ij} \mid Y_{i,j-1} = y)$$

to emphasize its dependence on $Y_{i,j-1}$. Then

$$\begin{aligned} \mu_{ij}(\boldsymbol{\beta}) &= \text{P}(Y_{ij} = 1 \mid \boldsymbol{x}_{ij}) \\ &= \text{P}(Y_{ij} = 1 \mid Y_{i,j-1} = 0, \boldsymbol{x}_{ij}, \boldsymbol{\gamma}, \Delta_{ij}) \, \text{P}(Y_{i,j-1} = 0 \mid \boldsymbol{x}_{i,j-1}, \boldsymbol{\beta}) \\ &\quad + \text{P}(Y_{ij} = 1 \mid Y_{i,j-1} = 1, \boldsymbol{x}_{ij}, \boldsymbol{\gamma}, \Delta_{ij}) \, \text{P}(Y_{i,j-1} = 1 \mid \boldsymbol{x}_{i,j-1}, \boldsymbol{\beta}) \\ &= \phi_{ij}(\boldsymbol{\gamma}, \Delta_{ij}, 0)\{1 - \mu_{i,j-1}(\boldsymbol{\beta})\} + \phi_{ij}(\boldsymbol{\gamma}, \Delta_{ij}, 1)\mu_{i,j-1}(\boldsymbol{\beta}). \quad (2.8) \end{aligned}$$

In Section 4.4 we use this model to analyze data from the CTQ I study (Section 1.4) and compare the results to those obtained with the logistic-normal random effects model described in Example 2.2. In Section 7.4, we also use this model to analyze data from the same study under MAR. $\qquad \square$

---

* In many cases, we implicitly assume $\boldsymbol{x}_{ij}$ includes relevant covariate history up to and including time j, obviating the need for overbar notation.

In practice, the MTM works well for first-order Markov dependence, but higher-order dependence structures can make computation of Δ_{ij} more difficult. Because the models are derived from log-linear specifications, they can be extended to handle ordinal or multinomial response (Lee and Daniels, 2007). MTMs are well-suited to temporally aligned data, but can be expanded to handle temporally misaligned observations through properly specified design matrices in both the mean and serial correlation. Su and Hogan (2007) develop a flexible approach using penalized splines.

When interest focuses only on the transition probabilities ϕ_{ij} and not the marginal mean, then the conditional part (2.7) of the MTM can be used for inference (it is simply a first-order Markov model). The intercept Δ_{ij} in (2.7) is replaced by $\boldsymbol{x}_{ij}\boldsymbol{\psi}$, and the parameters of interest are either the γ_j or $\boldsymbol{\psi}$.

Multivariate probit models

Another approach to handling longitudinal discrete data, the multivariate probit model, exploits the separation of mean and variance in the normal distribution by assuming the existence of a latent multivariate normal structure that gives rise to observed data. Probit models are widely used in psychometrics and econometrics, where justification for and interpretation of the latent scales can frequently be made on subject matter grounds (see Lancaster, 2004, Section 5.2, for example).

In epidemiology and public health applications, justification for the use of latent scales is less obvious. In any application, the data offer limited justification for the assumption that the latent variable follows a normal distribution, which has invited criticism of these models. Nevertheless, probit models afford considerable opportunity to parameterize covariance structures for discrete data and covariances between discrete and continuous data, a property that is difficult to overlook. Moreover, the assumed latent structure makes inference via posterior sampling rather easy to implement (see Chapter 3). The probit link is functionally quite similar to the logit link, so the fit is typically very similar between the two models. This example gives some details about specification. More details on the multivariate probit model, including methods for inference, are found in Chapters 3 and 6.

Example 2.6. *Multivariate probit model.*
The probit model based on an underlying multivariate normal model can be used to directly model the joint distribution of longitudinal binary data. The multivariate probit model assumes the existence of a vector of latent variables $\boldsymbol{Z}_i = (Z_{i1}, \ldots, Z_{iJ})^{\mathrm{T}}$ that follow a multivariate normal distribution

$$\boldsymbol{Z}_i \mid \boldsymbol{x}_i \sim N(\boldsymbol{\mu}_i, \boldsymbol{\Psi}_i),$$

where $\boldsymbol{\mu}_i = \boldsymbol{x}_i\boldsymbol{\beta}$, $\boldsymbol{\Psi}_i = \boldsymbol{\Psi}_i(\boldsymbol{\phi})$ is a $J \times J$ correlation matrix with elements $\{\psi_{kl}(\boldsymbol{\phi})\}$, and $\boldsymbol{\theta} = (\boldsymbol{\beta}, \boldsymbol{\phi})$. The diagonal elements ψ_{kk} are set to one for identifiability (Chib and Greenberg, 1998). Interpretation of regression parameters for probit models is most easily done on the latent scale: the coefficient β_p is the average difference in standard deviations of Z_{ij} corresponding to one-unit differences in x_{ijp}.

The binary observations Y_{ij} are viewed as manifestations of the underlying latent variables such that $Y_{ij} = I\{Z_{ij} > 0\}$. Marginally,

$$Y_{ij} \mid \boldsymbol{x}_{ij} \sim \text{Ber}(\mu_{ij}),$$

with $\mu_{ij} = \Phi(\boldsymbol{x}_{ij}\boldsymbol{\beta})$, where $\Phi(\cdot)$ is the cdf of a standard normal distribution. Within-subject dependence structure is modeled through $\boldsymbol{\Psi}(\boldsymbol{\phi})$. The underlying multivariate normal distribution of $\boldsymbol{Z}_i$ induces the correlation and leads to use of the normal cdf for the transformation of $\boldsymbol{x}_{ij}\boldsymbol{\beta}$ to the probability scale.

The multivariate distribution $p(\boldsymbol{y} \mid \boldsymbol{x}, \boldsymbol{\theta})$, and in particular its serial correlation structure, is obtained by integrating $p(\boldsymbol{z} \mid \boldsymbol{x}, \boldsymbol{\theta})$ over a subspace of $\mathbb{R}^J$ defined by the observed binary outcomes $\boldsymbol{Y}_i$; i.e.,

$$p(y_1, \ldots, y_J \mid \boldsymbol{x}, \boldsymbol{\theta}) = \int_{Q(\boldsymbol{y})} p(\boldsymbol{z} \mid \boldsymbol{x}, \boldsymbol{\theta}) \, d\boldsymbol{z},$$

where in this case $p(\boldsymbol{z} \mid \boldsymbol{x}, \boldsymbol{\theta})$ is a J-variate normal density function having mean and correlation parameters in $\boldsymbol{\theta}$, and $Q(\boldsymbol{y}) \subset \mathbb{R}^J$ defines the limits of integration over $\boldsymbol{z}$ such that the jth integral is taken over $(0, \infty)$ when $y_j = 1$ and over $(-\infty, 0]$ if $y_j = 0$. A probit model is used in Section 7.5 to illustrate auxiliary covariates. $\square$

2.6 Semiparametric and nonlinear regression

Our focus thus far has been on models where the regression function is specified as a linear function of covariates. Although it is possible, by transforming covariates, to specify models that are implicitly nonlinear, in practice linear models can be limiting in some settings. Models where the regression function is allowed to be explicitly nonlinear, either through specification of some known function or by allowing a flexible, unspecified function, offer a wide array of alternatives and significantly expand the capability of regression modeling. Our focus in this section is on models that allow nonlinearity in *time trends*, although in principle the models we discuss can be used to characterize nonlinear effects of any covariate.

In many models with nonlinear time trends, an appropriate functional form can be specified directly; the canonical reference for this line of work is the book by Davidian and Giltinan (1998) and a recent update by the same authors (Davidian and Giltinan, 2003). These models are especially useful in settings like pharmacokinetics, viral dynamics, and growth curve modeling, where processes like drug uptake, viral replication, and growth rate can be described by known but potentially complex functions of time (see Wakefield, 1996 and Wu and Wu, 2002, for examples).

The main focus of this section is on semiparametric models, particularly those in which covariate effects involving time can be captured with unspecified smooth functions. Here we divide the model types into two categories: generalized additive models (GAM) and varying coefficient models (VCM). In a GAM, the main effect of time is characterized by a smooth function $f(t)$; for longitudinal or repeated measures data, $f(t)$ can be inferred using random effects models where the smoothness parameter emerges as a variance component (Zhang et al., 1998; Ruppert et al., 2003); Example 2.7 below is used to illustrate. In the VCM, effects of a covariate (say x) are allowed to vary with time via $\beta(t)x$, where the regression coefficient $\beta(t)$ is a smooth function of time (Hoover et al., 1998; Chiang et al., 2001; Zhang, 2004). Example 2.8 below gives a formulation for continuous responses and a scalar covariate.

The review here is confined to likelihood-based methods, but readers should be aware of the sizable literature on moment-based methods for fitting semiparametric regression (see for example Lin and Carroll, 2001; Ruppert et al., 2003 and Wang et al., 2005). The emerging literature connecting *functional data analysis* with longitudinal repeated measures also is highly relevant; see for example Guo (2004), Rice (2004), and Zhao et al. (2004).

2.6.1 Generalized additive models based on regression splines

Consider the cross-sectional data model with scalar covariate

$$Y_i \mid x_i \sim N(\mu_i, \sigma^2),$$

with $\mu_i = f(x_i)$ for an unknown smooth function $f(\cdot)$. A natural cubic smoothing spline estimator of f is the value $\widehat{f}$ that minimizes a penalized sum of squares

$$\sum_{i=1}^{n}\{Y_i - f(X_i)\}^2 + \lambda \int \{f''(x)\}^2 \, dx, \qquad (2.9)$$

where f'' denotes the second derivative of f. For a suitably chosen $n \times n$ matrix $\boldsymbol{K}$, the penalty term can be rewritten as $\lambda \boldsymbol{f}^{\mathrm{T}} \boldsymbol{K} \boldsymbol{f}$, where $\boldsymbol{f} = (f(x_1), \ldots, f(x_n))^{\mathrm{T}}$; this term penalizes overfitting, and in particular the parameter λ governs smoothness of the fitted curve $\widehat{\boldsymbol{f}}$.

Regression splines for longitudinal data can be similarly constructed; when using splines to model time trends, longitudinal data affords the advantage of replicated curves across individuals. This structure can be exploited to permit automated smoothing by formulating models where λ is a variance component (Zhang et al., 1998; Ruppert et al., 2003), and permits fitting both GAM and VCM. The basic principle of formulating a semiparametric model as a linear regression model is illustrated in the context of penalized regression splines (called p-splines).

Example 2.7. *A regression spline for continuous longitudinal data.*
This example is designed to show that an unspecified time trend $f(t)$ can be
estimated within a linear model formulation. Assume $\boldsymbol{Y}_i = (Y_{i1}, \ldots, Y_{i,J_i})^{\mathrm{T}}$
is measured at times $\boldsymbol{t}_i = (t_{i1}, \ldots, t_{i,J_i})^{\mathrm{T}}$. The model is

$$\boldsymbol{Y}_i \mid \boldsymbol{t}_i \sim N(\boldsymbol{\mu}_i, \boldsymbol{\Sigma}_i),$$

with

$$\mu_{ij} = f(t_{ij}) \tag{2.10}$$

$$\sigma_{ijk}(\boldsymbol{\phi}) = \begin{cases} \sigma^2 & j = l \\ \sigma^2 \exp(-\phi|t_{ij} - t_{il}|) & j \neq l \end{cases}. \tag{2.11}$$

A *p-spline model* represents $f(t)$ in terms of an R-dimensional basis $\boldsymbol{B}(t) = \{B_1(t), \ldots, B_R(t)\}^{\mathrm{T}}$, where R is less than or equal to the number of unique
observation times. A common choice is based on pth-degree polynomial splines
with pre-selected knot points $s_1, \ldots, s_k$ along the time axis (Berry et al., 2002),

$$\boldsymbol{B}(t) = (1, t, t^2, \ldots, t^q, (t - s_1)_+^q, (t - s_2)_+^q, \ldots, (t - s_k)_+^q)^{\mathrm{T}}. \tag{2.12}$$

Here, $a_+ = aI\{a > 0\}$. This basis has dimension $R = q + k + 1$; the knots
can be chosen in a variety of ways; Berry et al. (2002) recommend using
percentiles of the unique observation times. In practice, a linear ($q = 1$) or
quadratic ($q = 2$) basis is often sufficient.

The linear model formulation replaces $f(t_{ij})$ in (2.10) with $\boldsymbol{B}(t_{ij})^{\mathrm{T}}\boldsymbol{\beta}$, where
$\boldsymbol{\beta}$ is $R \times 1$. Inference is then based on a likelihood having a penalty term
$\lambda\boldsymbol{\beta}^{\mathrm{T}}\boldsymbol{D}\boldsymbol{\beta}$, where $\boldsymbol{D}$ is a fixed $R \times R$ penalty matrix. The choice of basis function
will motivate the form of $\boldsymbol{D}$. If (2.12) is used, coefficients of the terms involving
knot points correspond to jumps in the qth derivative of $f(t)$ at each knot
point. To ensure smoothness, one could penalize the sum of squared jumps
at the knot points via a simple form of $\boldsymbol{D}$ that places 1's along the diagonal
corresponding to $\beta_{q+2}, \ldots, \beta_R$,

$$\boldsymbol{D} = \begin{bmatrix} \boldsymbol{0}_{(q+1)\times(q+1)} & \boldsymbol{0}_{(q+1\times k)} \\ & \\ \boldsymbol{0}_{k\times(q+1)} & \tau^2\boldsymbol{I}_k \end{bmatrix}$$

(Ruppert and Carroll, 2000). Given this form of $\boldsymbol{D}$, we partition the regression
parameters as

$$\boldsymbol{\beta} = (\alpha_0, \alpha_1, \ldots, \alpha_q, a_1, \ldots, a_k)^{\mathrm{T}} = (\boldsymbol{\alpha}^{\mathrm{T}}, \boldsymbol{a}^{\mathrm{T}})^{\mathrm{T}}.$$

The penalty can then be formulated as a (random effects) distribution on
$\boldsymbol{a} = (a_1, \ldots, a_k)^{\mathrm{T}}$,

$$\boldsymbol{a} \mid \tau \sim N(\boldsymbol{0}, \tau^2\boldsymbol{I}_k), \tag{2.13}$$

and fit using mixed models software. The original penalty parameter λ is
related to the variance component σ^2 in (2.11) and τ^2 in (2.13) via $\lambda = \sigma^2/\tau^2$.

In Section 7.6, we use this model to analyze longitudinal CD4 data from the
HER Study.   □

This approach can be used to estimate $f(t)$ via natural cubic smoothing
splines (Zhang et al., 1998), and applies to discrete data in a GLM framework;
see Lin and Zhang (1999) for a comprehensive account.

2.6.2 Varying coefficient models

Another type of semiparametric model for longitudinal data, similar to the
GAM, is the varying coefficient model. The model allows covariate effects to
change with time by allowing the coefficient $\beta(t)$ to be an unspecified smooth
function of time, a particularly useful feature in settings where covariate effects
are not time-constant and at the same time do not have an obvious functional
form. Although VCMs have a different specification than GAMs, the method
of estimation is similar. Recent work by Zhang (2004) shows that $\beta(t)$ can be
estimated by smoothing splines using a modified linear model similar to the
one described in Example 2.7.

Example 2.8. *Varying coefficient model for continuous responses.*
Let x_i represent a scalar baseline covariate, and let t denote time elapsed from
baseline. The following VCM allows the effect of x to vary over time through
an unspecified function for the regression coefficient. At the first level, we
assume

$$Y_{ij} \mid x_i, t_{ij} \sim N(\mu_{ij}, \sigma^2(t_{ij})),$$

where the correlation function from Example 2.7 applies. One can specify the
following VCM that allows an overall linear time trend, but an effect of x that
varies with time:

$$\mu_{ij} = \beta_0 + \beta_1 t_{ij} + \beta_2(t_{ij})x_i.$$

Here, β_0 and β_1 are unknown parameters, and $\beta_2(t)$ is a smooth function of t.
Hence

$$E(Y_{ij} \mid x_i, t = t^*) = \beta_0 + \beta_1 t^* + \beta_2(t^*)x$$

is the mean at some fixed time $t = t^*$. The model can be extended to allow
the intercept to depend on time via $\beta_0(t)$ (making the VCM a generalization
of the GAMs described above) and to models for discrete data (Zhang, 2004).

In Section 10.4, we use these models to handle continuous time dropout in
the Pediatric AIDS trial.   □

2.7 Interpreting covariate effects

Proper interpretation of covariate effects in models for longitudinal data requires careful attention to model specification and assumptions about covari-

ate processes. The use of time-varying covariates in particular allows investigators to study both within- and between-individual covariate effects, but incorporation of these covariates and interpretation of their effects must be done with care.

This section focuses on several aspects of model specification and interpretation, including assumptions about covariates, differentiation between longitudinal and cross-sectional effects of a time-varying covariate, and differentiation between marginal and conditional effects in random effects models. These topics are covered in considerable detail elsewhere; readers are referred to Chapters 1 and 12 of Diggle et al. (2002), Chapter 15 of Fitzmaurice et al. (2004), and Chapter 1 of Laird (2004).

2.7.1 Assumptions regarding time-varying covariates

Throughout the book, we assume covariates are exogenous and measured without error. Exogeneity is a standard assumption for nearly all regression models, and in its simplest form asserts independence between errors and covariates. In longitudinal models, exogeneity must be met for both time-constant and time-varying covariates. Time-varying covariates can be either deterministic or stochastic functions of time. Deterministic functions of time include calendar date or participant age; if they are exogenous at baseline, then their time-varying counterparts also will be.

Stochastic covariate processes can be more problematic and must be treated with caution. A stochastic time-varying covariate will usually be exogenous if it is completely 'external' to the measurement process (e.g., daily air pollution levels measured at the city level in a study where response is daily number of asthma events measured at the individual level). However, stochastic covariates measured at the same level as the response will tend to be endogenous because they may be systematically predicted by prior responses. In cohort studies of HIV infection, suppose that for measurement occasion j, X_j is exposure to antiviral treatment and Y_j is CD4 cell count. It is highly probable that Y_j is predictive of X_{j+1} because variations in CD4 count are clinically important to initiation or adjustment of treatment.

Consider another example where the objective is to characterize the effect of physical functioning (X_j) on depression (Y_j). It is easy to imagine that changes in physical functioning will be associated with subsequent changes in depression (i.e., X_j predicts Y_{j+1}), but the converse also may be true: changes in depression may also lead to changes in physical functioning (i.e., Y_j or some function of $Y_1, \ldots, Y_j$ predicts X_{j+1}).

Standard regression models are not usually suitable for characterizing the effects of time-varying stochastic covariates unless those covariates are *exogenous*. Effects of *endogenous* covariates are best summarized in terms of structural models. These require the use of instrumental variables or propensity score methods (Angrist et al., 1996; Hernán et al., 2001; Wooldridge, 2001; van der Laan and Robins, 2003). In other circumstances, it is more appropriate to treat $\{Y_j, X_j\}$ as a joint process and use a model to characterize its joint evolution and association structure. See Jones et al. (2003) and Liang et al. (2003) for applications in HIV.

Measurement error is another important concern and can introduce appreciable bias. The literature on this topic is deep and wide-ranging; see Fuller (1987) and Carroll et al. (1995) for a full treatment.

2.7.2 Longitudinal vs. cross-sectional effects

Longitudinal data affords researchers the opportunity to study both within- and between-subject effects for exogenous time-varying covariates (Ware, 1985; Diggle et al., 2002). In aging research, for example, there is interest in the average effect of age on variables like physical functioning and depression. Suppose that Y_{ij} represents level of depression and X_{ij} is age at assessment j (age is exogenous). The *cross-sectional* effect is a between-subject effect, and represents the difference in mean depression score between cohorts of individuals that differ by one unit in age. In the simplest formulation of the model, this difference is assumed to be time-constant. The *longitudinal effect* is a within-subject measure of association, and represents the average *change* in depression score per unit *change* in age, within an individual.

Clearly the longitudinal and cross-sectional effects measure different types of association, and frequently take different values. The linear model with identity link provides a convenient illustration The ideas here apply more generally to other models where the covariate is exogenous (see Laird, 2004, Section 1.4, for a general treatment). For the model

$$E(Y_{ij} \mid x_{ij}) = \beta_0 + \beta_1 x_{i1} + \beta_2(x_{ij} - x_{i1}),$$

β_1 measures cross-sectional (between-subject) association and β_2 captures longitudinal (within-subject) association. The choice of centering on x_{i1} is somewhat arbitrary; in this case, the cross-sectional effect is

$$\beta_1 = E(Y_{i1} \mid x_{i1} = x + 1) - E(Y_{i1} \mid x_{i1} = x),$$

and the longitudinal effect is

$$\beta_2 = E(Y_{ij} - Y_{ij'} \mid x_{ij} - x_{ij'} = 1).$$

More generally, we can write $E(Y_{ij} - Y_{ij'} \mid x_{ij}, x_{ij'}) = \beta_2(x_{ij} - x_{ij'})$. Besides the importance to interpretation of covariate effects, this decomposition is central to design of longitudinal studies (Diggle et al., 2002).

2.7.3 Marginal vs. conditional effects

As indicated earlier in this chapter, the regression coefficients have distinctly different interpretations in marginal and conditional models. The marginal mean is defined as the mean response, averaged over the population from which the sample is drawn. Regression coefficients in directly specified marginal models capture the effects of covariates between groups of individuals defined by differences in covariate values; these effects are sometimes referred to as 'population-averaged' effects (Zeger and Liang, 1992).

By contrast, regression coefficients in conditionally specified random effects models capture the effect of a covariate at the level of aggregation in the random effect. For example, in model (2.5), random effects capture individual-level variation; the regression coefficients β therefore represent covariate effects at the individual level. When the level of aggregation represented by random effects is an individual or subject, the conditional regression parameters are sometimes referred to as 'subject-specific' effects.

An important consideration in the interpretation of regression parameters from models of repeated measures is whether the covariate in question varies only between individuals, or whether it also varies within individuals (Fitzmaurice et al., 2004). Two examples serve to illustrate the point. Consider first the longitudinal studies of smoking cessation described in Section 1.4, where the binary response is cessation status Y_{ij} at measurement occasions $j = 1, \ldots, J$. Further suppose that each individual is enrolled in one of two treatment programs, denoted by $X_i \in \{0, 1\}$. A random effects logistic regression can be used to characterize a conditional or subject-specific effect of X,

$$\text{logit}\{E(Y_{ij} \mid x_i, b_i, \beta)\} = b_i + \beta_0 + \beta_1 x_i,$$

where (say) $b_i \sim N(0, \sigma^2)$. The random effect b_i hypothetically captures all observable variation in the mean of Y_{ij} that is not explained by x, and in that sense two individuals with the same value of b_i can be thought of as perfectly matched with respect to unobserved individual-level covariate effects. Hence the coefficient β_1 can be interpreted as the effect of treatment on smoking cessation for two individuals with the same unobserved covariate profile (i.e., the same value of b_i), or for two individuals whose predictors of smoking cessation, other than treatment, are exactly the same. Here, as in all random effects models, it is being assumed that unobserved characteristics can be captured with a scalar value b_i that follows a normal distribution. Both are strong assumptions, and the 'matching' interpretation hinges critically on correct specification of the model.

When X is a covariate that could potentially be manipulated (such as treatment or environmental exposure), and if the study design is appropriate, then β_1 sometimes can be interpreted as the causal effect of X on Y because it is the conditional effect of X, given a covariate measure that is technically unique to each individual and captures information from omitted individual-specific covariates. If X cannot be manipulated, for example as with gender or race at the individual level, then causal interpretations are not appropriate because they are not well-defined and information on within-subject variation is not available. See Chapter 13 in Fitzmaurice et al. (2004) for a discussion focused on appropriate use of random effects models for causal inference.

2.8 Further reading

Marginally specified multilevel models

Miglioretti and Heagerty (2004) specify marginalized models with Markov dependence and random effects in the setting of breast cancer screening. Schildcrout and Heagerty (2007) also propose marginalized models with Markov dependence and random effects but in settings of long series of binary responses. Ilk and Daniels (2007) specify a three level marginalized model to address serial correlation and multivariate dependence for multivariate longitudinal binary data.

Functional data analysis

Functional data analysis has important applications in longitudinal settings. For recent developments, see Zhao et al. (2004), Yao et al. (2005a) and Yao et al. (2005b). For a recent Bayesian treatment, see Rodriguez et al. (2007).

CHAPTER 3

Methods of Bayesian Inference

3.1 Overview

The Bayesian approach to modeling provides a natural framework for making inferences from incomplete data. Bayesian inference is characterized by specifying a model, then specifying prior distributions for the parameters of the models, and then lastly updating the prior information on the parameters using the model and the data to obtain the posterior distribution of the parameters. Analyses with missing data involve assumptions that cannot be verified; assumptions about the missing data (and our uncertainty about them) can be made explicit through prior distributions.

This chapter reviews some key ideas in Bayesian statistics for complete data, including specifying prior distributions, computation of the posterior distribution, and strategies for assessing model fit. Chapters 6, 8, and 9 will discuss extensions and modifications of these concepts in specific missing data settings. Many procedures discussed in this chapter can be implemented in WinBUGS software. We indicate those settings where special programming may be needed.

Many of our examples are based on models described in Chapter 2. In the following, we will mostly suppress conditioning on the model covariates x.

3.2 Likelihood and posterior distribution

3.2.1 Likelihood

Let $\boldsymbol{Y}_i = (Y_{i1}, \ldots, Y_{i,J_i})^{\mathrm{T}}$ denote observed response data for individual i, where $i = 1, \ldots, n$. Independence is assumed across subjects. For a parameter $\boldsymbol{\theta}$, the likelihood $L(\boldsymbol{\theta} \mid \boldsymbol{y})$ is any function proportional in $\boldsymbol{\theta}$ to the joint density evaluated at the observed sample,

$$
\begin{aligned}
L(\boldsymbol{\theta} \mid \boldsymbol{y}) \quad &\propto \quad p(\boldsymbol{y}_1, \ldots, \boldsymbol{y}_n \mid \boldsymbol{\theta}) \\
&= \quad \prod_{i=1}^{n} p(\boldsymbol{y}_i \mid \boldsymbol{\theta}).
\end{aligned}
$$

The function $p(\boldsymbol{y}_i \mid \boldsymbol{\theta})$ denotes a probability density (or mass) function for the J_i observations on subject i. The likelihood summarizes the information the data $\boldsymbol{y}$ has about the parameters $\boldsymbol{\theta}$. We will see in subsequent chapters that in incomplete data settings, the likelihood does *not* always provide information about all the components of $\boldsymbol{\theta}$. Likelihood for some of the models introduced in Chapter 2 are given next.

Example 3.1. *Likelihood for multivariate normal model for temporally aligned observations* (continuation of Example 2.3).
For the model with temporally aligned observations $J_i = J$, the likelihood takes the form

$$L(\boldsymbol{\theta} \mid \boldsymbol{y}) = \prod_{i=1}^{n} |\boldsymbol{\Sigma}(\boldsymbol{\phi})|^{-1/2} \exp\{-\tfrac{1}{2}\boldsymbol{e}_i(\boldsymbol{\beta})^{\mathrm{T}} \boldsymbol{\Sigma}(\boldsymbol{\phi})^{-1} \boldsymbol{e}_i(\boldsymbol{\beta})^{\mathrm{T}}\}, \qquad (3.1)$$

where $\boldsymbol{e}_i(\boldsymbol{\beta}) = \boldsymbol{y}_i - \boldsymbol{x}_i\boldsymbol{\beta}$ and $\boldsymbol{\theta} = (\boldsymbol{\beta}, \boldsymbol{\phi})$. □

Example 3.2. *Likelihood for multivariate normal model for temporally misaligned observations* (continuation of Example 2.4).
Suppose subject i has J_i observations at times $\{t_{i1}, \ldots, t_{i,J_i}\}$. In Example 3.1 each individual was assumed to have J observation times. Let $C(\cdot, \cdot; \boldsymbol{\phi})$ be a covariance function as defined in Example 2.4.

The likelihood contribution for subject i has the same form as (3.1), but with $\boldsymbol{\Sigma}(\boldsymbol{\phi})$ replaced with $\boldsymbol{\Sigma}_i$ having the (k, l) element parameterized via $C(t_{ik}, t_{il}; \boldsymbol{\phi})$. Here, we are assuming the covariance for observations at times t_k and t_l is the same across subjects. □

Example 3.3. *Likelihood for normal random effects model* (continuation of Example 2.1).
The marginal likelihood (after integrating out the random effects $\boldsymbol{b}_i$) is typically used for inference. It is given by

$$
\begin{aligned}
L(\boldsymbol{\theta} \mid \boldsymbol{y}) \;&\propto\; \prod_{i=1}^{n} \int |\boldsymbol{\Sigma}|^{-1/2} \exp\left\{-\tfrac{1}{2}\boldsymbol{e}_i(\boldsymbol{\beta}, \boldsymbol{b}_i)^{\mathrm{T}}(\boldsymbol{\Sigma}_i^b)^{-1}\boldsymbol{e}_i(\boldsymbol{\beta}, \boldsymbol{b}_i)\right\} p(\boldsymbol{b}_i \mid \boldsymbol{\Omega})d\boldsymbol{b}_i \\
&=\; \prod_{i=1}^{n} |\boldsymbol{A}_i|^{-1/2} \exp\left\{-\tfrac{1}{2}\boldsymbol{e}_i(\boldsymbol{\beta})^{\mathrm{T}}\boldsymbol{A}_i^{-1}\boldsymbol{e}_i(\boldsymbol{\beta})\right\},
\end{aligned}
$$

where

$$
\begin{aligned}
\boldsymbol{e}_i(\boldsymbol{\beta}, \boldsymbol{b}_i) &= \boldsymbol{y}_i - \boldsymbol{x}_i\boldsymbol{\beta} - \boldsymbol{w}_i\boldsymbol{b}_i, \\
\boldsymbol{e}_i(\boldsymbol{\beta}) &= \boldsymbol{y}_i - \boldsymbol{x}_i\boldsymbol{\beta}, \\
\boldsymbol{A}_i &= \boldsymbol{\Sigma}_i^b + \boldsymbol{w}_i\boldsymbol{\Omega}\boldsymbol{w}_i^{\mathrm{T}},
\end{aligned}
$$

$p(\boldsymbol{b} \mid \boldsymbol{\Omega})$ is a normal density with mean $\boldsymbol{0}$ and covariance matrix $\boldsymbol{\Omega}$, and $\boldsymbol{\theta} = (\boldsymbol{\beta}, \boldsymbol{\Sigma}, \boldsymbol{\Omega})$. □

Example 3.4. *Likelihood for marginalized transition model of order 1* (continuation of Example 2.5).

The likelihood takes the form

$$L(\boldsymbol{\theta} \mid \boldsymbol{y}) \;\; \propto \;\; \prod_{i=1}^{n} \left\{ \mu_{i1}^{y_{i1}} (1 - \mu_{i1})^{1-y_{i1}} \prod_{j=2}^{J} \phi_{ij}^{y_{ij}} (1 - \phi_{ij})^{1-y_{ij}} \right\},$$

where $\mu_{i1} = \mathrm{P}(Y_{i1} = 1 \mid \boldsymbol{x}_{i1}; \boldsymbol{\beta})$ and $\phi_{ij} = \mathrm{P}(Y_{ij} = 1 | y_{i,j-1}, \boldsymbol{x}_{ij}; \boldsymbol{\beta}, \boldsymbol{\alpha})$ for $j > 1$. The first term, $\mu_{i1}^{y_{i1}} (1 - \mu_{i1})^{1-y_{i1}}$, corresponds to the contribution from each subject at the initial time; the second, $\phi_{ij}^{y_{ij}} (1 - \phi_{ij})^{1-y_{ij}}$, each subject's contribution over the subsequent times. The forms of these probabilities are given in Example 2.5.

The likelihood for non-marginalized transition models takes the same form, but the conditional probabilities are not constrained by the model for the marginal mean (Diggle et al., 2002).   □

Example 3.5. *Likelihood for multivariate probit model* (continuation of Example 2.6).
The likelihood takes the form

$$L(\boldsymbol{\theta} \mid \boldsymbol{y}) \propto \prod_{i=1}^{n} \int_{Q(\boldsymbol{y}_i)} |\boldsymbol{\Psi}|^{-1/2} \exp \left\{ -\tfrac{1}{2} (\boldsymbol{z}_i - \boldsymbol{x}_i \boldsymbol{\beta})^{\mathrm{T}} \boldsymbol{\Psi}^{-1} (\boldsymbol{z}_i - \boldsymbol{x}_i \boldsymbol{\beta}) \right\} d\boldsymbol{z}_i,$$

where $Q(\boldsymbol{y}_i)$ determines the limits of integration of $\boldsymbol{z}_i$, and $\boldsymbol{\theta} = (\boldsymbol{\beta}, \boldsymbol{\Psi})$.   □

3.2.2 Score function and information matrix

Two important quantities related to the likelihood are the vector valued score function and the information matrix. Define $\ell(\boldsymbol{\theta} \mid \boldsymbol{y}) = \log L(\boldsymbol{\theta} \mid \boldsymbol{y})$. The score function, $\boldsymbol{S}(\boldsymbol{\theta} \mid \boldsymbol{y})$, is defined as

$$\boldsymbol{S}(\boldsymbol{\theta} \mid \boldsymbol{y}) = \frac{\partial \ell(\boldsymbol{\theta} \mid \boldsymbol{y})}{\partial \boldsymbol{\theta}}.$$

The equations given by $\boldsymbol{S}(\boldsymbol{\theta} \mid \boldsymbol{y}) = \boldsymbol{0}$ are called the score equations. The root $\widehat{\boldsymbol{\theta}}$ of the score equations is the maximum likelihood estimator (mle) for $\boldsymbol{\theta}$, and is typically used as the point estimator in frequentist inference.

A second quantity of interest is the information matrix $\boldsymbol{I}(\boldsymbol{\theta})$, defined as

$$\boldsymbol{I}(\boldsymbol{\theta}) = -E_{\boldsymbol{Y}} \left\{ \frac{\partial^2 \ell(\boldsymbol{\theta} \mid \boldsymbol{Y})}{\partial \boldsymbol{\theta} d\boldsymbol{\theta}^{\mathrm{T}}} \right\}.$$

This matrix quantifies the rate of curvature of the log likelihood at $\boldsymbol{\theta}$. In a frequentist setting, standard errors for $\widehat{\boldsymbol{\theta}}$ can be approximated using the diagonal elements of $\boldsymbol{I}(\widehat{\boldsymbol{\theta}})^{-1}$; i.e., the information evaluated at the mle. When $\boldsymbol{I}(\boldsymbol{\theta})$ does not take a closed form, as is often the case, the observed information

$$\boldsymbol{I}(\boldsymbol{\theta} \mid \boldsymbol{y}) = -\frac{\partial^2 \ell(\boldsymbol{\theta} \mid \boldsymbol{y})}{\partial \boldsymbol{\theta} d\boldsymbol{\theta}^{\mathrm{T}}} = -\frac{\partial \boldsymbol{S}(\boldsymbol{\theta} \mid \boldsymbol{y})}{\partial \boldsymbol{\theta}}$$

can be used.

3.2.3 The posterior distribution

Bayesian inference involves updating the prior distribution $p(\boldsymbol{\theta})$ using the information in the data as quantified by the likelihood $L(\boldsymbol{\theta} \mid \boldsymbol{y})$.

Definition 3.1. *Posterior distribution.*
Given a prior $p(\boldsymbol{\theta})$ and a likelihood $L(\boldsymbol{\theta} \mid \boldsymbol{y})$, the posterior distribution of $\boldsymbol{\theta}$ is

$$p(\boldsymbol{\theta} \mid \boldsymbol{y}) = \frac{L(\boldsymbol{\theta} \mid \boldsymbol{y})p(\boldsymbol{\theta})}{\int L(\boldsymbol{\theta} \mid \boldsymbol{y})p(\boldsymbol{\theta})d\boldsymbol{\theta}}. \tag{3.2}$$

The denominator is a normalizing constant, so as a function of $\boldsymbol{\theta}$, we have

$$p(\boldsymbol{\theta} \mid \boldsymbol{y}) \propto L(\boldsymbol{\theta} \mid \boldsymbol{y})p(\boldsymbol{\theta}).$$

$\square$

If $p(\boldsymbol{\theta})$ is proportional to a constant, then the posterior mode of $\boldsymbol{\theta}$ will be equivalent to the maximum likelihood estimator. However the posterior mode is not invariant to a re-parameterization of $\boldsymbol{\theta}$ because the prior distribution is not invariant to re-parameterizations.

In most cases, expressing the posterior distribution in closed form is complicated because the normalizing constant for the posterior is frequently an intractable high-dimensional integral (more on this later). The most common approach to inference using the posterior is to obtain a sample from the posterior distribution using techniques that do not require explicit evaluation of the denominator. Alternatively, the posterior distribution of $\boldsymbol{\theta}$ can be approximated with analogs of the score and information based on the likelihood *and* the prior (Tierney and Kadane, 1986); however, such approximations will prove more useful in implementing some of the computational approaches to sample from the posterior distribution (see Section 3.4.2). The posterior can also be viewed in a frequentist context, as we see below.

Definition 3.2. *Posterior consistency.*
We call the posterior distribution *consistent* if it approaches a point mass at the true value of the parameters as the sample size goes to infinity. $\square$

Posterior Summaries

The posterior is usually summarized with a few carefully chosen quantities. For point estimates, the posterior mean and median are common choices. Uncertainty is represented either by the standard deviation of the posterior (counterpart to the standard error in frequentist analysis) or by forming a credible interval based on the percentiles of the posterior distribution. For example, 95% credible intervals of a scalar parameter often are constructed using the 2.5th and 97.5th percentiles of the posterior distribution.

The choice of summary measures for the posterior distribution can be derived formally under a decision theoretic framework by minimizing a loss function, $\mathscr{L}(\theta, a)$, which is a function of the parameter θ and its estimator a. From a Bayesian decision-theoretic perspective, a is chosen to minimize the expectation of this loss with respect to the posterior distribution $p(\theta \mid \boldsymbol{y})$, and is called a Bayes estimator. For example, under $\mathscr{L}(\theta, a) = (\theta - a)^2$ (squared error loss), the Bayes estimator is the posterior mean. We refer the reader to Berger (1985) or Carlin and Louis (2000) for other loss functions and additional details.

Hypothesis testing can be conducted formally using Bayes Factors (Kass and Raftery, 1995) or more informally by examining relevant features of the posterior distribution. As an example of the latter, a point null hypothesis could be tested by examining whether the credible interval covers the null value. In addition, for a one-sided test, we could quantify the strength of evidence by computing, e.g., the posterior probability that the parameter is greater than the null value. We will illustrate these approaches in the data examples in Chapter 4. Hypothesis testing as it relates to comparing the fit of different models is discussed further in Section 3.5.

3.3 Prior Distributions

Prior distributions quantify *a priori* knowledge about $\boldsymbol{\theta}$. Prior information can be represented through a probability distribution, usually called an informative prior. In the absence of 'prior' information, vague prior beliefs can be reflected using diffuse probability distributions (often called default or noninformative priors).

We begin our review of priors by introducing conjugate and related priors, which are the most commonly used priors in Bayesian modeling. After reviewing common priors, we give recommendations on choosing priors for parameters in longitudinal models.

3.3.1 Conjugate priors

Conjugate priors are constructed such that the prior and posterior distribution are from the same family of distributions (e.g., both normal distributions). As a result, the posterior distribution can be expressed in closed form.

Example 3.6. *Conjugate priors for a normal linear regression model.* Consider a normal linear regression model

$$Y_i \mid \boldsymbol{x}_i, \boldsymbol{\beta}, \sigma^2 \sim N(\boldsymbol{x}_i\boldsymbol{\beta}, \sigma^2)$$

with σ^2 known. A conjugate prior on the regression coefficients $\boldsymbol{\beta}$ is

$$\boldsymbol{\beta} \mid \boldsymbol{\beta}_0, \boldsymbol{V}_\beta \sim N(\boldsymbol{\beta}_0, \boldsymbol{V}_\beta),$$

where β_0 and $\boldsymbol{V}_\beta$ are fixed. Parameters indexing the prior are called *hyperparameters*. The posterior distribution for $\boldsymbol{\beta}$ is $N(\widetilde{\boldsymbol{\beta}}, \widetilde{\boldsymbol{V}})$, where

$$\widetilde{\boldsymbol{\beta}} = \left(\sum_i \boldsymbol{x}_i \boldsymbol{x}_i^T / \sigma^2 + \boldsymbol{V}_\beta^{-1}\right)^{-1} \left\{\boldsymbol{V}_\beta^{-1}\widehat{\boldsymbol{\beta}} + \left(\sum_i \boldsymbol{x}_i \boldsymbol{x}_i^T / \sigma^2\right)\beta_0\right\}, \quad (3.3)$$

$$\widetilde{\boldsymbol{V}} = \left(\sum_i \boldsymbol{x}_i \boldsymbol{x}_i^T / \sigma^2 + \boldsymbol{V}_\beta^{-1}\right)^{-1}, \quad (3.4)$$

and $\widehat{\boldsymbol{\beta}}$ is the ordinary least squares estimator of $\boldsymbol{\beta}$. □

When conjugate priors cannot be found for the model of interest, priors that are *conditionally* conjugate are often specified to facilitate posterior sampling (see Section 3.4) using *full conditional distributions*.

Definition 3.3. *Full conditional distribution.*
The *full conditional distribution* for a parameter $\theta_j \in \boldsymbol{\theta}$ is the distribution of θ_j, given $(\{\theta_k : k \neq j\}, \boldsymbol{y})$. □

Definition 3.4. *Conditionally conjugate prior.*
A prior for a parameter (or set of parameters) is called *conditionally conjugate* when it is from the same family of distributions as its full conditional distribution. □

Here are some examples of conditionally conjugate priors; for a summary of (conditionally) conjugate priors, see Carlin and Louis (2000).

Example 3.7. *Conditionally conjugate priors for a normal linear regression model* (continuation of Example 3.6).
Consider again the normal linear regression model from Example 3.6, now with σ^2 unknown. A normal prior on $\boldsymbol{\beta}$ is a conjugate prior, *conditional* on σ^2, i.e., $p(\boldsymbol{\beta} \mid \boldsymbol{y}, \sigma^2)$ is normal with mean and variance given in (3.3) and (3.4), respectively. However, $p(\boldsymbol{\beta} \mid \boldsymbol{y})$ is not a normal distribution. If we place a Gamma(a, b) prior on $1/\sigma^2$, the full conditional distribution of $1/\sigma^2$ is

$$\text{Gamma}\left(n/2 + a, \left\{1/b + \sum_i (y_i - \boldsymbol{x}_i\boldsymbol{\beta})^2/2\right\}^{-1}\right).$$

The marginal posterior $p(1/\sigma^2 \mid \boldsymbol{y})$, however, is not a gamma distribution. These two full conditional distributions can be generated from to obtain a sample from the joint posterior distribution of $(\boldsymbol{\beta}, \sigma^2)$. Details follow in Section 3.4. □

Example 3.8. *Conditionally conjugate priors for a multivariate normal model with temporally aligned observations* (continuation of Example 2.4).
Recall the multivariate normal regression model

$$\boldsymbol{Y}_i \mid \boldsymbol{\beta}, \boldsymbol{\Sigma} \sim N(\boldsymbol{x}_i\boldsymbol{\beta}, \boldsymbol{\Sigma}).$$

In this model, the conditionally conjugate priors for $\boldsymbol{\beta}$ and $\boldsymbol{\Sigma}^{-1}$ are multivariate normal and Wishart, respectively (see Appendix). When $\boldsymbol{\Sigma}^{-1}$ is assumed to have a particular structure, conjugate priors of known form are often not available except in special cases.   □

Example 3.9. *Conditionally conjugate priors for normal random effects model* (continuation of Example 2.1).
Recall the normal random effects model,

$$\boldsymbol{Y}_i \mid \boldsymbol{b}_i, \boldsymbol{\beta}, \sigma^2 \;\sim\; N(\boldsymbol{x}_i\boldsymbol{\beta} + \boldsymbol{w}_i\boldsymbol{b}_i, \boldsymbol{\Sigma}_i^b)$$
$$\boldsymbol{b}_i \mid \boldsymbol{\Omega} \;\sim\; N(\boldsymbol{0}, \boldsymbol{\Omega}).$$

Here, we assume $\boldsymbol{\Sigma}_i^b = \sigma^2 \boldsymbol{I}$. The conditionally conjugate priors for this model are the same as Example 3.8 supplemented by a Wishart prior on $\boldsymbol{\Omega}^{-1}$.   □

Example 3.10. *Conditionally conjugate priors for Bayesian penalized splines* (continuation of Example 2.7).
Recall that the model is

$$\boldsymbol{Y}_i \mid \boldsymbol{t}_i \sim N(\boldsymbol{\mu}_i, \boldsymbol{\Sigma}_i), \tag{3.5}$$

with

$$\mu_{ij} = f(t_{ij}) \tag{3.6}$$
$$\sigma_{ijk}(\boldsymbol{\phi}) = \begin{cases} \sigma^2 & j = k \\ \sigma^2 \exp(-\phi|t_{ij} - t_{ik}|) & j \neq k \end{cases}.$$

The function $f(t)$ is represented in terms of the spline basis given in (2.12),

$$\boldsymbol{B}(t) = (1, t, \dots, t^q, (t - s_1)_+^q, \dots, (t - s_k)_+^q)^{\mathrm{T}}$$

having dimension $R = (q + 1) + k$ with $a_+ = aI\{a > 0\}$. The penalty has the form $\lambda \boldsymbol{\beta}^{\mathrm{T}} \boldsymbol{D} \boldsymbol{\beta}$, where $\boldsymbol{D}$ is a diagonal matrix having the first $q + 1$ diagonal elements equal to zero and the last k equal to one (Berry, Carroll, and Ruppert, 2002).

This penalty can be reformulated as a (conditionally conjugate) prior distribution on $\boldsymbol{\beta}$. To do this, we again partition $\boldsymbol{\beta}$ into $(\boldsymbol{\alpha}, \boldsymbol{a})$, where $\boldsymbol{\alpha}$ is $(q+1)$-dimensional coefficient vector corresponding to the first $(q + 1)$ components of the basis and $\boldsymbol{a}$ is k-dimensional coefficient vector corresponding to the last k components. The penalty described above corresponds to conditionally conjugate prior on $\boldsymbol{\alpha}$ (a flat prior on $\mathbb{R}^{p+1}$) and a normal prior on $\boldsymbol{a}$,

$$\boldsymbol{a} \sim N(\boldsymbol{0}, \tau^2 \boldsymbol{I}_k).$$

The smoothing parameter λ is the ratio of the two variances $\lambda = \sigma^2/\tau^2$. Conditionally conjugate priors for the inverse of the variance components (σ^2, τ^2) are Gamma priors (Craineceau, Ruppert, and Wand, 2005; Berry et al., 2002).   □

3.3.2 Noninformative priors

In many models, some components of $\boldsymbol{\theta}$ are not of primary interest (nuisance parameters) and/or may not have prior information available. Various approaches are available to express 'ignorance' via an appropriately specified prior distribution. These 'default' or noninformative priors are often constructed to satisfy some reasonable criterion such as invariance, admissibility, minimaxity, and/or favorable frequentist properties (Berger and Bernardo, 1992; Mukerjee and Ghosh, 1997). For example, the Jeffreys' prior $p(\boldsymbol{\theta}) \propto |\boldsymbol{I}(\boldsymbol{\theta})|^{1/2}$ is invariant to the parameterization of $\boldsymbol{\theta}$ (Jeffreys, 1961).

Example 3.11. *Jeffreys' prior for a normal linear regression model* (continuation of Example 3.6).
Again, consider the regression model in Example 3.6 with σ^2 known. Jeffreys' prior for the regression coefficients $\boldsymbol{\beta}$ is

$$p(\boldsymbol{\beta}) \propto 1.$$

The posterior for $\boldsymbol{\beta}$ is thus proportional to the likelihood; it is a normal distribution $N(\widetilde{\boldsymbol{\beta}}, \widetilde{\boldsymbol{V}})$, where

$$
\begin{aligned}
\widetilde{\boldsymbol{\beta}} &= \widehat{\boldsymbol{\beta}} \\
\widetilde{\boldsymbol{V}} &= \sigma^2 \left(\textstyle\sum_i \boldsymbol{x}_i \boldsymbol{x}_i^{\mathrm{T}} \right)^{-1}
\end{aligned}
$$

and $\widehat{\boldsymbol{\beta}}$ is the ordinary least squares estimator of $\boldsymbol{\beta}$.   □

Unfortunately, Jeffreys' prior tends to have suboptimal properties in higher dimensions (Berger and Bernardo, 1992). Since Jeffreys' work, there have been considerable developments in this area. The 'reference' priors of Berger and Bernardo (1992) overcome some of the flaws in Jeffreys' prior in multidimensions.

In one-dimensional problems with a continuous parameter, where Jeffreys' prior typically works well, the reference prior is the same as Jeffreys' prior (Bernardo, 2006). However, in multi-dimensional settings, the reference prior is superior (in terms of frequentist properties). Unfortunately, it is sometimes unclear how to 'order' the parameters to construct their joint distribution (different priors often result from different ordering of the parameters) and they can be difficult to derive (Bernardo, 2006). A variety of other default priors have been proposed in recent years, including the unit information priors of Kass and Wasserman (1995), probability matching priors (Datta and Ghosh, 1995; Mukerjee and Ghosh, 1997), and intrinsic priors (Berger and Pericchi, 1996). For a good review of non-informative priors, see Kass and Wasserman (1996) and Bernardo (2006).

Improper priors

Many vague or noninformative priors are not proper density functions; that is, $\int p(\boldsymbol{\theta})d\boldsymbol{\theta} = \infty$. In Jeffreys' prior from Example 3.11, $\int d\boldsymbol{\beta} = \infty$. When using improper priors, there is no guarantee that the posterior distribution will be proper in the sense that

$$\int L(\boldsymbol{\theta} \mid \boldsymbol{y}) \, p(\boldsymbol{\theta}) \, d\boldsymbol{\theta} < \infty.$$

The data analyst needs either to show that the posterior is proper or replace the improper priors with proper or 'just proper' priors.

To illustrate a 'just proper' prior, again consider the example of the prior on the regression coefficients in the normal linear regression model. Jeffreys' prior (and the reference prior) is $p(\boldsymbol{\beta}) \propto 1$; this can be replaced by $p(\boldsymbol{\beta}) = N(0, v\boldsymbol{I})$ where $v \gg 0$ and fixed. This prior is proper but as $v \to \infty$, it approaches the (improper) Jeffreys' prior. It also provides intuition behind the conjugacy of Jeffreys' prior seen in Example 3.11. In general, for means and regression coefficients, just proper normal priors are a convenient choice when little prior information is available. For WinBUGS, improper priors are not permitted, so just proper alternatives must be used.

Default priors for variances

Priors for variance (and covariances) have increased importance for incomplete longitudinal data. With complete data, variances and covariances are often important primarily in terms of efficiency, but with incomplete data, they have wider impact and when specified without care, can result in inconsistent estimation of mean parameters (see Chapter 6).

Common default choices for variance parameters in a normal model include $p(\sigma^2) \propto 1/\sigma^2$ (Jeffreys'), $p(\sigma^2) \propto I\{\sigma^2 > 0\}$ (flat), and $p(\sigma^2) \propto 1/(c + \sigma^2)^2$ (uniform shrinkage), where c is fixed (and corresponds to the prior median). The uniform shrinkage prior is often used when the variance σ^2 is at the 2nd level of a two-level model (see, e.g., the random effects variance); c can be chosen as a prior guess at the value of the first level variance. It is sensible when there is prior belief that the variance component might actually be zero, but little other prior information is available (Christensen and Morris, 1997; Daniels, 1999).

Jeffreys' prior tends to pull the variance toward zero and the flat prior pulls the variance away from zero. For a more detailed explanation and exploration of all these priors, including their frequentist properties, and further discussion of the choice of the constant c for the uniform shrinkage prior, see Daniels (1999).

The 'just proper' prior most frequently used in the literature is a Gamma prior (on $1/\sigma^2$) with parameters $(10^3, 10^{-3})$. Gelman (2006) recommends against using this prior and instead recommends bounded uniform or folded non-central t (or normal) priors on σ (not σ^2). All the proper priors for variances described above can be used in WinBUGS.

Priors for a covariance/correlation matrix

For the p–dimensional covariance matrix for a multivariate normal model (Example 2.3), the two most common default choices are Jeffreys' prior

$$p(\mathbf{\Sigma}) \propto |\mathbf{\Sigma}|^{-(p+1)/2}$$

and the 'just proper' Wishart distribution on $\mathbf{\Sigma}^{-1}$, which has degrees of freedom ν equal to p. The latter choice avoids concerns about the propriety of the posterior that arise when using improper priors (and can be used in Win-BUGS), but requires providing a value for the scale matrix $\mathbf{\Sigma}_0$. This choice is difficult and, as a default, is often chosen to be a diagonal matrix with marginal variances chosen to be roughly consistent with the variation in the data or chosen based on be the maximum likelihood estimator of $\mathbf{\Sigma}$. Unfortunately, this prior can only be made noninformative to a limited extent; the minimum value for ν is p, which can be thought of as a prior sample size. So, if p is large and the sample size is small, this prior may be more informative than desired (Daniels and Kass, 1999). Other default priors proposed in the literature include the reference prior of Yang and Berger (1994) which, though it has good frequentist properties, is improper and lacks the conditional conjugacy property of the Wishart.

For two-level models with a normal distribution on the random effects (for example, the models in Examples 2.1 and 2.2), common choices are the just proper Wishart prior (again) and the flat prior, $p(\mathbf{\Sigma}) \propto 1$ (Everson and Morris, 2000). Other priors proposed recently include a modification to the reference prior of Yang and Berger (Berger, Strawderman, and Tang, 2005) and the approximate uniform shrinkage prior (Natarajan and Kass, 2000).

Advantages of Jeffreys' and the flat prior as default choices are that they are both conditionally conjugate and do not require the choice of a scale matrix, $\mathbf{\Sigma}_0$. However, they are both improper and not available in WinBUGS.

Priors for the parameters in a structured covariance matrix, e.g., a first-order autoregressive covariance structure, have been examinined in Monahan (1983) and Ghosh and Heo (2003); specifying structure can be viewed as a very strong prior restricting $\mathbf{\Sigma}$ to be within a certain class of covariance models (Daniels, 2005); see Further Reading for more details.

Default priors for p–dimensional correlation matrices, which are needed for multivariate probit models (Example 2.6), have been proposed by Barnard et al. (2000), who consider either a joint prior that induces marginal uniform priors on the individual correlations in the correlation matrix or a uniform prior on the space of correlation matrices. Neither is (conditionally) conjugate, but both are proper. The latter is a uniform prior on the compact subspace of the $p(p-1)/2$ dimensional cube $[-1,1]^{p(p-1)/2}$ such that the correlation matrix is positive definite.

Recommendations for using default priors in longitudinal models

For regression coefficients $\boldsymbol{\beta}$ and unconstrained means $\boldsymbol{\mu}$, flat improper priors are a good choice if the posterior can be confirmed to be proper. Otherwise, the just proper normal priors should be used. In addition, if the WinBUGS software is being used, the just proper normal priors would be the only choice.

For variances, following Gelman's recommendations (Gelman, 2006), we suggest (truncated) uniform priors or folded normal priors on σ. For the 2nd-level variances (e.g., the variance of the random effects), uniform shrinkage priors can be considered (Daniels, 1999).

For covariance matrices, either the flat prior or the reference priors of Berger and colleagues would be recommended; unfortunately, both are improper and cannot be used in WinBUGS. As a result, the just proper Wishart is the most common choice. For covariance matrices with structure, it is hard to make a general recommendation as it will vary with how the structure is imposed. For a specific family of structured covariance matrices and recommended priors, see Section 6.4.1.

Most of the analyses in later chapters (Chapters 4, 7, 10) are done in WinBUGS, which will necessarily restrict the choice of priors.

3.3.3 Informative priors

Informative priors can be used in situations where prior information exists, whether in the form of historical information, expert opinion, or pilot data. With complete data, using strong priors has the effect of combining information from the prior, together with information from the data and model, to form the posterior. This can be illustrated very clearly in the normal linear regression model (Example 3.6). The posterior mean of $\boldsymbol{\beta}$,

$$E(\boldsymbol{\beta} \mid \boldsymbol{y}) = \left(\sum_i \boldsymbol{x}_i \boldsymbol{x}_i^{\mathrm{T}}/\sigma^2 + \boldsymbol{V}_\beta^{-1} \right)^{-1} \left\{ \boldsymbol{V}_\beta^{-1} \widehat{\boldsymbol{\beta}} + \left(\sum_i \boldsymbol{x}_i \boldsymbol{x}_i^{\mathrm{T}}/\sigma^2 \right) \boldsymbol{\beta}_0 \right\},$$

is a weighted average of the ordinary least squares estimator $\widehat{\boldsymbol{\beta}}$ (formed from the data and the model) and the prior mean $\boldsymbol{\beta}_0$, with weights depending on the data and the prior variance $\boldsymbol{V}_\beta$. The strength of the prior is (partially) determined by the magnitude of $\boldsymbol{V}_\beta$. The informativeness of this prior can be altered by appropriately modifying $\boldsymbol{V}_\beta$.

With incomplete data, the role of the prior takes on considerably added importance. When data are missing, information about some components of $\boldsymbol{\theta}$ often derives *only* from the prior. This places a premium on methods for obtaining and 'modeling' prior information.

Constructing informative priors that formalize expert opinion is difficult.

A common approach is to specify a particular family of distributions (often conditionally conjugate) and then to elicit information from the investigator to fill in values for the hyperparameters. A key component is determining an easy to comprehend scale to 'accurately' elicit information from an expert on the hyperparameters (or functions of them) (Chaloner, 1996). As an example, in linear regression models, Kadane et al. (1980), and later Laud and Ibrahim (1996), proposed constructing priors on regression coefficients based on elicited information about future data ($\boldsymbol{Y}$'s), and then deriving the induced hyperparameters on the regression parameters of interest; they viewed data predictions as a more understandable metric for the experts than the regression coefficients themselves.

The power priors of Ibrahim and Chen (2000) are a class of informative priors that provide a way to introduce prior information based on historical data. A prior is specified that is proportional to the likelihood of the historical data with a subjective choice of the power on this likelihood, i.e., how much prior 'weight' should be placed on the historical data.

We revisit informative priors in the context of sensitivity analysis in Chapter 9 and illustrate an elicited prior in Section 10.3.

3.3.4 Identifiability and incomplete data

In incomplete data settings, some components of $\boldsymbol{\theta}$ are well-identified by the data and others are not.

Definition 3.5. *Nonidentified parameter.*
A parameter θ_1 is called *nonidentified* when

$$p(\theta_1 \mid y) \equiv p(\theta_1)$$

for any choice of prior $p(\theta_1)$.   □

Note that for the component θ_1 of $\boldsymbol{\theta}$ to be nonidentified (as given by the above definition), the prior for $\boldsymbol{\theta}$ typically must be able to be factored as

$$p(\boldsymbol{\theta}) = p(\theta_1)p(\boldsymbol{\theta}_{-1}),$$

where $\boldsymbol{\theta}_{-1}$ is $\boldsymbol{\theta}$ with θ_1 removed. For a weakly identified parameter and for dependent priors, this equivalence will hold approximately; for a very careful discussion of identifiability, see Dawid (1979). We give a simple example next.

Example 3.12. *Nonidentifiability in a mixture of bivariate normals.*
Consider the following normal mixture model,

$$
\begin{aligned}
(Y_1, Y_2)^{\mathrm{T}} \mid R = 1 &\sim N(\boldsymbol{\mu}^{(1)}, \boldsymbol{\Sigma}) \\
(Y_1, Y_2)^{\mathrm{T}} \mid R = 0 &\sim N(\boldsymbol{\mu}^{(0)}, \boldsymbol{\Sigma}) \\
R \mid \phi &\sim \mathrm{Ber}(\phi)
\end{aligned}
$$

where in the first component of the mixture ($R = 1$), we observe $(Y_1, Y_2)^{\mathrm{T}}$; and for the second component ($R = 0$), we only observe Y_1. Suppose we specify independent priors on $\boldsymbol{\mu}^{(r)}$, $\boldsymbol{\Sigma}$, and ϕ. Let $\boldsymbol{\Sigma} = \{\sigma_{jk}\}$, and for $r = 0, 1$, let $\boldsymbol{\mu}^{(r)} = (\mu_1^{(r)}, \mu_2^{(r)})^{\mathrm{T}}$. Then one can show

$$
\begin{aligned}
p(\mu_2^{(0)} \mid \boldsymbol{y}, \boldsymbol{r}) \quad &\propto \quad p(\mu_1^{(1)}) \, p(\mu_2^{(1)}) \, p(\mu_1^{(0)}) \, p(\mu_2^{(0)}) \, p(\boldsymbol{\Sigma}) \\
&\quad \times \prod_{i=1}^{n} p(y_{i1}, y_{i2} \mid \mu_1^{(1)}, \mu_2^{(1)}, \boldsymbol{\Sigma}) \, p(y_{i1} \mid \mu_1^{(0)}, \sigma_{11}) \\
&\propto \quad \sigma_{11}^{-n_0/2} \exp \left\{ - \sum_{\{i : r_i = 0\}} (y_{i1} - \mu_1^{(0)})^2 / 2\sigma_{11} \right\} p(\mu_2^{(0)}) \\
&\propto \quad p(\mu_2^{(0)}),
\end{aligned}
$$

where n_0 is the number of subjects with Y_{i2} missing ($R_i = 0$). Since the likelihood does not contain $\mu_2^{(0)}$, the posterior for $\mu_2^{(0)}$ is proportional to its prior. □

In Chapters 5 and 8, we discuss various restrictions (implicitly, priors) that can be are used to identify $\mu_2^{(0)}$ in mixture models of this type.

In missing data problems, the Bayesian approach allows assumptions about nonidentified parameters, and our uncertainty about those assumptions, to be formalized through prior distributions. By contrast, in frequentist inference, nonidentified parameters are typically fixed at some value or constrained via untestable assumptions. We discuss this issue in more detail in Chapters 6, 8, and 9.

3.4 Computation of the posterior distribution

The 1990 paper by Gelfand and Smith (1990) marked the start of the revolution in Bayesian computations within the statistical community and popularized an approach to obtain exact (up to Monte Carlo error) inferences for complex Bayesian hierarchical models. The late 1990s saw the advent of software to do these computations, including WinBUGS. The seminal idea underlying all these approaches was to obtain a sample from the posterior distribution without the need for explicit evaluation of normalizing constant of the posterior by constructing a Markov chain having the posterior distribution of interest as its stationary distribution. Marginal posteriors and the posterior of functions of the parameters are easily obtained by doing Monte Carlo integration using the sample from the Markov chain. The approaches are often referred to as Markov chain Monte Carlo (MCMC) algorithms.

3.4.1 The Gibbs sampler

The Gibbs sampler (Geman and Geman, 1984) is the most common MCMC approach used in statistics. It is an algorithm that involves sampling a Markov chain whose kernel is the product of the sequentially updated full conditional distributions of the parameters and whose stationary distribution is the posterior.

To be more explicit, consider a q-dimensional parameter vector $\boldsymbol{\theta}$ with elements $(\theta_1, \theta_2, \ldots, \theta_q)$ and let $\theta_j^{(k)}$ correspond to the sample of the jth component of $\boldsymbol{\theta}$ at iteration k. We sample from the following distributions sequentially,

$$1. \qquad \theta_1^{(k)} \sim p(\theta_1^{(k)} \mid \theta_2^{(k-1)}, \theta_3^{(k-1)}, \ldots, \theta_q^{(k-1)}, \boldsymbol{y})$$

$$2. \qquad \theta_2^{(k)} \sim p(\theta_2^{(k)} \mid \theta_1^{(k)}, \theta_3^{(k-1)}, \ldots, \theta_q^{(k-1)}, \boldsymbol{y})$$

$$\vdots$$

$$q. \qquad \theta_q^{(k)} \sim p(\theta_q^{(k)} \mid \theta_1^{(k)}, \theta_2^{(k)}, \ldots, \theta_{q-1}^{(k)}, \boldsymbol{y}).$$

A common extension is block sampling (Roberts and Sahu, 1997), whereby subsets of $\boldsymbol{\theta}$ are sampled together. For example, we might sample from the joint distribution $p(\theta_1^{(k)}, \ldots, \theta_l^{(k)} \mid \theta_{l+1}^{(k-1)}, \ldots, \theta_q^{(k-1)}, \boldsymbol{y})$, instead of the componentwise distributions as above. We provide an example next to illustrate block sampling.

Example 3.13. *Block Gibbs sampling for normal random effects model* (continuation of Example 3.9).
Recall the normal random effects model

$$\boldsymbol{Y}_i \mid \boldsymbol{x}_i, \boldsymbol{b}_i \sim N(\boldsymbol{\mu}_i^b, \sigma^2 \boldsymbol{I}),$$

where

$$\boldsymbol{\mu}_i^b = \boldsymbol{x}_i \boldsymbol{\beta} + \boldsymbol{w}_i \boldsymbol{b}_i$$

and

$$\boldsymbol{b}_i \mid \boldsymbol{x}_i \sim N(\boldsymbol{0}, \boldsymbol{\Omega}).$$

Assume conditionally conjugate prior distributions for the parameters

$$\begin{aligned} \boldsymbol{\beta} \quad &\sim \quad N(\boldsymbol{\beta}_0, \boldsymbol{V}_\beta) \\ 1/\sigma^2 \quad &\sim \quad \text{Gamma}(a, b) \\ \boldsymbol{\Omega}^{-1} \quad &\sim \quad \text{Wishart}(p, \boldsymbol{\Omega}_0), \end{aligned}$$

where $p = \dim(\boldsymbol{\Omega})$. Here we sample the regression coefficients $\boldsymbol{\beta}$, the random effects $\boldsymbol{b}_i$, and the components of $\boldsymbol{\Omega}$ in blocks. The corresponding full conditional distributions are

$$\boldsymbol{\beta} \mid \boldsymbol{b}, \sigma^2, \boldsymbol{\Omega}, \boldsymbol{y}, \boldsymbol{x} \;\sim\; N\left(\boldsymbol{A}^{-1}\left\{\textstyle\sum_i \boldsymbol{x}_i^{\mathrm{T}}(\boldsymbol{y}_i - \boldsymbol{w}_i \boldsymbol{b}_i)/\sigma^2 + \boldsymbol{V}_\beta^{-1}\boldsymbol{\beta}_0\right\}, \boldsymbol{A}^{-1}\right)$$

$$\boldsymbol{b}_i \mid \boldsymbol{\beta}, \sigma^2, \boldsymbol{\Omega}, \boldsymbol{y}, \boldsymbol{x} \;\sim\; N\left(\boldsymbol{B}^{-1}\left\{\textstyle\sum_i \boldsymbol{w}_i^{\mathrm{T}}(\boldsymbol{y}_i - \boldsymbol{x}_i\boldsymbol{\beta})/\sigma^2 + \boldsymbol{\Omega}^{-1}\right\}, \boldsymbol{B}^{-1}\right)$$

$$\sigma^{-2} \mid \boldsymbol{\beta}, \boldsymbol{b}, \boldsymbol{\Omega}, \boldsymbol{y}, \boldsymbol{x} \;\sim\; \mathrm{Gamma}\left(n/2 + a, \left\{1/b + \textstyle\sum_i \boldsymbol{e}_i^{\mathrm{T}}\boldsymbol{e}_i/2\right\}^{-1}\right)$$

$$\boldsymbol{\Omega}^{-1} \mid \boldsymbol{\beta}, \boldsymbol{b}, \sigma^2, \boldsymbol{y}, \boldsymbol{x} \;\sim\; \mathrm{Wishart}\left(n + p, \left\{\boldsymbol{\Omega}_0^{-1} + \textstyle\sum_i \boldsymbol{b}_i\boldsymbol{b}_i^{\mathrm{T}}\right\}^{-1}\right),$$

where $\boldsymbol{e}_i = \boldsymbol{e}_i(\boldsymbol{\beta}, \boldsymbol{b}_i) = \boldsymbol{y}_i - \boldsymbol{x}_i\boldsymbol{\beta} - \boldsymbol{w}_i\boldsymbol{b}_i$ and

$$\boldsymbol{A} = \sum_{i=1}^n \boldsymbol{x}_i\boldsymbol{x}_i^{\mathrm{T}}/\sigma^2 + \boldsymbol{V}_\beta^{-1}$$

$$\boldsymbol{B} = \sum_{i=1}^n \boldsymbol{w}_i\boldsymbol{w}_i^{\mathrm{T}}/\sigma^2 + \boldsymbol{\Omega}^{-1}.$$

Each is available in closed form, demonstrating the computational advantage of using conditionally conjugate priors and the ease with which a posterior sample can be drawn. □

Having all the full conditional distributions available in closed form is atypical. We go through a more representative example next.

Example 3.14. *Gibbs sampling for logistic random effects model* (continuation of Example 2.2).
We specify priors for $\boldsymbol{\beta}$ and $\boldsymbol{\Omega}^{-1}$ as in Example 3.13. Recall the model is

$$\mathrm{logit}(\mu_{ij}^b) = \boldsymbol{x}_{ij}\boldsymbol{\beta} + \boldsymbol{w}_{ij}\boldsymbol{b}_i,$$

where $\boldsymbol{b}_i \sim N(0, \boldsymbol{\Omega})$. The full conditional distributions here are

$$\boldsymbol{\beta} \mid \boldsymbol{b}, \boldsymbol{\Sigma}, \boldsymbol{\Omega}, \boldsymbol{y}, \boldsymbol{x} \;\propto\; p(\boldsymbol{\beta}) \prod_{i=1}^n \prod_{j=1}^J (\mu_{ij}^b)^{y_{ij}}(1 - \mu_{ij}^b)^{1-y_{ij}}$$

$$\boldsymbol{b}_i \mid \boldsymbol{\beta}, \boldsymbol{\Sigma}, \boldsymbol{\Omega}, \boldsymbol{y}, \boldsymbol{x} \;\propto\; \exp\left(-\tfrac{1}{2}\boldsymbol{b}_i^{\mathrm{T}}\boldsymbol{\Omega}^{-1}\boldsymbol{b}_i\right) \prod_{j=1}^J (\mu_{ij}^b)^{y_{ij}}(1 - \mu_{ij}^b)^{1-y_{ij}}.$$

The full conditionals of $\boldsymbol{\beta}$ and $\boldsymbol{b}_i$ are not available in closed form. However, the full conditional distribution for $\boldsymbol{\Omega}^{-1}$ is a Wishart as in Example 3.13; i.e.,

$$\boldsymbol{\Omega}^{-1} \mid \boldsymbol{\beta}, \boldsymbol{b}, \boldsymbol{y}, \boldsymbol{x} \;\sim\; \mathrm{Wishart}(n + p, \{\boldsymbol{\Omega}_0^{-1} + \textstyle\sum_i \boldsymbol{b}_i\boldsymbol{b}_i^{\mathrm{T}}\}^{-1}).$$

□

As stated earlier, the inability to obtain all the full conditional distributions in closed form is not an uncommon occurrence in most models, in some cases due to the use of non-conjugate priors. For example, the binary data models in Examples 2.4 and 2.5 have at least one full conditional with an unknown form. Clearly, approaches are needed to sample from the full conditional distributions for such cases, and there are many techniques available to do this. We review one such approach in the next section.

3.4.2 The Metropolis-Hastings algorithm

The most commonly used approach for sampling from non-standard full conditional distributions is the Metropolis-Hastings algorithm (Hastings, 1970). We describe its implementation within a Gibbs sampling algorithm; this involves constructing another Markov chain (within the Gibbs sampler) that has as its stationary distribution the full conditional distribution of interest. For simplicity, denote the full conditional distribution of interest as

$$p(\theta_j^{(k)} \mid \{\theta_l^{(k)} : l < j\}, \{\theta_l^{(k-1)} : l > j\}, \boldsymbol{y}).$$

In the following, for clarity, we will remove the conditioning and write this full conditional as $p(\theta_j^{(k)})$.

At the kth iteration, $\theta_j^{(k)}$ is sampled from some candidate distribution, $q(\theta_j^{(k)} \mid \theta_j^{(k-1)})$, and the sampled value is accepted with probability $\alpha^\star$ given by

$$\alpha^\star = \min\left\{1, \frac{p(\theta_j^{(k)})q(\theta_j^{(k-1)} \mid \theta_j^{(k)})}{p(\theta_j^{(k-1)})q(\theta_j^{(k)} \mid \theta_j^{(k-1)})}\right\}.$$

This choice of acceptance probability ensures the stationary distribution of the Markov chain is the full conditional distribution of interest.

An important practical issue is the choice of candidate distribution q. Setting $q(\theta_j^{(k)} \mid \theta_j^{(k-1)}) = N(\theta_j^{(k-1)}, v)$, where v is a fixed variance, yields the *random walk Metropolis-Hastings algorithm*. For this choice, $q(\theta_j^{(k)} \mid \theta_j^{(k-1)}) = q(\theta_j^{(k-1)} \mid \theta_j^{(k)})$, and $\alpha^\star$ reduces to the ratio of the full conditionals evaluated at the proposal value and the previous value; i.e.,

$$\alpha^\star = \min\left\{1, \frac{p(\theta_j^{(k)})}{p(\theta_j^{(k-1)})}\right\}. \tag{3.7}$$

Optimal acceptance for this choice should be around $20 - 30\%$ (Brooks and Gelman, 1998). Higher acceptance rates will result in θ_j moving very slowly around the posterior distribution and a highly correlated sample of θ_j, creating an inefficient algorithm; these issues are discussed in detail in Section 3.4.4.

A common initial choice for v is some value proportional to the squared standard error (based on the information matrix), usually with proportionality constant less than one; this initial value can be adjusted by trial and error to attain an appropriate acceptance rate.

Another common choice for q is an approximation to the full conditional distribution $p(\cdot)$, independent of the previous value, $\theta_j^{(k-1)}$. For example, one might construct a normal approximation (Laplace approximation) to the full conditional distribution based on the modified score vector and information matrix as discussion in Section 3.2; this usually involves implementing an optimization algorithm (e.g., Newton-Raphson) at each iteration. In this case, (3.7) can be rewritten as the ratio of importance weights,

$$\alpha^\star = \min\left\{1, \frac{p(\theta_j^{(k)})/q(\theta_j^{(k)})}{p(\theta_j^{(k-1)})/q(\theta_j^{(k-1)})}\right\},$$

where $q(\cdot)$ is the approximation to the full conditional distribution. For this approach, the optimal acceptance rate is as close to 100% as possible. Poor choices of q will result in low acceptance often due to long runs of not accepting the candidate value. Heavier tailed candidates will often improve the efficiency (Chib and Greenberg, 1998; Daniels and Gatsonis, 1999). Choosing the candidate distribution based on approximating the full conditional distribution has been implemented with considerable success for sampling the fixed effects and random effects, β and b_i respectively, in generalized linear mixed models (Example 2.2). For some examples, see Daniels and Gatsonis (1999).

Recommendations

The random walk Metropolis-Hastings algorithm typically works best for sampling one parameter at a time and is computationally inexpensive (no need to do a maximization at each iteration). However, it does not usually generalize well to sampling blocks because of the difficulty in specifying an appropriate *covariance matrix* for the candidate distribution. In addition, sampling the parameter vector $\boldsymbol{\theta}$ componentwise can result in poor performance of the MCMC algorithm (more details in Section 3.4.4).

Using an approximation to the full conditional distribution for the candidate distribution in the Metropolis-Hastings algorithm often works well when it is not computationally expensive to do a maximization at each iteration (e.g., compute derivatives and do Newton-Raphson steps), and is a good approach when trying to sample more than one parameter simultaneously. Heavy-tailed approximations, e.g., a t-distribution instead of a normal distribution, will often increase efficiency (Chib and Greenberg, 1998). Other approaches to sample non-standard full conditionals are mentioned in Further Reading.

3.4.3 Data augmentation

As discussed previously, it is often the case that some of the full conditional distributions of $p(\boldsymbol{\theta} \mid \boldsymbol{y})$ are difficult to sample. An alternative to Metropolis-Hastings is to introduce latent data $\boldsymbol{Z}$ such that it is easy to sample $p(\boldsymbol{\theta} \mid$

z, y) and $p(z \mid \theta, y)$ and often improves the mixing properties of the sampler (Tanner and Wong, 1987; van dyk and Meng, 2001). This was coined *data augmentation* by Tanner and Wong. The approach proceeds in two steps:

1. Sample $p(z \mid \theta, y)$,
2. Sample $p(\theta \mid z, y)$ using Gibbs sampling.

This approach is well-suited to two (related) situations. First, complete data problems that inherently have latent structure based on known distributions, e.g., probit or random effects models; we illustrate this in Examples 3.15 and 3.16. Second, data augmentation is also a natural way to deal with incomplete data. For incomplete data problems, we often specify the model for the full data. As such, it is often easier to work with the full data posterior as opposed to the posterior based only on the observed data. In this case, Z would correspond to the missing data. Sampling $p(\theta \mid z, y)$ corresponds to sampling from the full data posterior; $p(z \mid \theta, y)$ to sampling from the posterior predictive distribution of the missing data. Chapters 6 and 8 provide details on data augmentation in this setting.

In the following, we provide details on data augmentation for some of the longitudinal models introduced in Chapter 2. Both correspond to complete data problems with latent structure, where data augmentation can make sampling from the posterior considerably easier.

Example 3.15. *Data augmentation for the probit model* (continuation of Example 2.6).
This is the multivariate probit formulation but with $J = 1$. The posterior for this model is

$$p(\beta \mid y) \propto \left\{ \prod_{i=1}^{n} \Phi(x_i\beta)^{y_i} \Phi(-x_i\beta)^{1-y_i} \right\} p(\beta), \qquad (3.8)$$

where $\Phi(\cdot)$ is the standard normal cdf and $p(\beta)$ is typically chosen to be a (multivariate) normal prior. Clearly, the posterior distribution of β is not available in closed form. However, we can use data augmentation here to simplify sampling from the posterior. We exploit the existence of the latent variables Z_i such that

$$L(\beta \mid y) \propto \int_{z_i \in Q(y_i)} p(z_i \mid \beta)dz_i,$$

where $Q(y_i) = (0, \infty)$ if $y_i = 1$ and $Q(y_i) = (-\infty, 0)$ if $y_i = 0$, and $p(z_i \mid \beta)$ is the pdf of a normal distribution with mean $x_i\beta$ and variance 1. With the introduction of the latent z_i, sampling from the posterior distribution of β can proceed (quite easily) in two steps:

1. Sample β from $p(\beta \mid z, y)$, a normal distribution (assuming the prior $p(\beta)$ is a normal distribution).
2. Sample z from $p(z \mid \beta, y)$, a truncated normal distribution.

Extensions to the multivariate model are conceptually straightforward in that they use the underlying latent structure as well. □

In some cases, t-distributions are used instead of normal distributions to obtain more robust inferences; the heavier tails of the t-distribution make it more robust to outliers (Lange, Little, and Taylor, 1989). In addition, they can be used to build multivariate logistic models for longitudinal binary data (Example 6.6.2). Unfortunately, regardless of the prior specification, the full conditional distributions of the location and scale parameters of the t-distribution are not available in closed form. However, we can use data augmentation to facilitate sampling from the location and scale parameters.

Example 3.16. *Data augmentation for the multivariate t-distribution* (continuation of Example 2.4).

To implement data augmentation for the t-distribution, we take advantage of the following relationship between the multivariate t-distribution with ν degrees of freedom and the multivariate normal distribution. If $\boldsymbol{Y}_i \mid \boldsymbol{\mu}, \boldsymbol{\Sigma} \sim \mathcal{T}_\nu(\boldsymbol{\mu}, \boldsymbol{\Sigma})$, then its density can be written as a gamma mixture of normals,

$$\int \frac{\tau_i^{J/2}}{(2\pi)^{J/2}|\boldsymbol{\Sigma}|^{1/2}} \exp\left\{-\tfrac{1}{2}(\boldsymbol{y}_i - \boldsymbol{\mu})^{\mathrm{T}}(\boldsymbol{\Sigma}/\tau_i)^{-1}(\boldsymbol{y}_i - \boldsymbol{\mu})\right\} p(\tau_i \mid \nu)d\tau_i, \quad (3.9)$$

where $p(\tau_i \mid \nu)$ is a Gamma density with parameters $(\nu/2, 2/\nu)$. Using this result, data augmentation approaches can be used to greatly simplify sampling. This result was first used in the context of Gibbs sampling by Chib and Albert (1993). By using the expanded parameter space with the latent variables τ_i instead of integrating them out as in (3.9), we can sample from the posterior of $(\boldsymbol{\mu}, \boldsymbol{\Sigma})$ using the following steps, assuming a multivariate normal prior on $\boldsymbol{\mu}$ and a Wishart prior on $\boldsymbol{\Sigma}^{-1}$:

1. Sample τ_i from $p(\tau_i|\boldsymbol{\mu}, \boldsymbol{\Sigma}, \boldsymbol{y})$, independent gamma distributions.
2. Sample $(\boldsymbol{\mu}, \boldsymbol{\Sigma}^{-1})$ from $p(\boldsymbol{\mu}, \boldsymbol{\Sigma} \mid \boldsymbol{\tau}, \boldsymbol{y})$ using Gibbs sampling:
   (a) Sample $\boldsymbol{\mu}$ from $p(\boldsymbol{\mu}|\boldsymbol{\Sigma}, \boldsymbol{\tau}, \boldsymbol{y})$, a multivariate normal distribution.
   (b) Sample $\boldsymbol{\Sigma}^{-1}$ from $p(\boldsymbol{\Sigma}^{-1}|\boldsymbol{\mu}, \boldsymbol{\tau}, \boldsymbol{y})$, a Wishart distribution. □

As a final example, we show that the Gibbs sampler for the normal random effects model in Example 3.13 implicitly uses data augmentation.

Example 3.17. *Data augmentation for the normal random effects model* (continuation of Example 3.13).

In the normal random effects model, the random effects $\boldsymbol{b}_i$ can be integrated out in closed form. Thus, the posterior distribution can be sampled by just using the full conditional distributions of $\boldsymbol{\beta}$, $\boldsymbol{\Omega}$, and σ^2 based on the integrated likelihood (3.2) and priors. However, the full conditional distributions of $\boldsymbol{\Omega}$ and σ^2 will not have known forms and in particular, the full conditional for $\boldsymbol{\Omega}$ will be very hard to sample (Daniels, 1998). On the other hand, if the random effects are not integrated out and are sampled from their full conditional distributions as well, the full conditional distribution of $\boldsymbol{\Omega}^{-1}$ will be a Wishart distribution (cf. Example 3.13) and $1/\sigma^2$ will be a Gamma distribution. Hence, from a computational perspective, we can view the random effects $\boldsymbol{b}_i$ as latent variables augmenting the model

$$\boldsymbol{Y}_i \sim N(\boldsymbol{x}_i\boldsymbol{\beta}, \boldsymbol{w}_i\boldsymbol{\Omega}\boldsymbol{w}_i^{\mathrm{T}} + \sigma^2\boldsymbol{I})$$

and the Gibbs sampler described in Example 3.13 can be viewed as a Gibbs sampler with data augmentation.   □

Now that we have reviewed the tools to obtain a sample from the posterior, we are ready to discuss how to use this sample for posterior inference.

3.4.4 Inference using the posterior sample

The sample generated by Gibbs sampling or other MCMC algorithms must be used carefully. Although WinBUGS will typically do the sampling described in Sections 3.4.1–3.4.3 automatically, we need to be sure we have obtained an accurate and correct sample using these approaches. As such, at least two issues need to be considered:

1. When do we consider the Markov chain to have reached the stationary distribution, i.e., the posterior distribution? (The time before reaching this is often referred to as the *burn-in*.)

2. How many additional samples are needed after the burn-in in order to make 'accurate' inferences? This can be tricky since MCMC approaches generate a *dependent* sample from the posterior distribution of the parameters.

Related to issue (1), it is often advocated to discard the first K samples as burn-in, giving the sampler time to find the stationary distribution, i.e., the posterior. To make sure the chain has actually converged to the correct place, it is recommended to run multiple chains with different starting values and make sure they all converge to the same region of the parameter space (Gelman and Rubin, 1992); Gelman and Rubin proposed a statistic to monitor convergence based on the variability within and between the multiple chains. We refer the reader to Cowles and Carlin (1996) for more details and other approaches to assess convergence. Figure 3.1 illustrates the concept; the sampler has reached the stationary distribution rather quickly after two or three iterations (i.e., when the four chains come together).

Once the sampler has converged, the issue then becomes how efficiently the sampler takes draws from the posterior. In most practical settings, the sequential MCMC draws are autocorrelated (dependent). Figure 3.2 is an autocorrelation plot showing lag-k correlations in a chain. It appears by lag 10 that the autocorrelation is negligible. If the chain moves slowly around the parameter space (high autocorrelation), more draws are needed for inference. For

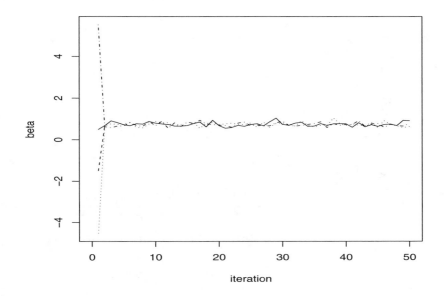

Figure 3.1 *Plot of multiple chains illustrating the concept of burn-in.*

example, the information in $10,000$ autocorrelated draws might be equivalent to only 500 independent draws; see Tierney (1994) for details.

Figure 3.3 shows one sampler that is mixing well and another that is mixing poorly. In the bottom plot, it is clear that the chain is moving slowly around the parameter space. Block Gibbs sampling or non-random walk Metropolis-Hastings approaches are often used to speed mixing by reducing autocorrelation.

Computation of functions of the posterior such as means, medians, and percentiles (credible intervals) can be done by just calculating the appropriate summary using the MCMC sample after throwing away the 'burn-in'. However, to compute other functions, such as the posterior standard deviation, care must be taken to account for the autocorrelation in the chain. One approach is 'thinning', whereby every kth MCMC draw is retained, with k chosen large enough so the lag-k correlation is small. Thinning yields an approximately independent sample. For example, based on Figure 3.2, we might choose $k = 10$.

Thinning can be inefficient because $(k-1)/k\%$ of the sample is discarded. Batching is usually a more efficient approach. It involves breaking the output

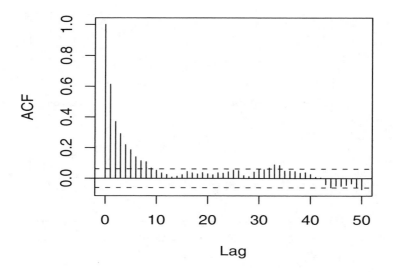

Figure 3.2 *Autocorrelation plot. Lag-k correlation is plotted vs. k.*

of the chain into m batches of length $m^\star$, such that $m^\star$ is large enough for the correlation between batch means to be negligible. To calculate the posterior variance, we use a suitably normalized corrected sum of squares of the batch means (see pp. 194–195 in Carlin and Louis, 2000). Other approaches, including some based on time series methodology, can be found in Chapter 5 of Carlin and Louis (2000).

Marginal posterior distributions

The MCMC sample obtained provides a (dependent) sample from the *joint* posterior distribution of interest. However, we are often interested in marginal posterior distributions; for example, in the multivariate normal model in Example 2.3, we may be interested in a specific regression coefficient, say β_j. A sample from the marginal posterior distribution of β_j (or, in general, any function of the parameters) is obtained by using only the sampled values of that parameter. To obtain the marginal posterior distribution for a function of the parameters, we can evaluate that function at each iteration, given the current values of the parameters, to obtain a sample from the marginal posterior of that function. See Section 4.4 for an illustration.

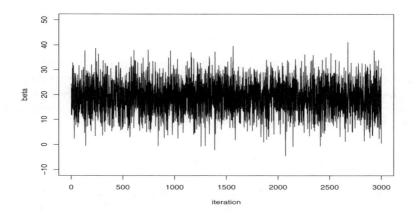

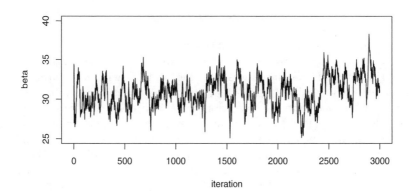

Figure 3.3 *Plot illustrating good mixing (top) and poor mixing (bottom).*

Reweighting

It will sometimes be the case that we have a sample from a distribution (e.g., an MCMC sample from the posterior) and we would like to use this sample to make inference based on a similar posterior (maybe with a slightly different likelihood and/or priors). We can avoid re-running a Gibbs sampler on this new model by appropriately reweighting the sample already obtained. In general, suppose we have a sample from some distribution, $p(\theta)$ and we want to make inference based on some different distribution, $p^\star(\theta)$. Then, we can reweight the sample using weights of the form

$$w = \frac{p(\theta)}{p^\star(\theta)}.$$

The reliability of this approach depends on the weights, w being stable (Peruggia, 1997). Reweighting will be useful for computing intractable likelihoods (e.g., the multivariate probit model in Example 2.6) that are sometimes needed for model selection criterion (see Section 3.5) and for computing several model selection criteria in the presence of incomplete data (see Chapters 6 and 8).

A note on improper priors and Gibbs sampling

We end this section on posterior sampling with a cautionary remark on using improper priors in Gibbs sampling (Hobert and Casella, 1996), illustrated by an example. Consider the normal random effects model in Example 2.1. Suppose we specify the following (improper) prior on $\boldsymbol{\Omega}$,

$$p(\boldsymbol{\Omega}) \quad \propto \quad |\boldsymbol{\Omega}|^{-(p+1)/2},$$

where $p = \dim(\boldsymbol{\Omega})$. Clearly, $\int p(\boldsymbol{\Omega})\, d\boldsymbol{\Omega} = \infty$.

The full conditional distribution of $\boldsymbol{\Omega}^{-1}$ will be a proper Wishart distribution. However, the posterior distribution of $\boldsymbol{\Omega}^{-1}$ will be *improper*. This phenomenon (when using improper priors) of all the full conditionals being proper distributions, but the posterior being improper, was first noticed in Hobert and Casella (1996). An even more problematic aspect from a practical perspective is that the sample from the improper posterior may *not* indicate any problems! So, when using improper priors, the propriety of the posterior distribution needs to be verified analytically. Otherwise, improper priors should *not* be used. If WinBUGS is being used, this is not a concern as it does not allow improper priors; however, for investigators writing their own code and using improper priors, this is an important issue.

3.5 Model comparisons and assessing model fit

When we fit a parametric model to a dataset, we should examine how well the model fits the observed data. A related issue is how to select among several plausible models (model selection), which tells us only about the fit of models relative to the others under consideration. We address model selection first. Two common criteria are the deviance information criterion (DIC) (Spiegelhalter et al., 2002) and posterior predictive loss (PPL) (Gelfand and Ghosh, 1998). Both take into account goodness of fit while penalizing models for overfitting (a complexity penalty).

3.5.1 Deviance Information Criterion (DIC)

The DIC is a *model-based criterion* composed of a goodness of fit term and a penalty term. The fit is measured by the deviance, a linear function of the log likelihood, given by

$$\text{Dev}(\boldsymbol{\theta}) = -2 \log L(\boldsymbol{\theta} \mid \boldsymbol{y}).$$

Larger values of the deviance indicate poorer fit.

The penalty term measures model complexity and is given by

$$p_D = E\{\text{Dev}(\boldsymbol{\theta}) \mid \boldsymbol{y}\} - \text{Dev}\{E(\boldsymbol{\theta} \mid \boldsymbol{y})\}. \tag{3.10}$$

The variable p_D is called the effective number of parameters. As the variability in the posterior of $\boldsymbol{\theta}$ decreases, $p_D \to 0$. How overfitting is penalized can best be understood by introducing the concept of the residual information in data $\boldsymbol{y}$ conditional on parameters $\boldsymbol{\theta}$, defined as $-2 \log\{p(\boldsymbol{y} \mid \boldsymbol{\theta})\}$ (Kullback and Liebler, 1951; Burham and Anderson, 1998). Recall that $L(\boldsymbol{\theta} \mid \boldsymbol{y}) \propto p(\boldsymbol{y} \mid \boldsymbol{\theta})$. Define $\widehat{\boldsymbol{\theta}}$ to be an estimator of $\boldsymbol{\theta}$ and $\boldsymbol{\theta}^\star$ to be the true parameter value. The difference between the residual information at the true parameter value and at the estimated parameter value is

$$-2 \log p(\boldsymbol{y} \mid \boldsymbol{\theta}^\star) + 2 \log p(\boldsymbol{y} \mid \widehat{\boldsymbol{\theta}}). \tag{3.11}$$

This can be interpreted as the degree of overfitting due to the influence of $\boldsymbol{y}$ on the estimator $\widehat{\boldsymbol{\theta}}$. In a Bayesian analysis, $\boldsymbol{\theta}$ is random and we can replace (3.11) with its posterior expectation, the effective number of parameters p_D given in (3.10).

The DIC itself is defined as

$$\text{DIC} = \text{Dev}\{E(\boldsymbol{\theta} \mid \boldsymbol{y})\} + 2p_D. \tag{3.12}$$

The first term measures goodness of fit and the second term is the complexity penalty. The form is very similar to the Akaike Information Criterion (AIC) (Akaike, 1973). Equivalently, the DIC can be written explicitly as a function of the log likelihood,

$$\text{DIC} = -4E\{\log L(\boldsymbol{\theta} \mid \boldsymbol{y}) \mid \boldsymbol{y}\} + 2 \log L\{E(\boldsymbol{\theta} \mid \boldsymbol{y}) \mid \boldsymbol{y}\}, \tag{3.13}$$

which will be a more convenient form for its development in the setting of incomplete data and for describing its computation in Chapters 6 and 8.

The DIC is easy to compute from a posterior sample; it requires calculating two quantities, $E\{\text{Dev}(\boldsymbol{\theta}) \mid \boldsymbol{y}\}$ and $\text{Dev}\{E(\boldsymbol{\theta} \mid \boldsymbol{y})\}$, using the output from MCMC approaches; WinBUGS will often calculate it automatically. Ease of implementation has contributed to its widespread use. In Section 4.2, we use the DIC to compare several multivariate normal models for the Growth Hormone data (described in Section 1.3)

An advantage of the DIC over approaches like AIC, where the user specifies the number of parameters, is that the (effective) number of parameters is counted automatically (see (3.10)). This is particularly helpful in multilevel models where the number of parameters is sometimes difficult to quantify. As an example, consider the normal random effects models in Example 2.1 with likelihood given by

$$L(\boldsymbol{\theta}, \boldsymbol{b}_i \mid \boldsymbol{y}) \propto \prod_{i=1}^{n} |\boldsymbol{\Sigma}|^{-1/2} \exp\left\{-\tfrac{1}{2} \boldsymbol{e}_i(\boldsymbol{\beta}, \boldsymbol{b}_i)^{\mathrm{T}} \boldsymbol{\Sigma}^{-1} \boldsymbol{e}_i(\boldsymbol{\beta}, \boldsymbol{b}_i)\right\}, \qquad (3.14)$$

where $\boldsymbol{e}_i(\boldsymbol{\beta}, \boldsymbol{b}_i) = \boldsymbol{y}_i - \boldsymbol{x}_i\boldsymbol{\beta} - \boldsymbol{w}_i\boldsymbol{b}_i$ and $\boldsymbol{\theta} = (\boldsymbol{\beta}, \boldsymbol{\Sigma})$. The random effects have not been integrated out and are now treated as parameters along with $\boldsymbol{\theta}$. On the surface, if we count the number of random effects (assume for simplicity they are one-dimensional), there are n. However, the effective number can be quite smaller because the random effects distribution $p(\boldsymbol{b}_i \mid \boldsymbol{\theta})$ shrinks the random effects to zero. As the variance of the random effects distribution goes to zero, there are fewer parameters; in fact, if the variance is zero, all the random effects are identically zero so there are in fact no parameters. On the other hand, as the variance increases, the number of parameters approaches n.

Despite its computational simplicity, the DIC does have drawbacks. The best model as determined by the DIC can change depending on the choice of 'likelihood' (see Trevisani and Gelfand, 2003); for example, again revisiting the normal random effects model (Example 2.1), the likelihood can take one of two forms: the integrated likelihood given in (3.2), or the likelihood without the random effects integrated out, given in (3.14).

In addition, the DIC is not invariant to the parameterization of $\boldsymbol{\theta}$. This occurs because the fit term $\mathrm{Dev}\{E(\boldsymbol{\theta} \mid \boldsymbol{y})\}$ in (3.12) involves a plug-in estimator for $\boldsymbol{\theta}$ based on the posterior, $E(\boldsymbol{\theta} \mid \boldsymbol{y})$; and in general, $E\{h(\boldsymbol{\theta}) \mid \boldsymbol{y}\} \neq h\{E(\boldsymbol{\theta} \mid \boldsymbol{y})\}$. For the multivariate normal model in Example 2.3, $\boldsymbol{\theta}$ could be defined as $(\boldsymbol{\beta}, \boldsymbol{\Sigma}^{-1})$ or $(\boldsymbol{\beta}, \boldsymbol{\Sigma})$. Using $\boldsymbol{\Sigma}$ vs. $\boldsymbol{\Sigma}^{-1}$ will result in different values for the DIC; see Section 4.2 for an illustration on the Growth Hormone data. For covariance matrices, Spiegelhalter et al. (2002) recommend using the inverse because its posterior mean is more stable.

Another limitation, common to all likelihood based criteria, is that for some models, the likelihood is not available in closed form (e.g., the multivariate probit model in Example 2.6). For many models, to evaluate the likelihood, it is possible to use Monte Carlo integration and reweighting. For example, in the multivariate probit model, we need to compute

$$E_{\boldsymbol{z}}\left(\prod_{j=1}^{J} I\{z_{ij} > 0\}^{y_{ij}} I\{z_{ij} < 0\}^{1-y_{ij}}\right)$$

given $(\boldsymbol{\beta}, \boldsymbol{\Sigma})$, where $\boldsymbol{z}_i$ follows a multivariate normal distribution with mean $\boldsymbol{x}_i\boldsymbol{\beta}$ and covariance matrix $\boldsymbol{\Sigma}$. We can sample from the distribution of $\boldsymbol{z}_i$ and compute the expectation by averaging the term in brackets over the samples.

However, it would be computationally prohibitive to do this for every sampled value of $\boldsymbol{\theta} = (\boldsymbol{\beta}, \boldsymbol{\Sigma})$ that is needed to compute $E\{\mathrm{Dev}(\boldsymbol{\theta})\}$ in the DIC.

A more practical approach is reweighting (as discussed in Section 3.4.4). To implement it here, we can take a likely value, say $\boldsymbol{\theta}^\star = E(\boldsymbol{\theta} \mid \boldsymbol{y})$, and sample L values, $\boldsymbol{z}^{(l)} : l = 1, \ldots, L$, from $Z_i \sim N(\boldsymbol{x}_i \boldsymbol{\beta}^\star, \boldsymbol{\Sigma}^\star)$, where $\boldsymbol{\theta}^\star = (\boldsymbol{\beta}^\star, \boldsymbol{\Sigma}^\star)$. Then, to compute the likelihood for other values of $\boldsymbol{\theta}$, we evaluate

$$\frac{\sum_l h(\boldsymbol{z}^{(l)}) w_l}{\sum_l w_l},$$

where $h(\boldsymbol{z}) = \prod_j I(z_{ij} > 0)^{y_{ij}} I(z_{ij} < 0)^{1-y_{ij}}$. The weights are given by

$$w_l = \frac{p(\boldsymbol{z}^{(l)} \mid \boldsymbol{\theta})}{p(\boldsymbol{z}^{(l)} \mid \boldsymbol{\theta}^\star)},$$

where $p(\cdot \mid \boldsymbol{\theta})$ is a multivariate normal distribution with parameters $\boldsymbol{\theta} = (\boldsymbol{\beta}, \boldsymbol{\Sigma})$ (Liu and Daniels, 2007). We illustrate this in Section 7.5. For further recommendations and discussion on the choice of likelihood and parameterization, we refer the reader to Spiegelhalter et al. (2002).

3.5.2 Posterior predictive loss

Posterior predictive loss (PPL) (Gelfand and Ghosh, 1998) is another model selection criterion. Before providing details, we first need to define the posterior predictive distribution.

Definition 3.6. *Posterior predictive distribution.*
The posterior predictive distribution is

$$p(\boldsymbol{y}_{\mathrm{rep}} \mid \boldsymbol{y}) = \int p(\boldsymbol{y}_{\mathrm{rep}} \mid \boldsymbol{\theta}, \boldsymbol{y}) p(\boldsymbol{\theta} \mid \boldsymbol{y}) d\boldsymbol{\theta}, \qquad (3.15)$$

where $p(\boldsymbol{y}_{\mathrm{rep}} \mid \boldsymbol{\theta}, \boldsymbol{y}) = p(\boldsymbol{y}_{\mathrm{rep}} \mid \boldsymbol{\theta})$. Samples from the posterior predictive distribution are replicates of the observed data generated by the model. □

PPL quantifies the fit of the model by comparing features of the (model-based) posterior predictive distribution to equivalent features of the observed data. The comparison is based on a user-chosen loss function. $\mathscr{L}(\boldsymbol{y}_{\mathrm{rep}}, a; \boldsymbol{y})$, where a is chosen to minimize the expectation of the loss with respect to the posterior predictive distribution $E\{\mathscr{L}(\boldsymbol{y}_{\mathrm{rep}}, a; \boldsymbol{y}) \mid \boldsymbol{y}\}$, i.e., the posterior predictive loss. For some choices of $\mathscr{L}$, the minimization has a closed form.

Gelfand and Ghosh consider loss functions of the form

$$\mathscr{L}_k(\boldsymbol{y}_{\mathrm{rep}}, a; \boldsymbol{y}) = \mathscr{L}(\boldsymbol{y}_{\mathrm{rep}}, a) + k \mathscr{L}(\boldsymbol{y}, a), \quad k \geq 0. \qquad (3.16)$$

For univariate y, if $\mathscr{L}$ is chosen as squared error loss, it can be shown that

$$\min_a E\left\{\mathscr{L}_k(\boldsymbol{y}_{\text{rep}}, a; \boldsymbol{y}) \mid \boldsymbol{y}\right\} = \sum_{i=1}^{n} \sigma_i^2 + \frac{k}{k+1} \sum_{i=1}^{n}(\mu_i - y_i)^2$$

$$= P + \frac{k}{k+1}G, \qquad (3.17)$$

where

$$\mu_i = E(Y_{i,\text{rep}} \mid \boldsymbol{y}) = \int y_{i,\text{rep}} \, p(y_{i,\text{rep}} \mid \boldsymbol{\theta}) \, p(\boldsymbol{\theta} \mid \boldsymbol{y}) \, d\boldsymbol{\theta} \, dy_{i,\text{rep}}$$

is the posterior predictive mean and $\sigma_i^2 = \text{var}(Y_{i,\text{rep}} \mid \boldsymbol{y})$ is the posterior predictive variance.

The first term in (3.17), $P = \sum_{i=1}^{n} \sigma_i^2$, is a penalty term. Overfitting the model will result in large predictive variances σ_i^2 and a large value for P. The second term, $G = \sum_{i=1}^{n}(\mu_i - y_i)^2$, is a goodness of fit term, which will decrease with model complexity. This statistic is easy to compute using samples from the posterior predictive distribution.

For other (smooth) choices of $\mathscr{L}(\cdot)$, the criterion can also be approximated in a similar form with a goodness of fit and a complexity term. Like the DIC, this criterion contains an 'automatic' penalty P. The choice of k determines how much weight is placed on the goodness of fit term relative to the penalty term. As $k \to \infty$, $k/(k+1) \to 1$. Unlike the DIC, which uses a non-invariant plug-in estimator for $\boldsymbol{\theta}$, PPL is based on the posterior predictive distribution and is invariant to the model parameterization.

The downsides of PPL are that it requires the choice of an appropriate loss function (which we do not specify in the course of most Bayesian analyses) and possibly nontrivial analytical calculations to obtain the criterion. Another issue is applying it to multivariate observations (e.g., longitudinal data), where we have to account for correlation when computing both the penalty and the fit terms. However, approaches such as using the log likelihood loss can account for correlation (Gelfand and Ghosh, 1998). Finally, extensions of this criterion to incomplete data are an area that needs further study.

A simple way to extend this approach to longitudinal (correlated) data, without using a loss function based on the likelihood, is to summarize each multivariate observation with a univariate measure $T_i = h(\boldsymbol{Y}_i)$, which is a function of the response vector for subject i, and then apply univariate methods (Hogan and Wang, 2001). The univariate summary T_i might be specified as a weighted average of the longitudinal responses, $T_i = \sum w_j Y_{ij}$ for some fixed set of weights. To emphasize fit based on the last observation time, we can set

$$w_l = \begin{cases} 0 & l = 1, \ldots, J-1 \\ 1 & l = J. \end{cases}$$

When emphasis is on a change from baseline, set

$$w_l = \begin{cases} 0 & l = 2, \ldots, J-1 \\ 1 & l = 1 \\ -1 & l = J. \end{cases}$$

Given the choice of T_i and using the Gelfand and Ghosh loss function (3.16) with squared error loss, the PPL criterion becomes

$$\text{PPL} = \sum_{i=1}^{n} \sigma_{i(T)}^2 + \frac{k}{k+1} \sum_{i=1}^{n} (\mu_{i(T)} - T_i)^2 \qquad (3.18)$$

$$= P + \frac{k}{k+1} G,$$

where $\mu_{i(T)} = E(T_{i,rep} \mid \boldsymbol{y})$ and $\sigma_{i(T)}^2 = \text{var}(T_{i,rep} \mid \boldsymbol{y})$, with $T_{i,\text{rep}} = h(\boldsymbol{Y}_{i,\text{rep}})$. In Section 4.4, we illustrate this approach on the CTQ I smoking cessation data (described in Section 1.4).

3.5.3 Posterior predictive checks

To determine how well the model fits the data in an absolute sense, posterior predictive checks can be used. They are a simple but versatile approach to determine whether particular aspects of the data are captured adequately by the model. They require sampling from the posterior predictive distribution given in (3.15). The draws from $p(\boldsymbol{y}_{\text{rep}} \mid \boldsymbol{y})$ can be made using the MCMC output. In particular, for each draw of $\boldsymbol{\theta}$ from the MCMC sample, we sample a set of replicated data from $p(\boldsymbol{y}_{\text{rep}} \mid \boldsymbol{\theta})$. This is easy to do in WinBUGS.

Model critique requires choosing an appropriate data summary $\boldsymbol{T}$, which may be a function of the parameters $\boldsymbol{\theta}$, and comparing its value based on the observed data, $\boldsymbol{T}(\boldsymbol{y}_{\text{obs}}; \boldsymbol{\theta})$, to its values based on the replicated data, $\boldsymbol{T}(\boldsymbol{y}_{\text{rep}}; \boldsymbol{\theta})$. We provide some examples relevant for longitudinal data next.

Gelman, Meng, and Stern (1996) proposed Pearson's χ^2 statistics as an overall measure of model fit (designed for independent data). For (temporally aligned) longitudinal data, we might use a multivariate version

$$T(\boldsymbol{y}; \boldsymbol{\theta}) = \sum_{i=1}^{n} Q_i(\boldsymbol{\theta}), \qquad (3.19)$$

where

$$Q_i(\boldsymbol{\theta}) = \{\boldsymbol{y}_i - E(\boldsymbol{y}_i \mid \boldsymbol{\theta})\}^{\mathrm{T}} \boldsymbol{\Sigma}(\boldsymbol{\theta})^{-1} \{\boldsymbol{y}_i - E(\boldsymbol{y}_i \mid \boldsymbol{\theta})\} \qquad (3.20)$$

and $\boldsymbol{\Sigma}(\boldsymbol{\theta}) = \text{var}(\boldsymbol{Y}_i \mid \boldsymbol{\theta})$. Another global measure is the empirical distribution, $\widehat{\boldsymbol{F}}$ of $Q_i(\boldsymbol{\theta})$,

$$\boldsymbol{T}(\boldsymbol{y}; \boldsymbol{\theta}) = \{\widehat{\boldsymbol{F}}\}. \qquad (3.21)$$

The distribution of residuals for each time point,

$$e_{it}(\boldsymbol{\theta}) = \frac{y_{it} - E(Y_{it} \mid \boldsymbol{\theta})}{\operatorname{var}(Y_{it} \mid \boldsymbol{\theta})^{1/2}},$$

might also be considered. Numerous other summaries can be chosen based on the application.

Posterior predictive probabilities are a means of quantifying the relationship between the statistics computed based on the observed data and the statistics computed based on the replicated data, and can be used to assess model fit.

Definition 3.7. *Posterior predictive probability.*
The posterior predictive probability based on data summary $T(\cdot; \boldsymbol{\theta})$ is defined as

$$\iint I\left\{ h(\boldsymbol{T}(\boldsymbol{y}_{\mathrm{obs}}; \boldsymbol{\theta}), \boldsymbol{T}(\boldsymbol{y}_{\mathrm{rep}}; \boldsymbol{\theta})) > c \right\} \, p(\boldsymbol{y}_{\mathrm{rep}} \mid \boldsymbol{\theta}, \boldsymbol{y}) \, p(\boldsymbol{\theta} \mid \boldsymbol{y}) \, d\boldsymbol{\theta} \, d\boldsymbol{y}_{\mathrm{rep}}$$

for some function $h(\cdot)$ and constant c. $\qquad\qquad\square$

For the multivariate version of Pearson's χ^2 (3.19), we might set $c = 0$ and

$$h\{T(\boldsymbol{y}_{\mathrm{obs}}; \boldsymbol{\theta}), T(\boldsymbol{y}_{\mathrm{rep}}; \boldsymbol{\theta})\} = T(\boldsymbol{y}_{\mathrm{obs}}; \boldsymbol{\theta}) - T(\boldsymbol{y}_{\mathrm{rep}}; \boldsymbol{\theta}).$$

For the empirical cdf (3.21), we might again set $c = 0$ and

$$h\{\boldsymbol{T}(\boldsymbol{y}_{\mathrm{obs}}; \boldsymbol{\theta}), \boldsymbol{T}(\boldsymbol{y}_{\mathrm{rep}}; \boldsymbol{\theta})\} = \mathrm{sign} \times \arg\max_{x} |\{\widehat{F}_{\mathrm{obs}}(x) - \widehat{F}_{\mathrm{rep}}(x)\}|, \qquad (3.22)$$

where $\widehat{F}_{\mathrm{obs}}(x)$ is the empirical cdf of $Q_i(\boldsymbol{\theta})$ based on $\boldsymbol{y}_{\mathrm{obs}}$, $\widehat{F}_{\mathrm{rep}}(x)$ is the empirical cdf of $Q_i(\boldsymbol{\theta})$ based on $\boldsymbol{y}_{\mathrm{rep}}$, and sign is the sign of this maximum deviation. Extreme probabilities, either close to 0 or close to 1, suggest lack of fit with respect to $\boldsymbol{T}(\cdot; \boldsymbol{\theta})$.

We illustrate these checks in our analyses of the Growth Hormone data in Section 4.2. We discuss modifications to these checks for incomplete data in Chapters 6 and 8.

3.6 Nonparametric Bayes

Nonparametric and semiparametric Bayesian approaches that weaken model assumptions have become much more common in the literature in recent years due to breakthroughs in computations. In the context of semiparametric regression, we have discussed spline approaches to model a trajectory over time nonparametrically in Examples 2.7 and Example 3.10. There are also a variety of approaches (see Further Reading) to specify distributions on the responses or random effects nonparametrically. Here we focus on mixtures of Dirichlet process models (Escobar, 1994; MacEachern, 1994) as a way to do this. This approach can often be implemented in WinBUGS (see Section 10.4 where it is used to specify the dropout distribution).

Consider univariate responses y_i with distribution $p(y_i \mid \theta_i)$. Assume the parameters θ_i follow some distribution $G(\theta_i)$. We specify a Dirichlet process prior on the distribution G as $G \sim DP(G_0, \alpha)$ with base measure αG_0 and mass parameter α. The distribution G_0 is often chosen as a simple parametric form (often what would be chosen for a fully parametric model). The mass parameter α provides a measure of how similar the nonparametric distribution G is to the parametric specification G_0. As $\alpha \to \infty$, $G \to G_0$.

As a simple nonparametric specification for a continuous response y_i, we can assume

$$
\begin{aligned}
y_i &\sim N(\mu_i, \sigma^2) \\
\mu_i &\sim G \\
G &\sim DP(G_0, \alpha),
\end{aligned}
$$

with G_0 a uniform (a, b) distribution.

These models can also be used for the random effects distribution. See for example Kleinman and Ibrahim (1998a, 1998b). For details on efficient computation for these models, see MacEachern and Muller (1998).

3.7 Further reading

Priors

In several recent papers, Daniels and others (Daniels and Kass, 1999, 2001; Daniels and Pourahmadi, 2002) have proposed approaches to shrink toward a parametric structure for the covariance matrix in longitudinal data; this offers both robustness to mis-specification of the structure and parsimony via the structure. This is well-suited to covariance matrices in longitudinal data models where parsimonious parametric models are often used to describe the covariance structure (see the multivariate normal models in Example 2.3 and Chapter 6), particularly Section 6.4.1. Gelman (2006) discusses the folded non-central t-distribution for a standard deviation parameter σ and shows it is conditionally conjugate for a normal model.

Some results on the propriety of the posterior for various longitudinal data models when using improper priors on regression coefficients, variances, and/or covariance matrices can be found in Natarajan and McCulloch (1995), Hobert and Casella (1996), Natarajan (2001), Berger, Strawderman, and Tang (2005), and Daniels (2006).

Prior elicitation and informative priors

For formulation and elicitation of informative priors, an alternative to conjugate priors is to use some sort of nonparametric prior, either constructed by just specifying quantiles (Berger and O'Hagan, 1988) or more formally using a modern Bayesian nonparametric approach (Oakley and O'Hagan, 2007). If multiple experts are to be used for elicitation, a major issue becomes how best to combine their opinions. Strategies can be found in Press (2003) and Garthwaite, Kadane, and O'Hagan (2005), with additional references therein.

For elicitation and informative priors for covariance matrices, we refer the reader to Brown, Le, and Zidek (1994), Garthwaite and Al-Awadhi (2001), and Daniels and Pourahmadi (2002); for correlation matrices, some recent work can be found in Zhang et al. (2006).

Other recent work on elicitation can be found in Kadane and Wolfson (1998), O'Hagan (1998), and Chen et al. (2003). Excellent reviews can be found in Chaloner (1996), Press (2003), and Garthwaite et al. (2005).

Computing

Slice sampling is an another approach to sample from unknown full conditional distributions (Damien, Wakefield, and Walker, 1999; Neal, 2003). The approach involves augmenting the parameter space with non-negative latent variables in a specific way. It is a special case of data augmentation. WinBUGS uses this approach for parameters with bounded domains.

Another approach that has been used to sample from unknown full conditionals is hybrid MC (Gustafson, 1997; Neal, 1996), which uses information from the first derivative of the log full conditional and has been implemented successfully in longitudinal models in Ilk and Daniels (2007) in the context of some extensions of MTM's for multivariate longitudinal binary data. For other candidate distributions for the Metropolis-Hastings algorithm, we refer the reader to Gustafson et al. (2004) who propose and review approaches that attempt to avoid the random walk behavior of the certain Metropolis-Hastings algorithm without necessarily doing an expensive numerical maximization at each iteration. This includes hybrid MC as a special case.

For settings where the logarithm of the full conditional distribution is log concave, Gilks and Wild (1992) have proposed easy-to-implement adaptive rejection sampling algorithms.

Parameter expansion algorithms to sample correlation matrices have recently been proposed (Liu, 2001; Liu and Daniels, 2006) that greatly simplify sampling by providing a conditionally conjugate structure by sampling a covariance matrix from the appropriate distribution and then transforming it back to a correlation matrix.

The efficiency of posterior summaries can be increased by using a technique called Rao-Blackwellization (Gelfand and Smith, 1990; Liu, Wong, and Kong, 1994). See Eberly and Casella (2003) for Rao-Blackwellization applied to credible intervals.

Additional extensions of data augmentation, termed *marginal, conditional,* and *joint,* that can further improve the efficiency and convergence of the MCMC algorithm can be found in van Dyk and Meng (2001).

Model comparison and model fit

For other diagnostics for assessing model fit, see Hodges (1998) and Gelfand, Dey, and Chang (1992).

Several authors have expressed concern that the posterior predictive probabilities, often called posterior predictive p-values, do not have a uniform distribution under the true model (Robins, Ventura, and van der Vaart, 2000). Given this concern, Hjort, Dahl, and Steinbakk (2006) recently proposed a way to appropriately calibrate these probabilities and more correctly call them p-values.

Semiparametric and nonparametric Bayes

For further discussion of priors in p-splines, see Berry et al. (2002) and Craineceau et al. (2005). Also, see Crainecau, Ruppert, and Carroll (2007) for recent developments in the longitudinal setting.

Most Bayesian approaches to regression splines with unknown number and location of knots use reversible jump MCMC methods (Green, 1995). See Denison, Mallick and Smith (1998) and DiMateo, Kass, and Genovese (2001) for methodology for a single longitudinal trajectory and Botts and Daniels (2007) for methodology for multiple trajectories (the typical longitudinal setting).

To model distributions nonparametrically, Dirichlet process priors (see Ferguson, 1982, and Sethuraman, 1994) and polya tree priors (Lavine, 1992, 1994) can be specified directly for the distribution of responses or for random effects.

CHAPTER 4

Worked Examples
using Complete Data

4.1 Overview

In this chapter we illustrate many of the ideas discussed in Chapters 2 and 3 by analyzing several of the datasets described in Chapter 1. For simplicity, we focus only on subjects with complete data. Missing data and dropout in these examples are addressed in a more definitive way in the analyses in Chapters 7 and 10. The sole purpose here is to illustrate using real data models from Chapter 2 and inferential methods described in Chapter 3.

4.2 Multivariate normal model: Growth Hormone study

We will illustrate aspects of model selection and inference for the multivariate normal model described in Example 2.3 using the data from the growth hormone trial described in Section 1.3. The primary outcome of interest is mean quadriceps strength (QS) at baseline, month 6, and month 12; the vector of outcomes for subject i is $\boldsymbol{Y}_i = (Y_{i1}, Y_{i2}, Y_{i3})^{\mathrm{T}}$. Subjects were randomized to one of four treatment groups: growth hormone plus exercise (EG), growth hormone (G), placebo plus exercise (EP), or placebo (P); we denote treatment group as Z_i, which takes values $\{1, 2, 3, 4\}$ for the four treatment groups, respectively. The main inferential objective for this data is to compare mean QS at 12 months in the four treatments.

4.2.1 Models

The general model is

$$\boldsymbol{Y}_i \mid Z_i = k \sim N(\boldsymbol{\mu}_k, \boldsymbol{\Sigma}_k), \tag{4.1}$$

with $\boldsymbol{\mu}_k = (\mu_{1k}, \mu_{2k}, \mu_{3k})^{\mathrm{T}}$ and $\boldsymbol{\Sigma}_k = \boldsymbol{\Sigma}(\boldsymbol{\phi}_k)$; $\boldsymbol{\phi}_k$ contains the six nonredundant parameters in the covariance matrix for treatment k. We compare models with $\boldsymbol{\phi}_k$ distinct for each treatment with reduced versions, including $\boldsymbol{\phi}_k = \boldsymbol{\phi}$.

Table 4.1 *Growth hormone trial: sample covariance matrices for each treatment. Elements below the diagonal are the pairwise correlations.*

	EG			G	
563	516	589	490	390	366
.68	1015	894	.85	429	364
.77	.87	1031	.83	.89	397

	EP			P	
567	511	422	545	380	292
.85	631	482	.81	403	312
.84	.91	442	.65	.81	367

4.2.2 Priors

Conditionally conjugate priors are specified as in Example 3.8 using diffuse choices for the hyperparameters,

$$\boldsymbol{\mu}_k \sim N(0, 10^6 \boldsymbol{I})$$
$$\boldsymbol{\Sigma}^{-1} \sim \text{Wishart}(\nu, \boldsymbol{A}^{-1}/\nu),$$

where $\nu = 3$ and $\boldsymbol{A} = \text{diag}\{(600, 600, 600)\}$. The scale matrix $\boldsymbol{A}$ is set so that the diagonal elements are roughly equal to those of the sample variances (Table 4.1).

4.2.3 MCMC details

For all the models, we ran four chains, each with $10,010$ iterations with a burn-in of 10 iterations (they all converged very quickly). The chains mixed well with minimal autocorrelation.

4.2.4 Model selection and fit

Based on an examination of sample covariance matrices in Table 4.1, we considered three models for the treatment specific covariance parameters:

(1) $\{\boldsymbol{\phi}_k : k = 1, \ldots, 4\}$
(2) $\{\boldsymbol{\phi}_k = \boldsymbol{\phi} : k = 1, \ldots, 4\}$,
(3) $\{\boldsymbol{\phi}_1, \boldsymbol{\phi}_k = \boldsymbol{\phi} : k = 2, 3, 4\}$.

We first used the DIC with $\boldsymbol{\theta} = \{\boldsymbol{\beta}, \boldsymbol{\Sigma}(\boldsymbol{\phi}_k)^{-1}\}$. Results appear in Table 4.2. The DIC results clearly favored covariance model (3). When we re-

Table 4.2 *Growth hormone trial: DIC for the three covariance models.*

	$\theta = \{\beta, \Sigma(\phi_k)^{-1}\}$			$\theta = \{\beta, \Sigma(\phi_k)\}$		
Model	DIC	$\text{Dev}(\overline{\theta})$	P_D	DIC	$\text{Dev}(\overline{\theta})$	P_D
(1)	2209	2139	35	2194	2153	20
(2)	2189	2152	18	2187	2154	17
(3)	2174	2127	24	2169	2132	19

parameterized the DIC using $\theta = (\beta, \Sigma(\phi_k))$, the difference between the DICs for models (1) and (2) decreased by over 60%, illustrating the sensitivity of the DIC to the parameterization of θ as discussed in Section 3.5.1.

We computed the multivariate version of Pearson's χ^2 statistic

$$T(y; \theta) = \sum_{i=1}^{n} (y_i - \mu_k)^{\mathrm{T}} \Sigma_k^{-1} (y_i - \mu_k) \tag{4.2}$$

as an overall measure of model fit based on the posterior predictive distribution of the residuals. The posterior predictive probability .11, which did not indicate a substantial departure of the model from the observed data.

4.2.5 Results

Based on the model selection results, we base inference on covariance model (3). Posterior means and 95% credible intervals for the mean parameters are given in Table 4.3. The posterior estimate of the baseline mean on the EG treatment differs substantially from the other arms, but our analysis is confined to completers. We can see the mean baseline quadriceps strength is very different from those who dropped out (cf. Table 1.2).

Histograms of the posterior distributions of the pairwise differences between the four treatment means at month 12 are given in Figure 4.1. We observe that for all the pairwise comparisons (except for G vs. P), most of the mass was either to the left (or right) of zero. We can further quantify this graphical determination by computing posterior probabilities that the month 12 mean was higher for pairs of treatments. As an example, we can quantify the evidence for the effect of growth hormone plus exercise over placebo plus exercise (EG vs. EP) using the posterior probability

$$P(\mu_{13} > \mu_{33} \mid y) = \int I\{(\mu_{13} - \mu_{33}) > 0\}\, p(\mu \mid y)\, d\mu.$$

This is an intractable integral, but it can be computed easily from the posterior

Table 4.3 *Growth hormone trial: posterior means and 95% credible intervals of the mean parameters for each treatment under covariance model (3).*

Treatment	Month		
	0	6	12
EG	78 (67, 89)	90 (76, 105)	88 (74, 103)
G	67 (59, 76)	64 (56, 72)	63 (56, 71)
EP	65 (57, 73)	81 (73, 89)	73 (65, 80)
P	67 (58, 76)	62 (54, 70)	63 (55, 71)

sample using

$$\frac{1}{K}\sum_{k=1}^{K} I\{\mu_{13}^{(k)} - \mu_{33}^{(k)} > 0\},$$

where k indexes draws from the MCMC sample of size K (after burn-in). For this comparison, the posterior probability was .99, indicating strong evidence that the 12-month mean on treatment EG is higher than treatment EP.

4.2.6 Conclusions

The analysis here suggests that the EG arm improves quadriceps strength more than the other arms. The DIC was used to choose the best fitting covariance model and posterior predictive checks suggested that this model fit adequately.

Our analysis has ignored dropouts, which can induce considerable bias in estimation of mean parameters. We saw this in particular by comparing the baseline mean on the EG treatment for the completers only vs. the full data given in Table 1.2. We discuss such issues, revisit this example, and do a more definitive analysis of this data in Section 7.2 under MAR and Section 10.2 using pattern mixture models that allow MNAR and sensitivity analyses.

4.3 Normal random effects model: Schizophrenia trial

We illustrate aspects of inference for the normal random effects model described in Example 2.1 using the data from the Schizophrenia Trial described in Section 1.2. The goal of this 6-week trial is to compare the mean change in schizophrenia severity (as measured by BPRS scores) from baseline to week 6 between four treatment groups. The treatment groups consist of three doses of a new treatment (low (L), medium (M), high(H)) and a standard dose (S) of an established treatment, halperidol.

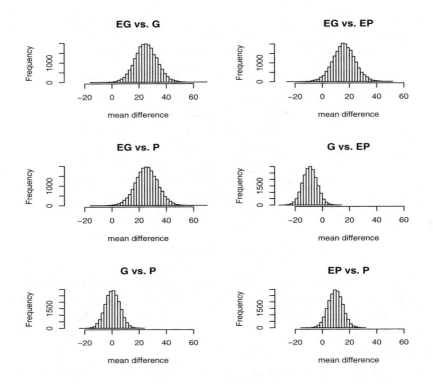

Figure 4.1 *Posterior distribution of pairwise differences of month 12 means for co-variance model (3) for the growth hormone trial. Reading across, the corresponding posterior probabilities that the pairwise difference in means are greater than zero are* $1.00, .97, 1.00, .04, .52, .96$.

The vector of outcomes for subject i is $\boldsymbol{Y}_i = (Y_{i1}, \ldots, Y_{i6})^{\mathrm{T}}$. We denote treatment group as Z_i, which takes values $\{1, 2, 3, 4\}$ for the four treatment groups, respectively.

4.3.1 Models

We use the normal random effects model described in Example 2.1,

$$
\begin{aligned}
\boldsymbol{Y}_i \mid \boldsymbol{b}_i, Z_i = k &\sim N(\boldsymbol{X}_i(\boldsymbol{\beta}_k + \boldsymbol{b}_i), \sigma^2 \boldsymbol{I}) \\
\boldsymbol{b}_i &\sim N(0, \boldsymbol{\Omega}),
\end{aligned}
$$

where the jth row of $\boldsymbol{X}_i$, $\boldsymbol{x}_{ij}$, is an orthogonal quadratic polynomial. Figure 1.1 provides justification for the quadratic trend. Using a random effects model here provides a parsimonious way to estimate var$(\boldsymbol{Y}_i \mid Z_i)$, which has 21 parameters, by reducing it to 7 parameters (6 in $\boldsymbol{\Omega}$ and the variance component σ^2).

4.3.2 Priors

Conditionally conjugate diffuse priors were specified as in Example 3.9 for $\boldsymbol{\beta}_k$ and $\boldsymbol{\Omega}^{-1}$,

$$\begin{aligned}
\boldsymbol{\beta}_k &\sim & N(\boldsymbol{0}, 10^6 \boldsymbol{I}_3), \\
\boldsymbol{\Omega}^{-1} &\sim & \text{Wishart}(\nu, \boldsymbol{A}^{-1}/\nu),
\end{aligned}$$

where $k = 1, 2, 3, 4$ indexes treatment group. For the Wishart prior, $\nu = 3$ and $\boldsymbol{A} = \text{diag}\{(120, 2, 2)\}$. Diagonal elements of $\boldsymbol{A}$ are chosen to be consistent with the observed variability (across subjects) of the orthogonal polynomial coefficients from fitting these to each subject individually. For the within-subject standard deviation σ, we use the bounded uniform prior

$$\sigma \sim \text{Unif}(0, 100).$$

4.3.3 MCMC details

For all the models, we ran four chains, each with $10,010$ iterations, with a burn-in of 10 iterations (similar to the Growth Hormone example, they all converged very quickly). The autocorrelation in the chains was higher than the growth hormone trial analysis in Section 4.2, but was negligible by iteration 25.

4.3.4 Results

Table 4.4 contains posterior means and 95% credible intervals for the treatment specific regression coefficients $\boldsymbol{\beta}_k$. Clearly the quadratic trend is necessary as the 95% credible interval for the quadratic coefficient for each treatment β_{2k} excluded zero.

Figure 4.2 plots the posterior means of the trajectories for each treatment over the 6 weeks of the trial; completers on the high dose appeared to do best, completers on the low dose did worst.

The last column of Table 4.4 gives the estimated change from baseline for all four treatments. All of the changes were negative, with credible intervals that excluded zero, indicating that all treatments reduced symptoms of schizophrenia severity. The smallest improvement was seen in the low dose arm, with a posterior mean of -12 and a 95% credible interval $(-19, -5)$.

Table 4.5 shows the estimated differences in the change from baseline between treatments. None of the changes from baseline were different between the four treatments as the credible intervals for all the differences covered zero.

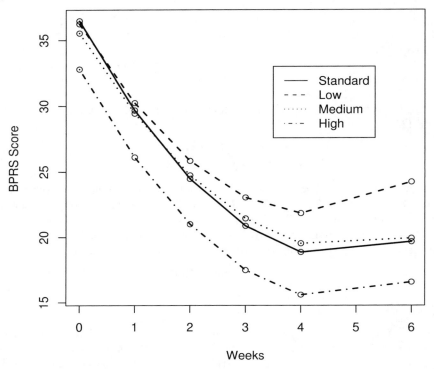

Figure 4.2 *Schizophrenia trial: posterior mean trajectories for each of the four treatments.*

Adequacy of the random effects covariance structure

Table 4.6 shows the sample covariance matrix and the posterior mean of the marginal covariance matrix under the random effects model $\Sigma = \sigma^2 I + w_i \Omega w_i^{\mathrm{T}}$. The random effects structure, with only seven parameters, appears to provide a good fit, capturing the form of the sample covariance matrix (unstructured). We do more formal covariance model selection for the schizophrenia data in Section 7.3.

4.3.5 Conclusions

Our analysis showed improvement in schizophrenia severity in all four treatment arms, but did not show significant differences in the change from baseline between the treatments. However, similar to the previous example with the growth hormone data, dropouts were ignored. Figure 1.1 suggests very differ-

Table 4.4 *Schizophrenia trial: posterior means and 95% credible intervals for the regression parameters and changes from baseline to week 6 for each treatment.*

Parameter	Treatment Group			
	Low	Medium	High	Standard
β_0	27	25	22	25
	(22, 32)	(22, 29)	(18, 26)	(21, 39)
β_1	−2.0	−2.6	−2.7	−2.8
	(−2.9, −1.1)	(−3.4, −1.9)	(−3.5, −2.0)	(−3.6, −2.1)
β_2	.8	.7	.8	.8
	(.5, 1.2)	(.4, 1.0)	(.5, 1.2)	(.5, 1.1)
Change	−12	−16	−16	−17
	(−17, −7)	(−20, −11)	(−21, −12)	(−21, −13)

Table 4.5 *Schizophrenia trial: posterior means and 95% credible intervals for the pairwise differences of the changes from baseline among the four treatments.*

L vs. M	L vs. H	L vs. S	M vs. H	M vs. S	H vs. S
4	4	5	0	1	1
(−3, 11)	(−3, 11)	(−2, 12)	(−6, 7)	(−5, 8)	(−6, 7)

ent BPRS scores between dropouts and completers. This example is revisited in Section 7.3 under an assumption of MAR using all the data.

4.4 Models for longitudinal binary data: CTQ I Study

We illustrate aspects of inference and model selection for longitudinal binary data models using data from the CTQ I smoking cessation trial, described in Section 1.4. The outcomes in this study are binary weekly quit status from weeks 1 to 12. The protocol called for women to quit at week 5. As such, the objective is to compare the time-averaged treatment effect from weeks 5 through 12. We fit both a logistic-normal random effects model (Example 2.2) and a marginalized transition model with first-order dependence (MTM(1), Example 2.5).

The vector of outcomes for woman i is $\boldsymbol{Y}_i = (Y_{i1}, \ldots, Y_{i,12})^{\mathrm{T}}$, binary indicators of weekly quit status for weeks 1 through 12. Treatment is denoted by $X_i = 1$ (exercise) or $X_i = 0$ (wellness).

Table 4.6 *Schizophrenia trial: sample covariance matrix and posterior mean of the marginal covariance matrix under the random effects structure.*

Sample Covariance					
126	104	81	85	75	74
.71	169	132	126	106	111
.58	.82	154	147	131	134
.57	.72	.89	179	149	151
.51	.62	.81	.85	171	162
.46	.59	.75	.78	.86	209

Posterior Mean Covariance					
153	111	99	90	86	86
.75	145	123	124	122	109
.62	.79	166	149	150	130
.53	.75	.84	189	169	150
.49	.71	.82	.86	204	168
.46	.60	.67	.73	.78	226

4.4.1 Models

(1) Logit normal random effects model

Our random effects specification uses a single random effect b_i, and assumes separate cessation rates before and after week 5, the designated quit week. Let $Z_j = I(j \geq 5)$. The treatment effect is the log odds ratio comparing time-averaged quit rate from week 5 to week 12 (i.e., over the period where $Z_j = 1$).

The model is given in two levels. At the first level, weekly quit indicators within individual are independent Bernoulli outcomes conditionally on b_i,

$$Y_{ij} \mid X_i, b_i, \sim \text{Ber}(\mu_{ij}^b),$$

with

$$\text{logit}(\mu_{ij}^b) = (1 - Z_j)\alpha + Z_j(b_i + \beta_0 + \beta_1 X_i).$$

Here, $\boldsymbol{\beta} = (\beta_0, \beta_1)^{\text{T}}$, and β_1 is the subject-specific log odds ratio capturing the treatment effect. At the second level, we assume

$$b_i \mid \tau^2 \sim N(0, \tau^2).$$

The marginal, or population-averaged, treatment effect is approximated by $\beta_1^m \approx \beta_1 K(\tau^2)$, where $K(\tau^2) = (1 + 0.346 \times \tau^2)^{-1/2}$ (Zeger and Liang, 1992).

(2) Marginalized transition model

We use a similar formulation to specify the MTM(1). Serial correlation is modeled only for the weeks following the quit date. The marginal mean follows

$$\text{logit}(\mu_{ij}) = (1 - Z_j)\alpha^m + Z_j(\beta_0^m + \beta_1^m X_i),$$

where superscript m denotes parameters from the marginal (population averaged) rather than conditional (subject-specific) distribution. For $j = 5, \ldots, 12$, serial correlation follows

$$\text{logit}(\phi_{ij}) = \Delta_{ij} + \gamma_i Y_{i,j-1},$$

where $\phi_{ij} = E(Y_{ij} \mid Y_{i,j-1})$. The correlation is allowed to differ by treatment group by setting

$$\gamma_i = \phi_0 + \phi_1 X_i.$$

4.4.2 Priors

Regression parameters in both models and serial correlation parameters in the MTM(1) are assumed independent with 'just proper' normal priors $N(0, 10^6)$. In the random effects model, the standard deviation τ of the random effects is given a Unif$(0, 100)$ prior.

4.4.3 MCMC details

For each model, inference is based on a posterior sample of size 30,000, generated by running three chains each with a burn-in comprising 1000 samples. The autocorrelation for some parameters in the logistic random effects model was quite large, not dying down until lag 30. For the MTM, the chains mixed very well with minimal autocorrelation.

4.4.4 Model selection

To select between the two models, we use posterior predictive loss (PPL). We specify the loss function given in (3.16). As a univariate summary of the multivariate response for each subject, we chose $T_i = \sum_{j=5}^{12} Y_{ij}$, the sum of the responses for weeks 5 to 12, which is proportional to the average quit rate for weeks 5 to 12. For these choices, the criterion takes the form

$$\text{PPL} = \sum_{i=1}^{n} \sigma_{i(T)}^2 + \frac{k}{k+1} \sum_{i=1}^{n} (\mu_{i(T)} - T_i)^2$$

$$= P + \frac{k}{k+1} G,$$

where

$$\mu_{i(T)} = E(T_{i,\text{rep}} \mid \boldsymbol{y}) = E\left(\sum_{j=5}^{12} Y_{i,\text{rep}}\right)$$

$$\sigma_{i(T)}^2 = \text{var}(T_{i,\text{rep}} \mid \boldsymbol{y}) = \text{var}\left(\sum_{j=5}^{12} Y_{i,\text{rep}}\right).$$

The choice of PPL here avoids the need to compute the integrated (over the random effects) likelihood necessary for the DIC. The PPL results appear in Table 4.7. The PPL for the MTM is considerably smaller than the logistic random effects model. The fit term G is almost ten times smaller; in fact, the fit term for the random effects models here is about the same as fitting an independence model (results not shown). In addition, the complexity term P is also much smaller under the MTM; the large complexity term for the random effects model is directly related to the large random effects variance (given in Table 4.8). Clearly, inference should be based on the MTM.

Table 4.7 *CTQ I: posterior predictive loss (PPL) for the marginalized transition model (MTM) and logistic random effects (LRE) models.*

Model	PPL	G	P
LRE	4346	2168	2178
MTM	303	166	137

4.4.5 Results

Table 4.8 summarizes the key parameters. Turning first to the MTM(1), we can compute the posterior mean and credible intervals of the smoking rates in the control and treatment conditions using the MCMC draws of β_0^{m} and β_1^{m}. The rates are given by

$$h_1(\beta_0^{\text{m}}) = \exp(\beta_0^{\text{m}})/\{1 + \exp(\beta_0^{\text{m}})\}$$
$$h_2(\beta_0^{\text{m}}, \beta_1^{\text{m}}) = \exp(\beta_0^{\text{m}} + \beta_1^{\text{m}})/\{1 + \exp(\beta_0^{\text{m}} + \beta_1^{\text{m}})\}.$$

We can compute their posterior means and credible intervals by evaluating these functions at each iteration of the MCMC sample to obtain a posterior sample of these parameters, giving posterior means (credible intervals) of .30 (.24, .37) and .40 (.32, .48), respectively.

In the MTM, the posterior mean of the population-averaged treatment log odds ratio is $\beta_1^{\mathrm{m}} = .43$ (odds ratio roughly 1.5). Its posterior credible interval $(-.04, .88)$ suggests those on exercise have higher cessation rates. Within-subject correlation in the control group is very high, as indicated by the posterior distribution of ϕ_0; the posterior of ϕ_1 does not suggest any difference in correlation between treatment groups (credible interval covers zero).

In the random effects model, the between subject variance τ^2 is very large, leading to a substantial difference between the subject-specific effect β_1 (posterior mean 1.8) and its model-derived population-averaged effect β_1^{m} (posterior mean .50). The mean of the population-averaged effect is similar between the two models.

Table 4.8 *CTQ I: posterior estimates of key model parameters from logistic-normal random effects (LRE) model and first-order marginalized transition model (MTM).*

Model	Parameter	Posterior Mean	95% Credible Interval
LRE	β_1	1.8	$(-.12, 3.7)$
	τ^2	6.0	$(4.7, 7.5)$
	β_1^{m}	.50	$(-.03, 1.0)$
MTM	β_0^{m}	$-.85$	$(-1.2, -.53)$
	β_1^{m}	.43	$(-.04, .88)$
	ϕ_0	4.4	$(3.9, 4.9)$
	ϕ_1	.79	$(.00, 1.6)$

4.4.6 Conclusions

Both models suggest a positive effect of exercise on quit rates. The population averaged treatment effect was very different than the subject specific effect in the random effects model due to the large variability of the random effect in the population. PPL strongly indicated that the MTM fit the data better than the logistic random effects model.

Figure 1.3 shows large difference in the observed quit rates between completers and dropouts. In Section 7.4, we do a more appropriate analysis under an MAR assumption using all the observed data.

4.5 Summary

We have used several of the data examples introduced in Chapter 1 to illustrate some features of the regression models introduced in Chapter 2 using the inferential methods described in Chapter 3. To maintain focus on key ideas, we only used subjects with complete data. Thus, the conclusions should not be viewed as valid given that a sensible approach to handle missing data was not undertaken. These analyses only serve to illustrate the models and inferential methods on real data.

In the rest of the book, we discuss ways to more appropriately handle missing data, starting in the next chapter with a careful definition of different types of missing data and dropout. All the approaches proposed to handle dropout in the remainder of the book will use all the observed data (not just completers).

CHAPTER 5

Missing Data Mechanisms and Longitudinal Data

5.1 Introduction

This chapter lays out the definitions and assumptions that are commonly used to make inference about a full-data distribution from incompletely observed data. Some specific examples help to motivate the discussion.

Example 5.1. *Dropout in a longitudinal clinical trial.*
In the Growth Hormone study described in Section 1.3, individuals were scheduled for measurement at baseline, month 6, and month 12. The target of inference is mean difference in quadriceps strength between the two treatment groups, at month 12, among everyone who began follow up. This is a parameter indexing the distribution of *full data*; i.e., the data that would have been observed had everyone completed the study. However several individuals dropped out of the study before completing follow up.   □

Example 5.2. *Dropout and mortality in a cohort study.*
The HER Study (Section 1.5) followed 871 HIV-infected women for up to 6 years, recording clinical and behavioral outcomes every 6 months. All women were scheduled for 12 follow-up measurements. About 10 percent died from AIDS and other causes, and about another 20 percent withdrew or were lost to follow-up. For a specific outcome of interest, like CD4 count, the full data could be defined in several ways. One definition is all CD4 counts that were scheduled to be observed (12 for each woman). Because of the difficulty in conceptualizing CD4 count for someone who has died, an alternate definition is CD4 counts taken while still alive.   □

A key theme of this chapter is the distinction, for purposes of model specification and inference, between *full data* and *observed data*. Inferences about full data that are based on incomplete observations must rely on assumptions about the distribution of missing responses. We demonstrate here that these assumptions are always encoded in a full data model by some combination of modeling assumptions and constraints on the parameter space. This is even true — actually it is especially true — for commonly used assumptions such as MAR and ignorability.

Section 5.2 provides definitions for full data and characterizes some common missing data processes such as dropout and monotone missingness. Section 5.3

85

gives a general conceptualization of the full-data model, and Section 5.4 describes missing data mechanisms such as MAR in the context of a full-data model. In Section 5.5 we show how the MAR assumption applies to dropout in longitudinal studies. The remainder of the chapter is devoted to model specification and interpretation under various assumptions for the missing data mechanism. A number of issues are highlighted for further reading.

5.2 Full vs. observed data

5.2.1 Overview

When drawing inference from incomplete data, it is necessary to expand the context of the modeling problem to characterizing some set of 'full' — but incompletely observed — data. In this section we differentiate *full data*, *observed data*, and *full data response*. The *full data* refers to all elements of the data that are observed or are intended to be observed, usually including some response or dependent variable, covariates of direct interest, auxiliary variables, and missing data indicators. Observed data comprise the observed subset of the full data. Finally, the full-data response refers to the dependent variable of primary interest. While missing data indicators are part of the full data, their distribution or relation to covariates typically is not of primary interest to the modeler.

As an example, consider the schizophrenia trial (Section 1.2), which compares standard treatment to one of several doses of a new therapy. The design calls for symptom severity to be recorded weekly for 6 weeks, but many patients drop out of the study before their follow-up is complete. Here, the full data consists of

- Symptom severity scores for all weeks, regardless of whether they actually were recorded;

- A binary random variable for each week, indicating whether the symptom severity was recorded (sometimes called missing data indicators);

- The covariate of interest, namely treatment group;

- Other baseline covariates that may have been recorded.

The full *response* data comprise only the first item. The *observed* data comprise the observed elements of symptom score, the missing data indicators, and the covariates.

The remainder of this section lays out notation and definitions; in Section 5.3 we define parameters that can be used to characterize various aspects of the full-data distribution.

5.2.2 Data structures

Consider first the case of bivariate response, where the first response is always observed but the second one may be missing. The *full data* for individual i consists of a response vector $\boldsymbol{Y}_i = (Y_{i1}, Y_{i2})^{\mathrm{T}}$, a $2 \times p$ matrix $\boldsymbol{X}_i$ of model covariates, a $2 \times q$ matrix $\boldsymbol{V}_i$ of auxiliary covariates, and the missing data indicator R_i. Throughout the book, we assume the process that causes response data to be missing is stochastic (as opposed to being part of a study design); hence R_i (and its generalizations below) are always treated as random variables.

The *observed* data for individual i are $\boldsymbol{O}_i = (\boldsymbol{Y}_{i,\text{obs}}, \boldsymbol{X}_i, \boldsymbol{V}_i, R_i)$, where

$$\boldsymbol{Y}_{i,\text{obs}} = \begin{cases} (Y_{i1}, Y_{i2})^{\mathrm{T}} & \text{if } R_i = 1 \\ Y_{i1} & \text{if } R_i = 0 \end{cases}.$$

For more general patterns of longitudinal data, similar structures apply. Consider first the case of temporally aligned observations. The full data response vector is $\boldsymbol{Y}_i = (Y_{i1}, \ldots, Y_{iJ})^{\mathrm{T}}$. The vector $\boldsymbol{R}_i = (R_{i1}, \ldots, R_{iJ})^{\mathrm{T}}$ indicates which components are observed, with $R_{ij} = 1$ if Y_{ij} is observed, and $R_{ij} = 0$ otherwise. Let $J_i = \sum_j R_{ij}$ denote the number of full-data components of $\boldsymbol{Y}_i$ that are observed. A useful way to represent the observed and missing components of the full data is the (possibly temporally unordered) partition $(\boldsymbol{Y}_{i,\text{obs}}^{\mathrm{T}}, \boldsymbol{Y}_{i,\text{mis}}^{\mathrm{T}})^{\mathrm{T}}$. The subvector $\boldsymbol{Y}_{i,\text{obs}}^{\mathrm{T}}$ is $J_i \times 1$, with elements $\{Y_{ij} : R_{ij} = 1\}$; similarly, $\boldsymbol{Y}_{i,\text{mis}}$ is $(J - J_i) \times 1$, with elements $\{Y_{ij} : R_{ij} = 0\}$.

For temporally misaligned data, the situation is somewhat different; the definition of full data is not always obvious because the timing of measurements is itself a random variable. For example, the full response data could be all values of a stochastic process $\{Y_i(t)\}$ over a fixed range of t, and the observed data is the subset $\{Y_i(t_{i1}), \ldots, Y_i(t_{iJ_i})\}$. Here the observation times $t_{i1}, \ldots, t_{iJ_i}$ might arise from a marked counting process $N_i(t)$, with observations of $Y_i(t)$ taken at points where $dN_i(t) = 1$; i.e., where the counting process jumps (see Yao et al., 1998; Lin and Ying, 2001; and Lin et al., 2004b for examples). While our focus is primarily on full data with fixed (as opposed to random) observation times, many of the topics we discuss can be applied to settings where observation times are random. For instance, a key consideration in modeling longitudinal data is whether the missing data process defined by $\boldsymbol{R}$ is independent of the responses $\boldsymbol{Y}$; for continuous-time processes, the consideration is whether $N(t)$ is in some sense independent of $Y(t)$. For a detailed treatment, see Lin and Ying (2001) and Tsiatis and Davidian (2004).

5.2.3 Dropout and other processes leading to missing responses

Any number of events can lead to missing data. Commonly encountered examples include

- Missed visits, either at random or for reasons related to response, such as when a patient in a study of depression fails to show up when he is experiencing symptoms;

- Withdrawal from a study, decided either by the participant or by the investigator conducting the study; common examples in pharmacologic trials are withdrawal due to side effects, toxicity, or lack of efficacy;

- Loss to follow-up, distinguished from withdrawal because the reasons are not reported;

- Death or disabling event, possibly related to the outcome but sometimes not; for example, in a longitudinal study of HIV, accidental death is not outcome related;

- Missingness by design, as in a longitudinal survey where only a subset of individuals is selected for follow-up.

This certainly is not exhaustive but covers many reasons for missing data in longitudinal studies. Adding to the complexity of handling missing data is that missingness may have different causes, and in most cases should be treated differently. Withdrawal for lack of efficacy is a different process than withdrawal for toxicity; outcome-related mortality must be treated differently than death by other causes.

Our focus throughout the book is primarily on dropout, and for simplicity we begin with the assumption that dropout — and its relation to the response process — can be captured using a single random variable. When there are distinct types of dropout such that they are related to outcome in different ways, then it is straightforward to introduce a multinomial version of the missing data indicator, and many of the same ideas discussed here will apply directly (see Rotnitzky et al., 2001, for example). Also, see Section 8.4.5.

Before moving forward to describing models for incomplete data, it is necessary to define formally two important terms: *dropout* and *monotone missing data pattern*. For many of the models we discuss in this and subsequent chapters, monotone missingness is a key requirement.

Definition 5.1. *Dropout process.*
For full-data responses $Y_1, \ldots, Y_J$ scheduled to be recorded at times $t_1, \ldots, t_J$, let $R_1, \ldots, R_J$ denote the missing data indicators, with $R_j = 1$ if Y_j is observed and $R_j = 0$ if missing. A missing data process is a dropout process if for some j such that $1 < j < J$, $R_j = 0 \Rightarrow R_{j+k} = 0$ for all $1 < k \leq J - j$; that is, there exists a measurement occasion j such that a missing response at time j implies all subsequent observations are missing.   □

Missingness that does not lead to dropout usually is called *intermittent* missingness because rather than truncating the longitudinal process, it creates gaps. Dropout that occurs in the absence of intermittent missingness leads to a monotone pattern for the responses

Definition 5.2. *Monotone missing data pattern.*
A missing data pattern is monotone if, for each individual, there exists a measurement occasion j such that $R_1 = \cdots = R_{j-1} = 1$ and $R_j = R_{j+1} = \cdots = R_J = 0$; that is, all responses are observed through time $j - 1$, and no responses are observed thereafter.   □

Throughout the book, we will use time-to-event variables S and U to characterize follow-up time and dropout time, respectively.

Definition 5.3. *Follow-up time.*
Assuming discrete time measurement at times times $t_1, \ldots, t_J$, the *follow-up time* is the time of the last observed measurement. It is denoted by t_S, where $S = \max\{j : R_j = 1\}$. When missingness is monotone, $S = \sum_j R_j$.   □

Definition 5.4. *Dropout time.*
Again assuming discrete-time measurement at times $t_1, \ldots, t_J$, the *dropout time* is the first scheduled measurement time after the follow-up time (i.e., after the last observed response). It is denoted by t_U, where $U = 1 + S$ and by convention, $U = 1 + J$ for those with complete follow-up. When missingness is monotone, $U = 1 + \sum_j R_j$.   □

Clearly S and U carry equivalent information. The distinction is made because in some situations it is more convenient to define the missing data process in terms of one or the other.

5.3 Full-data models and missing data mechanisms

5.3.1 Targets of inference

In almost all practical settings, the analyst is interested in drawing inference about a function of the parameter $\boldsymbol{\theta}$ that indexes the full data *response* model

$$p(\boldsymbol{y} \mid \boldsymbol{x}, \boldsymbol{\theta}) = p(y_1, y_2, \ldots, y_J \mid \boldsymbol{x}, \boldsymbol{\theta}).$$

Examples include the full-data mean $\boldsymbol{\mu}(\boldsymbol{\theta}) = E(\boldsymbol{Y} \mid \boldsymbol{\theta})$ and coefficients $\boldsymbol{\beta} = \boldsymbol{\beta}(\boldsymbol{\theta})$ in a regression model for $E(\boldsymbol{Y} \mid \boldsymbol{X}, \boldsymbol{\beta})$.

When data are completely observed, $p(\boldsymbol{y} \mid \boldsymbol{x}, \boldsymbol{\theta})$ can be specified directly. If responses are not fully observed, $\boldsymbol{\theta}$ is a function of the parameter $\boldsymbol{\omega}$ indexing a larger model $p(\boldsymbol{y}, \boldsymbol{r} \mid \boldsymbol{x}, \boldsymbol{\omega})$ of the full data $(\boldsymbol{Y}, \boldsymbol{R}, \boldsymbol{X})$; hence $\boldsymbol{\theta} = \boldsymbol{\theta}(\boldsymbol{\omega})$. The form of the full-data response model $p(\boldsymbol{y} \mid \boldsymbol{x}, \boldsymbol{\theta}(\boldsymbol{\omega}))$ will therefore depend on specifications of and assumptions about the full-data model $p(\boldsymbol{y}, \boldsymbol{r} \mid \boldsymbol{x}, \boldsymbol{\omega})$. When information on auxiliary covariates $\boldsymbol{V}$ is used, the full-data model is $p(\boldsymbol{y}, \boldsymbol{r}, \boldsymbol{v} \mid \boldsymbol{x}, \boldsymbol{\omega})$. We defer discussion of auxiliary covariates to Section 5.4.4.

Definition 5.5. *Full-data model.*
Let $\boldsymbol{Y}$ denote the full-data response vector for an individual, let $\boldsymbol{R}$ denote the associated vector of missingness indicators, and let $\boldsymbol{X}$ represent covariates of interest. The *full-data model* $p(\boldsymbol{y}, \boldsymbol{r} \mid \boldsymbol{x}, \boldsymbol{\omega})$ describes the joint distribution of $\boldsymbol{Y}$

and $\boldsymbol{R}$, conditionally on covariates $\boldsymbol{X}$, and is indexed by a finite-dimensional parameter $\boldsymbol{\omega}$.   □

Definition 5.6. *Full-data response model.*
The full-data response model characterizes the distribution of $\boldsymbol{Y}$ conditionally on covariates $\boldsymbol{X}$. It is indexed by a parameter $\boldsymbol{\theta} = \boldsymbol{\theta}(\boldsymbol{\omega})$ that is a subset or function of the full-data parameter $\boldsymbol{\omega}$, and is related to the full-data model via

$$p(\boldsymbol{y} \mid \boldsymbol{x}, \boldsymbol{\theta}(\boldsymbol{\omega})) = \sum_{r \in \mathscr{R}} p(\boldsymbol{y}, \boldsymbol{r} \mid \boldsymbol{x}, \boldsymbol{\omega}),$$

where $\mathscr{R}$ is the sample space of $\boldsymbol{R}$.   □

A given full-data model will give rise to only one observed-data model, but the converse of course is not true. Our motivation for starting with a full-data model is to force the analyst to specify, through appropriate subject-matter motivations, a model for the data generating process and its relationship to the missingness indicators.

Inference about $\boldsymbol{\theta}$ will depend crucially on choices made in specifying the full-data model $p(\boldsymbol{y}, \boldsymbol{r} \mid \boldsymbol{x}, \boldsymbol{\omega})$; however, observed data offer no information to validate these choices as they relate to the distribution of missing responses $\boldsymbol{Y}_{\mathrm{mis}}$. Assumptions made by the data analyst will generally exert considerable influence over final inferences, even for very large samples of observed data. This point can sometimes be lost when using popular models based on assumptions like missing at random (MAR) and implemented in standard software packages. In many cases, this approach yields valid inference about a particular full-data model under specific assumptions about the joint distribution of $\boldsymbol{Y}$ and $\boldsymbol{R}$ and/or constraints on the parameter $\boldsymbol{\omega}$.

As an example, many software packages now fit random effects models to incomplete data using maximum likelihood, under the assumption that both the random effects and the residual error distributions are normal (e.g. PROC MIXED in SAS or xtreg in Stata). When the data are incomplete, the analyst is implicitly assuming that the model specified in the software routine applies to the full-data response, and that missingness is MAR (or more specifically, *ignorable*; see Definition 5.12).

5.3.2 Missing data mechanisms

In general terms, the missing data mechanism is the stochastic mechanism leading to missingness among elements of $\boldsymbol{Y}$. Formally, we will use the term *missing data mechanism* to refer to the conditional distribution of missing data indicators $\boldsymbol{R}$ given the full-data response $\boldsymbol{Y}$ and covariates $\boldsymbol{X}$, denoted by $p(\boldsymbol{r} \mid \boldsymbol{y}, \boldsymbol{x}, \boldsymbol{\omega})$. Generally speaking, specification of $p(\boldsymbol{y}, \boldsymbol{r} \mid \boldsymbol{x}, \boldsymbol{\omega})$ will imply a missing data mechanism; in practice, the modeler often begins with a working assumption about the missing data mechanism and uses it to specify the full-data model.

Definition 5.7. *Missing data mechanism.*
The missing data mechanism is the model for the joint distribution of missing data indicators $\boldsymbol{R}$ as a function of $\boldsymbol{Y}$ and $\boldsymbol{X}$; it is indexed by a finite-dimensional parameter $\boldsymbol{\psi} = \boldsymbol{\psi}(\boldsymbol{\omega})$ and written as $p(\boldsymbol{r} \mid \boldsymbol{y}, \boldsymbol{x}, \boldsymbol{\psi}(\boldsymbol{\omega}))$. $\qquad\square$

Any full-data model can therefore be factored as the product of a full-data response model and the associated missing data mechanism,

$$p(\boldsymbol{y}, \boldsymbol{r} \mid \boldsymbol{x}, \boldsymbol{\omega}) = p(\boldsymbol{y} \mid \boldsymbol{x}, \boldsymbol{\theta}(\boldsymbol{\omega})) \, p(\boldsymbol{r} \mid \boldsymbol{y}, \boldsymbol{x}, \boldsymbol{\psi}(\boldsymbol{\omega})). \tag{5.1}$$

Being able to characterize the missing data mechanism does not depend on whether the full-data model has been specified according to (5.1), which is commonly referred to as a selection model factorization. The implied missing data mechanism can, in principle, be derived from any specification of the full-data model, but it may not always take a closed form.

5.4 Common assumptions about the missing data mechanism

Restrictions on the missing data mechanism can be classified as missing completely at random (MCAR), missing at random (MAR), or missing not at random (MNAR); these progressively weaker assumptions delineate the dependence of the missing data indicators $\boldsymbol{R}$ on the observed and missing parts of the full-data response vector $\boldsymbol{Y}$ (conditionally on $\boldsymbol{X}$ and possibly also on $\boldsymbol{V}$). The taxonomy and its terminology were developed by Rubin (1976), and generalized using the 'coarsening' framework by Heitjan and Rubin (1991). Robins et al. (1995), Jacobsen and Keiding (1995), Gill and Robins (1997), and Gill and Robins (2001) develop formalizations for stochastic process and longitudinal data settings. Recent surveys on the characterization of missing data mechanisms in longitudinal studies, from a variety of different perspectives, can be found in Little (1995), Little and Rubin (2002), van der Laan and Robins (2003), Hogan et al. (2004b), Tsiatis (2006), and Molenberghs and Kenward (2007).

To define missing data assumptions, it is useful to rewrite the missing data mechanism as

$$p(\boldsymbol{r} \mid \boldsymbol{y}, \boldsymbol{x}, \boldsymbol{\psi}) = p(\boldsymbol{r} \mid \boldsymbol{y}_{\text{obs}}, \boldsymbol{y}_{\text{mis}}, \boldsymbol{x}, \boldsymbol{\psi})$$

to emphasize dependence of $\boldsymbol{R}$ on the observed and missing components of $\boldsymbol{Y}$. Although we write $\boldsymbol{\psi}$, it is assumed that $\boldsymbol{\psi} = \boldsymbol{\psi}(\boldsymbol{\omega})$ unless stated otherwise.

5.4.1 Missing completely at random (MCAR)

Missingness completely at random occurs when data are missing for reasons wholly unrelated to either the observed or missing parts of $\boldsymbol{Y}$, conditionally on $\boldsymbol{X}$. With repeated measurements, the data can be used to provide evidence against the MCAR assumption through the association between $\boldsymbol{R}$ and $\boldsymbol{Y}_{\text{obs}}$; e.g., by regressing R_j on one or more observed components of $(Y_1, \ldots, Y_{j-1})^{\text{T}}$ (see also Chen and Little, 1999). In general, however, analyses under the weaker MAR assumption will provide valid inference when MCAR holds. We include a description here for the purposes of completeness and to introduce MAR and MNAR.

Definition 5.8. *Missing completely at random.*
Missing responses are missing completely at random (MCAR) if, for all $\boldsymbol{x}$ and $\boldsymbol{\psi}$,

$$p(\boldsymbol{r} \mid \boldsymbol{y}, \boldsymbol{x}, \boldsymbol{\psi}) = p(\boldsymbol{r} \mid \boldsymbol{x}, \boldsymbol{\psi});$$

i.e., if $p(\boldsymbol{r} \mid \boldsymbol{y}, \boldsymbol{x}, \boldsymbol{\psi})$ is a constant function of $\boldsymbol{y}$.   □

One implication of MCAR is that missingness can be fully explained by covariates $\boldsymbol{X}$ that are included in the full-data model. Another is that the full-data distribution can be factored as

$$p(\boldsymbol{y}, \boldsymbol{r} \mid \boldsymbol{x}, \boldsymbol{\omega}) = p(\boldsymbol{y} \mid \boldsymbol{x}, \boldsymbol{\omega})\, p(\boldsymbol{r} \mid \boldsymbol{x}, \boldsymbol{\omega}),$$

meaning that the observed and missing response data have the same distribution, conditionally on $\boldsymbol{X}$. Under MCAR, valid inference about the full-data response distribution can be based solely on those with complete response data. Even though this is a valid approach, it discards data on those without complete response and therefore does not make optimal use of the available data; the posterior variance of $\boldsymbol{\theta}$ may be higher than necessary.

Example 5.3. *Missing completely at random without covariates.*
A longitudinal cohort study of school performance will record test scores each year for 2 years of secondary school; these are Y_1 and Y_2. In the first year, 1000 students are sampled and their test scores are recorded. In the second year, budget constraints force the investigators to reduce the sample size to 800. Data on Y_2 are collected for a random subsample of the original 1000 students. Because the subsample is randomly drawn, missing responses on Y_2 are missing completely at random.   □

Example 5.4. *Missing completely at random with covariates.*
In the same study, suppose the investigators are interested in comparing test scores between boys and girls, and let X denote gender. The distribution of interest is $p(y_1, y_2 \mid x, \boldsymbol{\theta})$. Suppose further that the random subsample at time 2 oversamples girls, so that girls are more likely to have Y_2 observed. Because gender is a model covariate, the missing data on Y_2 are missing completely at random.   □

The MCAR assumption usually is not realistic for longitudinal studies because unplanned missingness is so common. Consider the first CTQ study of smoking cessation described in Section 1.4, where $\boldsymbol{Y} = (Y_1, \ldots, Y_{12})^{\mathrm{T}}$ are the cessation indicators measured weekly over the course of the study, X is the binary indicator of treatment, and $\boldsymbol{R} = (R_1, \ldots, R_{12})^{\mathrm{T}}$ are the binary missingness indicators. The full-data model of interest is $p(\boldsymbol{y} \mid x, \boldsymbol{\theta})$, which characterizes the joint distribution of smoking outcomes conditionally on treatment group X. The MCAR assumption states that $p(\boldsymbol{r} \mid \boldsymbol{y}_{\mathrm{obs}}, \boldsymbol{y}_{\mathrm{mis}}, x, \boldsymbol{\psi}) = p(\boldsymbol{r} \mid x, \boldsymbol{\psi})$; in this context, it means that probability of nonresponse can depend on treatment group, but within treatment group, nonresponse is completely independent of cessation outcomes. Under the plausible (and testable) scenario that within treatment group, participants observed to be heavier smokers are more likely to drop out, the MCAR assumption would not hold.

5.4.2 Missing at random (MAR)

A more realistic condition for many longitudinal studies is missing at random (MAR), which requires that missingness is independent of missing responses $\boldsymbol{Y}_{\mathrm{mis}}$, conditionally on observed responses $\boldsymbol{Y}_{\mathrm{obs}}$ and model covariates $\boldsymbol{X}$.

Definition 5.9. *Missing at random.*
Missing responses are missing at random (MAR) if, for all $\boldsymbol{y}_{\mathrm{obs}}$, $\boldsymbol{x}$ and $\boldsymbol{\psi}$,

$$p(\boldsymbol{r} \mid \boldsymbol{y}_{\mathrm{obs}}, \boldsymbol{y}_{\mathrm{mis}}, \boldsymbol{x}, \boldsymbol{\psi}) = p(\boldsymbol{r} \mid \boldsymbol{y}_{\mathrm{obs}}, \boldsymbol{x}, \boldsymbol{\psi});$$

i.e., if $p(\boldsymbol{r} \mid \boldsymbol{y}_{\mathrm{obs}}, \boldsymbol{y}_{\mathrm{mis}}, \boldsymbol{x}, \boldsymbol{\psi})$ is a constant function of $\boldsymbol{y}_{\mathrm{mis}}$. $\square$
The MAR assumption is one component of the *ignorability* condition, which allows valid inference about $\boldsymbol{\theta}$ to be based on the likelihood function for $\boldsymbol{Y}_{\mathrm{obs}}$. Ignorability is discussed in detail in Section 5.7.

Example 5.5. *A missing at random mechanism.*
Continuing with Examples 5.3 and 5.4, if those with lower values of the first test score Y_1 are less likely to sit for the second test, then missingness in Y_2 depends on Y_1. If missingness further depends on X, but does not depend on any other variable including Y_1, then the missing data mechanism is MAR. $\square$

Application of the MAR condition for dropout mechanisms requires some more development and is discussed in more detail in Section 5.5.

5.4.3 Missing not at random (MNAR)

Although the MAR assumption is fairly general in that it allows the missing data mechanism (and sometimes the missing data itself) to be explained by observables, there are cases where MAR may fail to hold. MAR will not be valid, for example, when the probability of missingness depends on the value of the missing response — or on other unobservables — even after conditioning on observed data. We refer to mechanisms of this type as missing not at random (MNAR).

Definition 5.10. *Missing not at random.*

Missing responses are missing not at random (MNAR) if, for some $\boldsymbol{y}_{\mathrm{mis}} \neq \boldsymbol{y}'_{\mathrm{mis}}$,

$$p(\boldsymbol{r} \mid \boldsymbol{y}_{\mathrm{obs}}, \boldsymbol{y}_{\mathrm{mis}}, \boldsymbol{x}, \boldsymbol{\psi}) \neq p(\boldsymbol{r} \mid \boldsymbol{y}_{\mathrm{obs}}, \boldsymbol{y}'_{\mathrm{mis}}, \boldsymbol{x}, \boldsymbol{\psi});$$

i.e., if $\boldsymbol{R}$ depends on some part of $\boldsymbol{Y}_{\mathrm{mis}}$, even after conditioning on $\boldsymbol{Y}_{\mathrm{obs}}$ and $\boldsymbol{X}$. $\square$

Example 5.6. *Missing not at random mechanism: dropout in a smoking cessation study.*

Returning to the smoking cessation trial, dropout is an inevitable complication. Consider a simplified version where the full data consist of cessation outcomes (Y_1, Y_2), treatment indicator X, and missing data indicator R. Under MAR,

$$P(R = 1 \mid X, Y_1, Y_2 = 1) = P(R = 1 \mid X, Y_1, Y_2 = 0),$$

which implies that, conditional on treatment group and time-1 smoking status Y_1, the dropout probability is the same for both smokers and nonsmokers at time 2. In smoking cessation studies, this assumption will not hold if the dropout probability differs by Y_2 (or equivalently, if smoking status Y_2 differs between dropouts and non-dropouts having the same (X, Y_1) profile). Lichtenstein and Glasgow (1992) cite empirical evidence that in smoking cessation studies, participants followed up after dropout are more likely than not to have experienced relapse, raising the possibility that MAR is not plausible here. $\square$

5.4.4 Auxiliary variables

Recall that the full-data response model is $p(\boldsymbol{y} \mid \boldsymbol{x}, \boldsymbol{\theta}(\boldsymbol{\omega}))$, and primary interest is in parameters governing this distribution. We have implicitly assumed up to this point that $(\boldsymbol{Y}, \boldsymbol{R}, \boldsymbol{X})$ constitutes the only data that potentially are available.

In many cases, however, investigators may have access to a set of *auxiliary variables*, denoted by $\boldsymbol{V}$, that may be correlated with one or more of $(\boldsymbol{Y}, \boldsymbol{R}, \boldsymbol{X})$. Examples of auxiliary variables include covariates that are observed but not of direct interest, or other longitudinal responses that are measured concurrently with $\boldsymbol{Y}$. In a clinical trial, the model covariates $\boldsymbol{X}$ would include only treatment indicators and possibly stratification variables; however, other covariates might be available. In an HIV cohort study where longitudinal CD4 counts are the response of interest, the investigator may also have access to viral load measurements taken at the same time points.

If the primary responses are completely observed, the modeler generally has little or nothing to gain by using these auxiliary variables to estimate covariate effects or other parameters; however, when data are incomplete, auxiliary variables can be helpful to the extent that they explain variation in the joint distribution of Y and R. Missingness mechanisms can be defined conditionally on auxiliary variables as well.

Definition 5.11. *Auxiliary variable MAR (A-MAR).*
Missing responses are A-MAR if MAR holds conditionally on auxiliary variables V; i.e., if, for all y_{obs}, x, z, and ψ,

$$p(r \mid y_{\mathrm{obs}}, y_{\mathrm{mis}}, x, v, \psi) = p(r \mid y_{\mathrm{obs}}, x, v, \psi).$$

$\square$

Clearly, A-MAR does not imply MAR. Consequently, when primary interest is in the full-data response model $p(y \mid x, \theta)$ — which does not involve v — it is necessary to specify a full-data model conditional on v and integrate it out to obtain the unconditional model. Specifically, the full-data model of $(Y, R \mid X)$ is

$$p(y, r \mid x, \omega) = \int p(y, r, v \mid x, \omega) \, dv. \qquad (5.2)$$

Returning to the longitudinal HIV example, suppose Y is longitudinal CD4 count and V is longitudinal viral load, and that there is appreciable missingness in CD4. The MAR condition requires the analyst to assume that conditional on observed CD4 history, missingness is unrelated to the CD4 count that would have been measured; this may be unrealistic. Further suppose that the investigator can confidently specify a joint model for CD4 count and viral load (e.g., based on knowledge of disease progression dynamics), and that viral load is thought to explain sufficient variability in CD4 count that one could use it to predict missing responses. In other words, the analyst is willing to assume A-MAR, or MAR conditional on viral load.

This approach requires specifying the joint distribution of $(Y, R, V \mid X)$ under the integral sign in (5.2). One possibility is to write

$$p(y, r, v \mid x, \omega) = p(y, v \mid x, \omega) \, p(r \mid y, v, x, \omega),$$

where the first factor is the joint model for CD4 and viral load, the second is the missing data mechanism. As we will see in Chapter 7, incorporating auxiliary covariates can sometimes be simpler than it would appear. Under A-MAR (and assuming ignorability, see Section 5.7), the missing data mechanism can be left unspecified, although the joint model for Y and V does have to be specified. Section 6.6 describes some specific models that can be used for this purpose; in Section 7.5, we illustrate by analyzing data from the CTQ II trial, where the responses are smoking cessation outcomes measured weekly for 8 weeks, and the auxiliary process is individual weight, recorded concurrently.

5.5 Missing at random applied to dropout processes

The definitions of MCAR, MAR, and MNAR given in the previous section were general. In this section we describe the implications of the MAR assumption for handling monotone missing data patterns caused by dropout. Although the MAR condition is rather simple to state and interpret for cross-sectional data, it is somewhat less transparent when applied to longitudinal data, particularly if missingness patterns are non-monotone. When missingness follows a monotone pattern, as when dropout is the sole cause of missing data, it turns out that MAR has an intuitive representation in terms of the hazard of dropout, with obvious analogies to the cross-sectional setting.

Our discussion assumes a discrete set of measurement times within individual, and assumes dropout happens at one of the scheduled times, $t_1, \ldots, t_J$. Recall that the missingness indicator at time t_j is R_j, with $R_j = I(Y_j$ observed). Hence dropout time, denoted by U, is $U = 1 + \sum_j R_j$ (see also Definition 5.4). Further define, for each individual,

$$\overline{\boldsymbol{Y}}_j = \{Y_1, \ldots, Y_j\}$$

to be the subset of the full-data vector $\boldsymbol{Y}$ that would be observed up to and including measurement time j (i.e., the response history up to and including t_j).

Under monotone dropout, the MAR condition states that

$$P(U = j \mid \overline{\boldsymbol{Y}}_J) = P(U = j \mid \overline{\boldsymbol{Y}}_{j-1}),$$

or that the marginal probability of dropping out at t_j cannot depend on measurements at or beyond t_j (Molenberghs et al., 1998). In practical settings, this version of MAR can be somewhat difficult to interpret because it makes reference to the marginal probability of dropping out at t_j, $P(U = j)$, rather than the more intuitive hazard function, $P(U = j \mid U \geq j)$. The marginal dropout rate is the probability of dropping out at t_j among all those who begin follow-up, while the hazard rate is the probability of dropping out at t_j among those still eligible to be measured at t_j.

At time t_j and for any given response history $\overline{\boldsymbol{Y}}_k = \{Y_1, \ldots, Y_k\}$, let

$$h(t_j \mid \overline{\boldsymbol{Y}}_k) = P(U = j \mid U \geq j, \overline{\boldsymbol{Y}}_k)$$

denote the hazard of dropout at t_j conditional on $\overline{\boldsymbol{Y}}_k$. It turns out that monotone missingness caused by dropout is MAR if and only if the hazard of dropout at t_j depends *only* on observed response history $\overline{\boldsymbol{Y}}_{j-1}$, and conditionally on $\overline{\boldsymbol{Y}}_{j-1}$ is independent of present and future outcomes $\{Y_j, Y_{j+1}, \ldots, Y_J\}$.

Proposition 1. *Under monotone dropout, missingness is MAR if and only if for all j, the hazard of dropout at t_j is independent of current and future responses $Y_j, \ldots, Y_J$ conditionally on past responses $Y_1, \ldots, Y_{j-1}$.*

Proof: Recall that under MAR, $P(U = j \mid \overline{\boldsymbol{Y}}_J) = P(U = j \mid \overline{\boldsymbol{Y}}_{j-1})$. By definition, the hazard can be written in terms of the marginal dropout probabilities

$$h(t_j \mid \overline{\boldsymbol{Y}}_k) = \frac{P(U = j \mid \overline{\boldsymbol{Y}}_k)}{1 - \sum_{l=1}^{j-1} P(U = l \mid \overline{\boldsymbol{Y}}_k)},$$

for any (j, k) such that $1 \leq j \leq J$ and $1 \leq k \leq J$. Hence, under MAR, we have

$$
\begin{aligned}
h(t_j \mid \overline{\boldsymbol{Y}}_J) &= \frac{P(U = j \mid \overline{\boldsymbol{Y}}_J)}{1 - \sum_{l=1}^{j-1} P(U = l \mid \overline{\boldsymbol{Y}}_J)} \\
&\overset{\text{MAR}}{=} \frac{P(U = j \mid \overline{\boldsymbol{Y}}_{j-1})}{1 - \sum_{l=1}^{j-1} P(U = l \mid \overline{\boldsymbol{Y}}_{l-1})} \\
&= \text{function of } Y_1, \ldots, Y_{j-1} \text{ only.}
\end{aligned}
$$

Conversely, suppose the hazard of dropout at t_j given $\overline{\boldsymbol{Y}}_J$ depends only on $\overline{\boldsymbol{Y}}_{j-1}$; that is,

$$
\begin{aligned}
P(R_j = 0 \mid R_1 = \cdots = R_{j-1} = 1, \overline{\boldsymbol{Y}}_J) \\
= P(R_j = 0 \mid R_1 = \cdots = R_{j-1} = 1, \overline{\boldsymbol{Y}}_{j-1}).
\end{aligned}
$$

To show this implies MAR, we need to demonstrate that $P(U = j \mid \overline{\boldsymbol{Y}}_J)$ is a constant with respect to $Y_j, \ldots, Y_J$, or equivalently, is a function of $Y_1, \ldots, Y_{j-1}$ only. Writing it in terms of the hazard, we have

$$
\begin{aligned}
P(U = j \mid \overline{\boldsymbol{Y}}_J) &= P(R_1 = \cdots = R_{j-1} = 1, R_j = 0 \mid \overline{\boldsymbol{Y}}_J) \\
&= P(R_j = 0 \mid R_1 = \cdots = R_{j-1} = 1, \overline{\boldsymbol{Y}}_J) \\
&\quad \times P(R_{j-1} = 1 \mid R_{j-2} = \cdots = R_1 = 1, \overline{\boldsymbol{Y}}_J) \\
&\quad \times \cdots \times P(R_1 = 1 \mid \overline{\boldsymbol{Y}}_J) \\
&= h(t_j \mid \overline{\boldsymbol{Y}}_J) \prod_{k=1}^{j-1} \{1 - h(t_k \mid \overline{\boldsymbol{Y}}_J)\}. \qquad (5.3)
\end{aligned}
$$

But if $h(t_j \mid \overline{\boldsymbol{Y}}_J) = h(t_j \mid \overline{\boldsymbol{Y}}_{j-1})$, then the probability of dropout at j, given by (5.3), reduces to

$$
\begin{aligned}
P(U = j \mid \overline{\boldsymbol{Y}}_J) &= h(t_j \mid \overline{\boldsymbol{Y}}_{j-1}) \prod_{k=1}^{j-1} \{1 - h(t_k \mid \overline{\boldsymbol{Y}}_{k-1})\} \\
&= \text{function of } Y_1, \ldots, Y_{j-1} \text{ only.}
\end{aligned}
$$

$\square$

This result is useful for making the MAR assumption more transparent; dropout is really an event process, and it is more intuitive to think about imposing conditions on hazard of dropout rather than on its marginal probability. It also is helpful in writing down the observed data likelihood under MAR, which we discuss in Section 5.7

In the foregoing, we have not discussed the role of covariates, and in particular time-varying covariates. These results will be expected to hold when exogenous covariates are included in the full-data model; i.e., covariates meeting the following conditions (see also Section 2.7):

1. Time-invariant covariates observed at baseline;

2. Time-varying covariates that are a deterministic function of baseline value (e.g., age);

3. Stochastic time-varying covariates that are external to the response process, and can be expected to be observed throughout the measurement period (e.g., when air pollution level measured from a monitoring station is a covariate, and health status of an individual is the response process; the individual may drop out of the study, but air pollution will continue to be measured.).

5.6 Observed data posterior of full-data parameters

Thus far, we have described processes that cause data to be missing, distinguished full from observed data, described the Rubin taxonomy for missing data mechanisms, and showed how it applies to monotone missingness induced by dropout. In particular we showed that in monotone patterns, the MAR condition is equivalent to assuming the hazard of dropout at some time t_j depends only on observed response data up to and including time t_{j-1}. We are now ready to describe formal methods for drawing posterior inference about full-data parameters from observed (incomplete) data.

When using incomplete data to draw posterior inference about aspects of the full data, constraints or model specifications must be imposed on a full-data model and on priors for parameters indexing that model. The full-data model $p(\boldsymbol{y}, \boldsymbol{r} \mid \boldsymbol{\omega})$, including its assumptions and constraints, will encode both the full-data response model $p(\boldsymbol{y} \mid \boldsymbol{\theta}(\boldsymbol{\omega}))$ and the missing data mechanism $p(\boldsymbol{r} \mid \boldsymbol{y}, \boldsymbol{\psi}(\boldsymbol{\omega}))$. Because all components of $\boldsymbol{\omega}$ may not be identified, the prior distribution $p(\boldsymbol{\omega})$ will directly inform certain aspects of the posterior distribution $p(\boldsymbol{\omega} \mid \boldsymbol{y}_{\mathrm{obs}}, \boldsymbol{r})$. We do not confine ourselves in this discussion to a particular factorization or specification of the full-data model; the setup described here applies to *any* full-data model. Posterior inference under a variety of full-data model specifications is described and discussed throughout the remainder of this chapter.

By definition, the posterior distribution of the full-data parameter $\boldsymbol{\omega}$, given observed data $(\boldsymbol{y}_{\mathrm{obs}}, \boldsymbol{r})$, is

$$
\begin{aligned}
p(\boldsymbol{\omega} \mid \boldsymbol{y}_{\mathrm{obs}}, \boldsymbol{r}) &= \frac{p(\boldsymbol{y}_{\mathrm{obs}}, \boldsymbol{r} \mid \boldsymbol{\omega})\, p(\boldsymbol{\omega})}{\int p(\boldsymbol{y}_{\mathrm{obs}}, \boldsymbol{r} \mid \boldsymbol{\omega})\, p(\boldsymbol{\omega}) d\boldsymbol{\omega}} \\
&= \frac{p(\boldsymbol{y}_{\mathrm{obs}}, \boldsymbol{r} \mid \boldsymbol{\omega})\, p(\boldsymbol{\omega})}{p(\boldsymbol{y}_{\mathrm{obs}}, \boldsymbol{r})}.
\end{aligned}
\tag{5.4}
$$

(For clarity we will drop covariate dependence, but it is implied throughout.) The *observed-data likelihood*, proportional in ω to $p(\boldsymbol{y}_{\mathrm{obs}}, \boldsymbol{r} \mid \omega)$, is obtained by averaging the full-data model over all possible realizations of the missing data,

$$L(\omega \mid \boldsymbol{y}_{\mathrm{obs}}, \boldsymbol{r}) \quad \propto \quad p(\boldsymbol{y}_{\mathrm{obs}}, \boldsymbol{r} \mid \omega)$$

$$= \int p(\boldsymbol{y}_{\mathrm{obs}}, \boldsymbol{y}_{\mathrm{mis}}, \boldsymbol{r} \mid \omega) \, d\boldsymbol{y}_{\mathrm{mis}}.$$

Because the marginal distribution $p(\boldsymbol{y}_{\mathrm{obs}}, \boldsymbol{r})$ in the denominator of (5.4) is a constant with respect to ω, the observed-data posterior is proportional in ω to the product of the full data prior $p(\omega)$ and the observed-data likelihood,

$$p(\omega \mid \boldsymbol{y}_{\mathrm{obs}}, \boldsymbol{r}) \quad \propto \quad p(\omega) \, L(\omega \mid \boldsymbol{y}_{\mathrm{obs}}, \boldsymbol{r}).$$

In the following sections we describe specific strategies for observed-data posterior inference about full-data parameters. In general the missing data distribution is not identifiable, and assumptions are needed for the full-data model and informative priors need to be specified. When missingness is MAR, the ignorability assumption provides considerable simplification of the observed-data posterior. We define this assumption in the next section.

5.7 The ignorability assumption

5.7.1 Likelihood and posterior under ignorability

The ignorability condition, first described by Rubin (1976), can be used to facilitate posterior inference without having to specify the missing data mechanism. An ignorable missing data mechanism meets three conditions, given here in its definition.

Definition 5.12. *Ignorable missing data mechanism (Little and Rubin, 2002).* A missing data mechanism is said to be *ignorable* for the purposes of posterior inference if

1. The missing data mechanism is MAR.
2. The full data parameter ω can be decomposed as $\omega = (\boldsymbol{\theta}, \boldsymbol{\psi})$, where $\boldsymbol{\theta}$ indexes the full-data response model $p(\boldsymbol{y} \mid \boldsymbol{\theta})$ and $\boldsymbol{\psi}$ indexes the missing data mechanism $p(\boldsymbol{r} \mid \boldsymbol{y}, \boldsymbol{\psi})$.
3. The parameters $\boldsymbol{\theta}$ and $\boldsymbol{\psi}$ are *a priori* independent; i.e.,

$$p(\boldsymbol{\theta}, \boldsymbol{\psi}) = p(\boldsymbol{\theta})p(\boldsymbol{\psi}). \qquad \qquad \square$$

An important consequence of ignorability is that inference about $\boldsymbol{\theta}$ — typically of direct interest — can be based on a likelihood function that is proportional in $\boldsymbol{\theta}$ to the observed-data response model,

$$p(\boldsymbol{y}_{\mathrm{obs}} \mid \boldsymbol{\theta}) = \int p(\boldsymbol{y}_{\mathrm{obs}}, \boldsymbol{y}_{\mathrm{mis}} \mid \boldsymbol{\theta}) \, d\boldsymbol{y}_{\mathrm{mis}}.$$

This can be shown as follows. Recall that the observed data posterior for $\boldsymbol{\omega}$ is given by

$$p(\boldsymbol{\omega} \mid \boldsymbol{y}_{\text{obs}}, \boldsymbol{r}) \quad \propto \quad p(\boldsymbol{\omega}) \, L(\boldsymbol{\omega} \mid \boldsymbol{y}_{\text{obs}}, \boldsymbol{r}). \tag{5.5}$$

By condition (2) of the ignorability assumption, we can write

$$L(\boldsymbol{\omega} \mid \boldsymbol{y}_{\text{obs}}, \boldsymbol{r}) = L(\boldsymbol{\theta}, \boldsymbol{\psi} \mid \boldsymbol{y}_{\text{obs}}, \boldsymbol{r}).$$

Conditions (1) and (3) — MAR and prior independence between $\boldsymbol{\theta}$ and $\boldsymbol{\psi}$ — lead to the further simplification

$$
\begin{aligned}
L(\boldsymbol{\theta}, \boldsymbol{\psi} \mid \boldsymbol{y}_{\text{obs}}, \boldsymbol{r}) \quad &\propto \quad p(\boldsymbol{r}, \boldsymbol{y}_{\text{obs}} \mid \boldsymbol{\theta}, \boldsymbol{\psi}) \\
&= \quad \int p(\boldsymbol{r} \mid \boldsymbol{y}_{\text{obs}}, \boldsymbol{y}_{\text{mis}}, \boldsymbol{\psi}) \, p(\boldsymbol{y}_{\text{obs}}, \boldsymbol{y}_{\text{mis}} \mid \boldsymbol{\theta}) \, d\boldsymbol{y}_{\text{mis}} \\
&\stackrel{\text{MAR}}{=} \quad p(\boldsymbol{r} \mid \boldsymbol{y}_{\text{obs}}, \boldsymbol{\psi}) \int p(\boldsymbol{y}_{\text{obs}}, \boldsymbol{y}_{\text{mis}} \mid \boldsymbol{\theta}) \, d\boldsymbol{y}_{\text{mis}} \\
&= \quad p(\boldsymbol{r} \mid \boldsymbol{y}_{\text{obs}}, \boldsymbol{\psi}) \, p(\boldsymbol{y}_{\text{obs}} \mid \boldsymbol{\theta}) \\
&= \quad L_1(\boldsymbol{\psi} \mid \boldsymbol{r}, \boldsymbol{y}_{\text{obs}}) \, L_2(\boldsymbol{\theta} \mid \boldsymbol{y}_{\text{obs}}).
\end{aligned}
\tag{5.6}
$$

Conditions (2) and (3) from the definition of ignorability imply that $p(\boldsymbol{\omega}) = p(\boldsymbol{\theta})p(\boldsymbol{\psi})$. Substituting the simplified likelihood (5.6) into the observed-data posterior (5.5) and using the factored prior, we have

$$
\begin{aligned}
p(\boldsymbol{\omega} \mid \boldsymbol{y}_{\text{obs}}, \boldsymbol{r}) \quad &\propto \quad p(\boldsymbol{\omega}) \, L(\boldsymbol{\omega} \mid \boldsymbol{y}_{\text{obs}}, \boldsymbol{r}) \\
&= \quad \{p(\boldsymbol{\psi})L_1(\boldsymbol{\psi} \mid \boldsymbol{r}, \boldsymbol{y}_{\text{obs}})\} \, \{p(\boldsymbol{\theta})L_2(\boldsymbol{\theta} \mid \boldsymbol{y}_{\text{obs}})\}
\end{aligned}
$$

(Little and Rubin, 2002). Because the posterior factors over $\boldsymbol{\psi}$ and $\boldsymbol{\theta}$, the ignorability condition implies

$$p(\boldsymbol{\theta} \mid \boldsymbol{y}_{\text{obs}}) \propto p(\boldsymbol{\theta}) \, L_2(\boldsymbol{\theta} \mid \boldsymbol{y}_{\text{obs}}), \tag{5.7}$$

which does not involve the part of the likelihood corresponding to $p(\boldsymbol{r} \mid \boldsymbol{y}, \boldsymbol{\psi})$; hence the missing data mechanism is 'ignored' for the purposes of drawing posterior inference about $\boldsymbol{\theta}$. This should not be interpreted to mean that observations with missing data can be ignored or discarded; what is meant by 'ignoring' the missing data mechanism is that $p(\boldsymbol{r} \mid \boldsymbol{y}, \boldsymbol{\psi})$ does not have to be specified (Laird, 1988).

We illustrate with a derivation for the case where $J = 2$. For those with missing Y_2, the observed data are $(Y_1, R = 0)$, and the contribution to the likelihood is proportional to $p(y_1, r \mid \boldsymbol{\theta}, \boldsymbol{\psi})$, evaluated at $r = 0$. The expression for this term must be derived by integrating over the distribution of missing responses — in this case y_2 — for the assumed full-data model,

$$p(y_1, r \mid \boldsymbol{\theta}, \boldsymbol{\psi}) \quad = \quad \int p(y_1, y_2 \mid \boldsymbol{\theta}) \, p(r \mid y_1, y_2, \boldsymbol{\psi}) \, dy_2. \tag{5.8}$$

If missingness in Y_2 is MAR, then $p(r \mid y_1, y_2, \boldsymbol{\psi}) = p(r \mid y_1, \boldsymbol{\psi})$ and (5.8) reduces to

$$
\begin{aligned}
p(y_1, r \mid \boldsymbol{\theta}, \boldsymbol{\psi}) &= \int p(y_1, y_2 \mid \boldsymbol{\theta})\, p(r \mid y_1, \boldsymbol{\psi})\, dy_2 \\
&= p(r \mid y_1, \boldsymbol{\psi}) \int p(y_1, y_2 \mid \boldsymbol{\theta})\, dy_2 \\
&= p(r \mid y_1, \boldsymbol{\psi}) p(y_1 \mid \boldsymbol{\theta}).
\end{aligned}
$$

Hence the observed-data likelihood contribution for individual i factors over $\boldsymbol{\psi}$ and $\boldsymbol{\theta}$ as

$$
\begin{aligned}
L(\boldsymbol{\theta}, \boldsymbol{\psi} \mid \boldsymbol{y}_{i,\text{obs}}, r_i) &\propto \{p(y_{i1}, y_{i2} \mid \boldsymbol{\theta})\}^{r_i} \{p(y_{i1} \mid \boldsymbol{\theta})\}^{(1-r_i)} \\
&\quad \times p(r_i \mid y_{i1}, \boldsymbol{\psi}) \\
&= L_1(\boldsymbol{\theta} \mid \boldsymbol{y}_{i,\text{obs}})\, L_2(\boldsymbol{\psi} \mid r_i, y_{i1}) \qquad (5.9)
\end{aligned}
$$

An additional implication of ignorability is that missing responses $\boldsymbol{Y}_{\text{mis}}$ can be imputed or extrapolated from observed data $\boldsymbol{Y}_{\text{obs}}$ using only the full-data response model (ignoring the missing data mechanism); this can be seen as follows:

$$
\begin{aligned}
p(\boldsymbol{y}_{\text{mis}} \mid \boldsymbol{y}_{\text{obs}}, \boldsymbol{r}, \boldsymbol{\omega}) &= \frac{p(\boldsymbol{y}_{\text{mis}}, \boldsymbol{y}_{\text{obs}}, \boldsymbol{r} \mid \boldsymbol{\omega})}{p(\boldsymbol{y}_{\text{obs}}, \boldsymbol{r} \mid \boldsymbol{\omega})} \\
&= \frac{p(\boldsymbol{r} \mid \boldsymbol{y}_{\text{obs}}, \boldsymbol{y}_{\text{mis}}, \boldsymbol{\psi})\, p(\boldsymbol{y}_{\text{obs}}, \boldsymbol{y}_{\text{mis}} \mid \boldsymbol{\theta})}{p(\boldsymbol{r} \mid \boldsymbol{y}_{\text{obs}}, \boldsymbol{\psi})\, p(\boldsymbol{y}_{\text{obs}} \mid \boldsymbol{\theta})} \qquad (5.10) \\
&= \frac{p(\boldsymbol{y}_{\text{obs}}, \boldsymbol{y}_{\text{mis}} \mid \boldsymbol{\theta})}{p(\boldsymbol{y}_{\text{obs}} \mid \boldsymbol{\theta})} \qquad (5.11) \\
&= p(\boldsymbol{y}_{\text{mis}} \mid \boldsymbol{y}_{\text{obs}}, \boldsymbol{\theta}),
\end{aligned}
$$

where (5.10) follows from condition (2) of ignorability, and (5.11) is a direct consequence of MAR (condition (1) of ignorability). Applied to the simple case of bivariate response data where Y_1 is always observed but Y_2 may be missing, we see that

$$
p(y_2 \mid y_1, r = 0, \boldsymbol{\theta}) = p(y_2 \mid y_1, r = 1, \boldsymbol{\theta});
$$

hence missing values of Y_2 can be imputed from a model of $p(y_2 \mid y_1, r = 1, \boldsymbol{\theta})$ (e.g., from a regression of Y_2 on Y_1 among those with complete data). This consequence of the ignorability assumption figures prominently when using data augmentation techniques for posterior inference under MAR (Chapter 6).

5.7.2 Factored likelihood with monotone ignorable missingness

When the missing data pattern is monotone — as in the previous example — the method of factored likelihood can be used to simplify computations for posterior sampling based on observed-data likelihoods for longitudinal data (Little and Rubin, 2002). The joint distribution of observed data in (5.9) can be rewritten as

$$p(\boldsymbol{y}_{i,\mathrm{obs}}, r_i \mid \boldsymbol{\theta}, \boldsymbol{\psi}) \;=\; p(y_{i2} \mid y_{i1}, \boldsymbol{\theta})^{r_i}\, p(y_{i1} \mid \boldsymbol{\theta})\, p(r_i \mid y_{i1}, \boldsymbol{\psi}).$$

If there exists an invertible mapping $g(\boldsymbol{\theta}) = \boldsymbol{\phi} = (\boldsymbol{\phi}_{2|1}, \boldsymbol{\phi}_1)$ such that

$$p(y_2, y_1 \mid \boldsymbol{\theta}) \;=\; p(y_2 \mid y_1, \boldsymbol{\phi}_{2|1})\, p(y_1 \mid \boldsymbol{\phi}_1),$$

then (5.9) can be factored over the parameters $\boldsymbol{\phi}_{2|1}$ and $\boldsymbol{\phi}_1$ indexing 'complete-data' likelihood terms,

$$\begin{aligned} L(\boldsymbol{\phi}, \boldsymbol{\psi} \mid \boldsymbol{y}_{i,\mathrm{obs}}, r_i) \;&\propto\; p(y_{i2} \mid y_{i1}, \boldsymbol{\phi}_{2|1})^{r_i}\, p(y_{i1} \mid \boldsymbol{\phi}_1)\, p(r_i \mid y_{i1}, \boldsymbol{\psi}) \\ &=\; L_1(\boldsymbol{\phi}_{2|1} \mid \boldsymbol{y}_{i,\mathrm{obs}})\, L_2(\boldsymbol{\phi}_1 \mid \boldsymbol{y}_{i,\mathrm{obs}})\, L_3(\boldsymbol{\psi} \mid r_i, y_{i1}). \end{aligned}$$

For distributions where this factorization and transformation of the parameter space yields simple models for the conditional distributions (e.g., of Y_2 given Y_1) and for priors specified directly on $\boldsymbol{\phi}$, posterior sampling is simplified because each term of the likelihood behaves like a 'complete-data' model, using observable data.

The general form of a factored likelihood for observed data under ignorable monotone missingness is

$$L(\boldsymbol{\phi} \mid \boldsymbol{y}_{i,\mathrm{obs}}) = p(y_{i1} \mid \boldsymbol{\phi}_1) \prod_{j=2}^{J} p(y_{ij} \mid y_{i,j-1}, \ldots, y_{i1}, \boldsymbol{\phi}_{j|\{1,\ldots,j-1\}})^{r_{ij}}, \qquad (5.12)$$

where $\boldsymbol{\phi}_{j|\{1,\ldots,j-1\}}$ denotes parameters indexing the conditional distribution of $\boldsymbol{Y}_j$ given $\{Y_1, \ldots, Y_{j-1}\}$.

For some models used in applied settings, generating a factored likelihood is sometimes possible and sometimes not. The multivariate normal distribution without covariates can be factored easily, and is further illustrated in Example 5.7. For other models, such as the marginalized transition model in Example 5.8, or most models having time-constant covariate effects, the likelihood may not factor easily over separable components of $\boldsymbol{\phi}$.

5.7.3 The practical meaning of 'ignorability'

Calling the missing data mechanism 'ignorable' can be deceptive; if we view the process of inference as starting from a full-data model and moving forward with a series of assumptions, then plainly we have not 'ignored' the missing data mechanism at all; indeed, we have made explicit constraints on it by placing the MAR assumption on $p(\boldsymbol{r} \mid \boldsymbol{y}, \boldsymbol{\psi})$, by partitioning the parameter space of $\boldsymbol{\omega}$ into $(\boldsymbol{\theta}, \boldsymbol{\psi})$, and by assuming *a priori* independence between $\boldsymbol{\theta}$ and $\boldsymbol{\psi}$. It is possible to form qualitative justifications of these assumptions, but they cannot be empirically verified, and therefore should not be viewed as benign. The term 'ignorable' means that once these assumptions and parameter constraints have been imposed, and assuming the full-data model has been correctly specified, the functional form of the missing data mechanism $p(r \mid y, \psi)$ can be ignored when making inference about the full-data response model parameter $\theta = \theta(\omega)$ from observed data.

5.8 Examples of full-data models under MAR

We are now ready to illustrate applications of MAR and ignorability in constructing models from some specific distributions. Here we focus on construction of the observed data likelihood, identification of full-data parameters, and the predictive distribution of Y_{mis} given Y_{obs}.

Example 5.7. *Bivariate normal with missing Y_2.*
Consider a study, such as a pretest-posttest design, where measurements Y_1 are taken at baseline and Y_2 at follow-up. All individuals have their outcome recorded at baseline, but some are missing at time 2. Further assume that primary interest is in $E(Y_2)$, the mean of Y_2.

The full data for each individual is (Y_1, Y_2, R), where $R = 1$ if Y_2 is observed and $R = 0$ if not. In general terms, the full-data model is $p(y_1, y_2, r \mid \omega)$. If we wish to characterize the full-data response distribution $p(y_1, y_2 \mid \theta(\omega))$ using a bivariate normal model under ignorability, we must make the following assumptions:

1. The full-data parameter ω can be decomposed as
$$p(y_1, y_2, r \mid \omega) \quad = \quad p(y_1, y_2 \mid \theta)\, p(r \mid y_1, y_2, \psi),$$
   where $\omega = (\theta, \psi)$;

2. The full-data response model $p(y_1, y_2 \mid \theta)$ is the $N(\mu, \Sigma)$ density function, where $\Sigma = [\{\sigma_{ij}\}]$ and $\theta = (\mu_1, \mu_2, \sigma_{11}, \sigma_{12}, \sigma_{22})$;

3. The missing data mechanism is MAR, which implies
$$p(r \mid y_1, y_2, \psi) \quad = \quad p(r \mid y_1, \psi);$$

4. The prior distribution factors as $p(\theta, \psi) = p(\theta)\, p(\psi)$.

Let $\sigma = (\sigma_{11}, \sigma_{12}, \sigma_{22})$, so that $\theta = (\mu, \sigma)$. From our bivariate example (5.9), the observed-data likelihood contribution for individual i is

$$
\begin{aligned}
L(\theta \mid y_{\text{obs},i}) \quad &\propto \quad p(y_{i1}, y_{i2} \mid \mu, \sigma)^{r_i} \left\{ \int p(y_{i1}, y_2 \mid \mu, \sigma)\, dy_2 \right\}^{(1-r_i)} \\
&= \quad p(y_{i1}, y_{i2} \mid \mu, \sigma)^{r_i}\, p(y_{i1} \mid \mu_1, \sigma_{11})^{(1-r_i)}. \quad (5.13)
\end{aligned}
$$

A more convenient parameterization for posterior inference directly on the observed data posterior uses a factored likelihood. If $(Y_1, Y_2)^{\text{T}} \sim N(\mu, \Sigma)$, then

$$Y_2 \mid Y_1 \; \sim \; N(\beta_0 + \beta_1 Y_1, \tau^2)$$
$$Y_1 \; \sim \; N(\mu_1, \sigma_{11}^2).$$

Using standard results from the normal distribution, there exists an invertible mapping $g(\boldsymbol{\theta}) = \boldsymbol{\phi}$ such that $\boldsymbol{\phi} = (\beta_0, \beta_1, \tau, \mu_1, \sigma_{11})$ (see also Little and Rubin, 2002). The parameter of interest, $\mu_2 = E(Y_2)$, is simply a function of $\boldsymbol{\phi}$,

$$E(Y_2) = E_{Y_1}\{E(Y_2 \mid Y_1)\} = \beta_0 + \beta_1 \mu_1.$$

The observed-data likelihood (5.13) is equivalent to

$$L(\boldsymbol{\phi} \mid \boldsymbol{y}_{i,\text{obs}}) \; \propto \; p(y_{i2} \mid y_{i1}, \beta_0, \beta_1, \tau)^{r_i} \, p(y_{i1} \mid \mu_1, \sigma_{11}).$$

This likelihood factors over separable components of $\boldsymbol{\phi}$; hence posterior computations can be based directly on the observed data-posterior. Moreover, the model is more naturally parameterized for sensitivity analysis, a topic that is taken up in considerable detail in Chapters 9 and 10. In Section 10.2, we use a trivariate normal model with similar parameterization to analyze data from the Growth Hormone Study. □

Posterior computations can also be based on the full-data model directly using data augmentation (see Section 3.4.3), which is described in the setting of ignorable missingness in Section 6.3. This approach is more convenient when we cannot use the factored likelihood or priors are specified directly on the full-data response model parameters. The following example shows an instance where the missing data mechanism is ignorable but the factored likelihood approach cannot be used because of specific parameter constraints. Here, the effect of covariates on the marginal mean is assumed constant over time.

Example 5.8. *Marginalized transition model under ignorability.*
For longitudinal responses $Y_1, \ldots, Y_J$ and covariates $\boldsymbol{X}_j$, the full-data MTM(1) model assumes the marginal regression

$$g(\pi_j) = \boldsymbol{x}_j \boldsymbol{\beta},$$

where $\pi_j = E(Y_j \mid \boldsymbol{x}_j)$. The full-data distribution is specified as

$$Y_1 \mid \boldsymbol{x}_1 \; \sim \; \text{Ber}(\pi_1)$$
$$Y_j \mid Y_{j-1}, \boldsymbol{x}_j \; \sim \; \text{Ber}(\phi_j) \quad j = 2, \ldots, J,$$

where

$$h(\phi_j) \;\; = \;\; \Delta_{ij} + y_{j-1}\gamma.$$

Hence $\boldsymbol{\theta} = (\boldsymbol{\beta}, \gamma)$. Given $\boldsymbol{x}_j$, $\boldsymbol{\beta}$, and γ, the parameter $\Delta_j = \Delta(\boldsymbol{x}_j)$ is determined by the marginal regression constraints given in (2.8).

Under ignorability and monotone missingness, there is no need to specify

$p(\boldsymbol{r} \mid \boldsymbol{y})$. To write the observed-data likelihood, let $\pi_{i1}(\boldsymbol{\beta}) = g^{-1}(\boldsymbol{x}_{i1}\boldsymbol{\beta})$, $\Delta_{ij} = \Delta(\boldsymbol{x}_{ij})$, and $\phi_{ij}(\Delta_{ij}, \gamma) = h^{-1}(\Delta_{ij} + y_{i,j-1}\gamma)$. Then

$$L(\boldsymbol{\beta}, \gamma \mid \boldsymbol{y}_{i,\text{obs}}) = p(y_{i1} \mid \boldsymbol{x}_{i1}, \boldsymbol{\beta}) \prod_{j=2}^{J} p(y_{ij} \mid y_{i,j-1}, \boldsymbol{x}_{ij}, \Delta_{ij}, \gamma), \quad (5.14)$$

where

$$p(y_{i1} \mid \boldsymbol{x}_{i1}, \boldsymbol{\beta}) = \{\pi_{i1}(\boldsymbol{\beta})\}^{y_{i1}} \{1 - \pi_{i1}(\boldsymbol{\beta})\}^{1-y_{i1}}$$
$$p(y_{ij} \mid y_{i,j-1}, \boldsymbol{x}_{ij}, \Delta_{ij}, \gamma) = \{\phi_{ij}(\Delta_{ij}, \gamma)\}^{y_{ij}} \{1 - \phi_{ij}(\Delta_{ij}, \gamma)\}^{1-y_{ij}}.$$

Comparing (5.14) to the general version of the factored likelihood (5.12), we see that the observed-data likelihood for this model does not factor over $(\boldsymbol{\beta}, \gamma)$ because γ and $\boldsymbol{\beta}$ are common to each term under the product sign in (5.14) through Δ. In Section 7.4, this model is used to analyze data from the first CTQ trial, using an ignorability assumption. □

Although MAR is needed for ignorability, not all MAR mechanisms lead to a model where the missing data is ignorable. This is particularly true for mixture specifications, evident from this example.

Example 5.9. *Nonignorability under MAR: mixture of bivariate normals.* The two-component mixture of bivariate normal distributions having common variance is given by

$$(Y_1, Y_2)^{\mathrm{T}} \mid R = 1 \sim N(\boldsymbol{\mu}^{(1)}, \boldsymbol{\Sigma})$$
$$(Y_1, Y_2)^{\mathrm{T}} \mid R = 0 \sim N(\boldsymbol{\mu}^{(0)}, \boldsymbol{\Sigma})$$
$$R \sim \text{Ber}(\phi),$$

where $\boldsymbol{\Sigma}$ has elements $\boldsymbol{\sigma} = (\sigma_{11}, \sigma_{12}, \sigma_{22})$; hence $\boldsymbol{\omega} = (\boldsymbol{\mu}^{(0)}, \boldsymbol{\mu}^{(1)}, \boldsymbol{\sigma}, \phi)$. In this model, the components are defined by whether Y_2 is observed ($R = 1$). The missing data mechanism is the discriminant function (e.g., Anderson, 1984, Chapter 6)

$$\log \left\{ \frac{P(R = 1 \mid \boldsymbol{y}, \boldsymbol{\omega})}{P(R = 0 \mid \boldsymbol{y}, \boldsymbol{\omega})} \right\} = \psi_0 + \psi_1 y_1 + \psi_2 y_2,$$

where

$$\psi_0 = \log \frac{\phi}{1 - \phi} + (\boldsymbol{\mu}^{(0)})^{\mathrm{T}} \boldsymbol{\Sigma}^{-1} \boldsymbol{\mu}^{(0)} - (\boldsymbol{\mu}^{(1)})^{\mathrm{T}} \boldsymbol{\Sigma}^{-1} \boldsymbol{\mu}^{(1)}$$

$$\psi_1 = \frac{\Delta_1 \sigma_{22} - \Delta_2 \sigma_{12}}{\sigma_{11}\sigma_{22} - \sigma_{12}^2} \quad (5.15)$$

$$\psi_2 = \frac{\Delta_2 \sigma_{11} - \Delta_1 \sigma_{12}}{\sigma_{11}\sigma_{22} - \sigma_{12}^2}, \quad (5.16)$$

with $\Delta_j = \mu_j^{(1)} - \mu_j^{(0)}$ denoting between-pattern difference in means at measurement time j. It can be shown that when the pattern-specific variances are equal, MAR holds if and only if

$$E(Y_2 \mid Y_1, R = 0, \boldsymbol{\omega}) = E(Y_2 \mid Y_1, R = 1, \boldsymbol{\omega})$$

for all possible realizations of Y_1; therefore, in the mixture of normals model, MAR holds if and only if

$$\mu_2^{(0)} - \frac{\sigma_{12}}{\sigma_{22}}\mu_1^{(0)} + \frac{\sigma_{12}}{\sigma_{22}}Y_1 \;\; = \;\; \mu_2^{(1)} - \frac{\sigma_{12}}{\sigma_{22}}\mu_1^{(1)} + \frac{\sigma_{12}}{\sigma_{22}}Y_1$$

for all Y_1. Hence MAR holds if and only if

$$\Delta_2 = \frac{\sigma_{12}}{\sigma_{11}}\Delta_1.$$

From (5.15) and (5.16) we see that MAR corresponds to $\psi_1 = \Delta_1/\sigma_{11}$ and $\psi_2 = 0$; however, Δ_1 and σ_{11} index both the missing data mechanism (through ψ_1) and the full-data response model

$$p(\boldsymbol{y} \mid \boldsymbol{\omega}) \;\; = \;\; \phi\, p(\boldsymbol{y} \mid r = 1, \boldsymbol{\mu}^{(1)}, \boldsymbol{\sigma}) + (1 - \phi)\, p(\boldsymbol{y} \mid r = 0, \boldsymbol{\mu}^{(0)}, \boldsymbol{\sigma}),$$

violating condition (2) in the definition of ignorability (Definition 5.12). Hence the missing data mechanism is MAR but not ignorable.   □

5.9 Full-data models under MNAR

The steps involved in drawing inference from incomplete data are specification of a full data model, specification of the priors, and then sampling from the posterior distribution of full-data parameters, given the observed data $\boldsymbol{Y}_{\text{obs}}$, $\boldsymbol{R}$ and $\boldsymbol{X}$. In Section 5.7 we showed the form of the posterior distribution for the full-data response parameter $\boldsymbol{\theta}$ when the missing data mechanism is ignorable.

As we saw in the previous section, identification of a full-data response model — particularly the part involving $\boldsymbol{Y}_{\text{mis}}$ — requires making unverifiable assumptions about the full-data model $f(\boldsymbol{y}, \boldsymbol{r} \mid \boldsymbol{x}, \boldsymbol{\omega})$. In the previous section, we relied on the ignorability assumption to identify the full-data response model $f(\boldsymbol{y} \mid \boldsymbol{x}, \boldsymbol{\theta})$.

When ignorability is believed not to be a suitable assumption, one can use a more general class of models that allows missing data indicators to depend on missing responses themselves. Essentially these models allow one to parameterize the conditional dependence between $\boldsymbol{R}$ and $\boldsymbol{Y}_{\text{mis}}$, given $\boldsymbol{Y}_{\text{obs}}$ and $\boldsymbol{X}$. Without the benefit of untestable assumptions, this association structure cannot be identified from observed data for the simple reason that $\boldsymbol{Y}_{\text{mis}}$ cannot be observed. Inference therefore depends on some combination of (a) unverifiable parametric assumptions and (b) informative prior distributions. This can be viewed either as a strength or a weakness. The goal of this section is to illustrate, using some simple examples, the construction, identification, and practical utility of some commonly used 'nonignorable' models.

5.9.1 Selection models

The selection model (SM) approach factors the full-data distribution as

$$p(\boldsymbol{y}, \boldsymbol{r} \mid \boldsymbol{x}, \boldsymbol{\omega}) = p(\boldsymbol{y} \mid \boldsymbol{x}, \boldsymbol{\omega}) p(\boldsymbol{r} \mid \boldsymbol{y}, \boldsymbol{x}, \boldsymbol{\omega}),$$

so that the full-data response model $p(\boldsymbol{y} \mid \boldsymbol{x}, \boldsymbol{\omega})$ and the missing data mechanism $p(\boldsymbol{r} \mid \boldsymbol{y}, \boldsymbol{x}, \boldsymbol{\omega})$ must be specified by the analyst. We have written the model so that both factors are indexed by a common parameter $\boldsymbol{\omega}$; it is common but not necessary to assume that $\boldsymbol{\omega}$ can be decomposed as $(\boldsymbol{\theta}, \boldsymbol{\psi})$, separate parameters for each factor.

Selection models can be attractive for several reasons. First, the analyst is usually interested in the full-data response distribution $p(\boldsymbol{y} \mid \boldsymbol{x})$, which in the SM is specified directly. Second, the SM factorization appeals to the missing data taxonomy described in Section 5.3, enabling easy characterization of the missing data mechanism. Third, when the missing data pattern is monotone, the missing data mechanism can be formulated as a hazard function, where hazard of dropout at some time t can depend on parts of the full-data vector $\boldsymbol{Y}$. Under monotone missingness, assumptions about the hazard function translate directly into the MCAR-MAR-MNAR taxonomy.

Two potential downsides to selection models are their sensitivity to model specification and the sometimes opaque nature of identifiability conditions. Several features of selection models can be illustrated using a simple example involving the bivariate normal distribution.

Example 5.10. *Selection model for bivariate normal data.*
Consider a situation where Y_1 is always observed but Y_2 may be missing, and as usual define $R = I(Y_2 \text{ observed})$. For simplicity we assume no covariates. A selection model factors the full-data distribution as

$$p(y_1, y_2, r \mid \boldsymbol{\omega}) = p(y_1, y_2 \mid \boldsymbol{\theta}) \, p(r \mid y_1, y_2, \boldsymbol{\psi}),$$

where we assume $\boldsymbol{\omega} = (\boldsymbol{\theta}, \boldsymbol{\psi})$. Suppose we specify $p(y_1, y_2 \mid \boldsymbol{\theta})$ as a bivariate normal density $N(\boldsymbol{\mu}, \boldsymbol{\Sigma})$. The distribution $p(r \mid y_1, y_2, \boldsymbol{\psi})$ is a Bernoulli distribution with some regression structure for the mean; e.g.,

$$R_i \mid y_{i1}, y_{i2} \sim \text{Ber}\{\pi_i(\boldsymbol{\psi})\},$$

where

$$g(\pi_i) = \psi_0 + \psi_1 y_{i1} + \psi_2 y_{i2}. \tag{5.17}$$

Setting $g(\cdot)$ equal to the inverse normal cdf $\Phi^{-1}(\cdot)$ yields the Heckman probit selection model (Heckman, 1976). For further details on this model, see Section 8.3.3. □

The model generalizes easily to longitudinal data by using a multivariate normal distribution for $\boldsymbol{Y}_i = (Y_{i1}, \ldots, Y_{iJ})^{\text{T}}$ and replacing π_i in (5.17) with a discrete-time hazard function for dropout

$$h(t_j \mid \overline{\boldsymbol{Y}}_{ij}) \;\; = \;\; P(R_{ij} = 0 \mid R_{i,j-1} = 1, Y_{i1}, \dots, Y_{ij}).$$

Diggle and Kenward (1994) use the logit function to model the discrete-time hazard in terms of observed response history $\overline{\boldsymbol{Y}}_{j-1}$ and the current but possibly missing Y_{ij}. Referring back to (5.17), if $\psi_2 = 0$, we have MAR. If, in addition, $p(\boldsymbol{\mu}, \boldsymbol{\Sigma}, \boldsymbol{\psi}) = p(\boldsymbol{\mu}, \boldsymbol{\Sigma})\, p(\boldsymbol{\psi})$, then we have ignorability.

Even though the parameter ψ_2 in (5.17) characterizes the association between R and the incompletely observed Y_2, the parametric assumptions being made in this example will identify ψ_2, even in the absence of prior information; that is, the observed-data likelihood is a function of ψ_2.

The contribution of individual i to the observed-data likelihood for the selection model in Example 5.10 can be written as

$$L(\boldsymbol{\mu}, \boldsymbol{\Sigma}, \boldsymbol{\psi} \mid \boldsymbol{y}_{i,\text{obs}}, r_i) \;\; \propto \;\; \{ p(y_{i1}, y_{i2} \mid \boldsymbol{\mu}, \boldsymbol{\Sigma})\, \pi(y_{i1}, y_{i2}, \boldsymbol{\psi}) \}^{r_i}$$

$$\times \left[\int p(y_{i1}, y_2 \mid \boldsymbol{\mu}, \boldsymbol{\Sigma})\, \{ 1 - \pi(y_{i1}, y_2, \boldsymbol{\psi}) \}\, dy_2 \right]^{(1-r_i)},$$

where $\pi(y_1, y_2, \boldsymbol{\psi}) = \text{pr}(R = 1 \mid y_1, y_2, \boldsymbol{\psi}) = g^{-1}(\psi_0 + \psi_1 y_1 + \psi_2 y_2)$, For common choices of $g(\cdot)$, such as probit and logit, the observed data likelihood generally will be a function of the parameter ψ_2.

Moreover, although the parameter ψ_2 does *not* index the full-data response model $p(y_1, y_2 \mid \boldsymbol{\mu}, \boldsymbol{\Sigma})$, it *does* index the joint distribution of observables $\boldsymbol{Y}_{\text{obs}}$ and R, and in general is identified by observed data. This can be seen by writing

$$p(\boldsymbol{y}_{\text{obs}}, r) \;\; = \;\; \int p(\boldsymbol{y}_{\text{obs}}, \boldsymbol{y}_{\text{mis}}, \boldsymbol{\mu}, \boldsymbol{\Sigma})\, p(r \mid \boldsymbol{y}_{\text{obs}}, \boldsymbol{y}_{\text{mis}}, \boldsymbol{\psi})\, d\boldsymbol{y}_{\text{mis}}$$

$$= \;\; \text{function of } (\boldsymbol{\mu}, \boldsymbol{\Sigma}, \boldsymbol{\psi})$$

(see Heckman, 1976 for a complete treatment). Although there is nothing counterintuitive about this — the distribution of the selected sample is a function of the selection parameter — this property of parametric selection models make them ill-suited to assessing sensitivity to assumptions about the missing data mechanism. This point is taken up in more detail in Chapters 8 and 9.

To summarize, selection models represent a natural generalization from ignorability to nonignorability by expanding $p(\boldsymbol{r} \mid \boldsymbol{y}_{\text{obs}}, \boldsymbol{y}_{\text{mis}})$ to depend on $\boldsymbol{y}_{\text{mis}}$. The structure of selection models appeals directly to the MAR-MNAR taxonomy developed by Rubin (1976), and in some cases the models can even be fit using standard software. On the other hand, identification of the missing data distribution is accomplished primarily through parametric assumptions about the full-data response model $p(\boldsymbol{y} \mid \boldsymbol{\theta})$ and the explicit form of the missing data mechanism (e.g., linear in $\boldsymbol{y}$). For the simple case where the full-data response model is bivariate normal and the selection model is linear in y, the parameter ψ_2 indexing association between R and the partially observed Y_2 is identifiable from observed data. This feature of the model places considerable importance on assumptions that cannot be verified, and has the potential to make sensitivity analysis problematic.

5.9.2 Mixture models

Mixture models represent another way to factor the full-data model so that $\boldsymbol{R}$ depends on $\boldsymbol{Y}_{\mathrm{mis}}$. In contrast to selection models, the MM approach uses the factorization

$$p(\boldsymbol{y}, \boldsymbol{r} \mid \boldsymbol{x}, \boldsymbol{\omega}) = p(\boldsymbol{y} \mid \boldsymbol{r}, \boldsymbol{x}, \boldsymbol{\omega}) \, p(\boldsymbol{r} \mid \boldsymbol{x}, \boldsymbol{\omega}). \qquad (5.18)$$

As with the selection model, it is sometimes useful to partition the parameter space as $\boldsymbol{\omega} = (\boldsymbol{\alpha}, \boldsymbol{\phi})$, where $\boldsymbol{\alpha}$ and $\boldsymbol{\phi}$, respectively, index the two factors in (5.18). (Recall that the decomposition $\boldsymbol{\omega} = (\boldsymbol{\theta}, \boldsymbol{\phi})$ required for ignorability implicitly refers to a selection model factorization of the full-data model, so the partition $\boldsymbol{\omega} = (\boldsymbol{\alpha}, \boldsymbol{\phi})$ is not equivalent.) The full-data response model is a mixture

$$p(\boldsymbol{y} \mid \boldsymbol{x}, \boldsymbol{\alpha}, \boldsymbol{\phi}) = \sum_{r \in \mathscr{R}} p(\boldsymbol{y} \mid \boldsymbol{r}, \boldsymbol{x}, \boldsymbol{\alpha}) \, p(\boldsymbol{r} \mid \boldsymbol{x}, \boldsymbol{\phi}).$$

The missing data mechanism can be derived using Bayes' rule,

$$p(\boldsymbol{r} \mid \boldsymbol{y}, \boldsymbol{x}, \boldsymbol{\alpha}, \boldsymbol{\phi}) = \frac{p(\boldsymbol{y} \mid \boldsymbol{r}, \boldsymbol{x}, \boldsymbol{\alpha}) \, p(\boldsymbol{r} \mid \boldsymbol{x}, \boldsymbol{\phi})}{p(\boldsymbol{y} \mid \boldsymbol{x}, \boldsymbol{\alpha}, \boldsymbol{\phi})}. \qquad (5.19)$$

Despite its seeming intractability, in some circumstances the missing data mechanism implied by a mixture model actually takes a closed form, as in the mixture of normals model in Example 5.9.

One criticism of mixture models, not readily apparent when studying the case of bivariate data with missingness in Y_2 only, is that the hazard of missingness at time t_j can depend not only on the potentially missing observation Y_j, but also on future measurements $Y_{j+1}, \ldots, Y_J$. In one sense this is a reasonable criticism because in a stochastic process it does not seem sensible for the future to predict the past. However, the construction of mixture models asks for specification of the full-data distribution conditional on different dropout times or patterns. It seems entirely sensible, in considering the association between dropout and response, to specify whether and how the distribution of (say) the vector $(Y_1, Y_2, Y_3)^{\mathrm{T}}$ differs between those who drop out after one measurement and those who record complete follow-up. If desired, mixture models can be constrained such that, conditionally on past responses, hazard of dropout at time t_j depends only on the past and current (but possibly missing) response, and conditionally on those, does not depend on future responses (Kenward et al., 2003). Further discussion of this point is provided in Section 8.4.2, and an illustrative comparison of mixture models where hazard of dropout does and does not depend on future observations is given in our analysis of the Growth Hormone trial in Section 10.2.

From the point of view of modeling, mixture models treat dropout or missingness as a source of variation in the full data distribution. Specifying a different distribution for each dropout time or missing data pattern may seem cumbersome, but it frequently has the advantage of making explicit the parameters that cannot be identified by observed data. This can be seen as follows (again leaving aside covariates). Recall the mixture model specification is

$$p(\boldsymbol{y}_{\mathrm{obs}}, \boldsymbol{y}_{\mathrm{mis}}, \boldsymbol{r} \mid \boldsymbol{\omega}) \;\; = \;\; p(\boldsymbol{y}_{\mathrm{mis}}, \boldsymbol{y}_{\mathrm{obs}} \mid \boldsymbol{r}, \boldsymbol{\alpha}) \, p(\boldsymbol{r} \mid \boldsymbol{\phi}).$$

Further decompose the first term on the right-hand side as

$$p(\boldsymbol{y}_{\mathrm{obs}}, \boldsymbol{y}_{\mathrm{mis}} \mid \boldsymbol{r}, \boldsymbol{\alpha}) \;\; = \;\; p(\boldsymbol{y}_{\mathrm{mis}} \mid \boldsymbol{y}_{\mathrm{obs}}, \boldsymbol{r}, \boldsymbol{\alpha}_{\mathrm{E}}) \, p(\boldsymbol{y}_{\mathrm{obs}} \mid \boldsymbol{r}, \boldsymbol{\alpha}_{\mathrm{O}}), \quad (5.20)$$

where $\boldsymbol{\alpha}_{\mathrm{E}} = \boldsymbol{\lambda}_1(\boldsymbol{\alpha})$, $\boldsymbol{\alpha}_{\mathrm{O}} = \boldsymbol{\lambda}_2(\boldsymbol{\alpha})$ are functions of the mixture component parameter $\boldsymbol{\alpha}$. The parameters $\boldsymbol{\alpha}_{\mathrm{E}}$ and $\boldsymbol{\alpha}_{\mathrm{O}}$ may overlap. The subscript E indicates that $\boldsymbol{\alpha}_{\mathrm{E}}$ indexes an *extrapolation distribution*; i.e., the distribution of missing responses given the observables under the assumed full-data model. The subscript O indicates that $\boldsymbol{\alpha}_{\mathrm{O}}$ indexes a model for observables. In general, $\boldsymbol{\alpha}_{\mathrm{E}}$ cannot be fully identified from data alone.

Now suppose there exists a partition $(\boldsymbol{\alpha}_{\mathrm{E:I}}, \boldsymbol{\alpha}_{\mathrm{E:NI}})$ of $\boldsymbol{\alpha}_{\mathrm{E}}$ such that the observed-data likelihood is a function of $\boldsymbol{\alpha}_{\mathrm{E:I}}$ but not of $\boldsymbol{\alpha}_{\mathrm{E:NI}}$ (i.e., $\boldsymbol{\alpha}_{\mathrm{E:I}}$ is identified from data but $\boldsymbol{\alpha}_{\mathrm{E:NI}}$ is not). A parameter $\boldsymbol{\alpha}_{\mathrm{E:NI}}$ satisfying this condition makes a suitable basis for formulating sensitivity analysis or for encoding the conditional distribution of missing responses using informative prior distributions; in general, mixture models lend themselves well to parameterizations having these properties. To illustrate, we revisit Example 5.9.

Example 5.11. *Mixture of bivariate normals with missingness in Y_2.*
We generalize the specification of Example 5.9 to allow separate variance matrices by pattern of missingness; specifically let $\boldsymbol{Y} = (Y_1, Y_2)^{\mathrm{T}}$, with

$$\boldsymbol{Y} \mid R = 1 \;\; \sim \;\; N(\boldsymbol{\mu}^{(1)}, \boldsymbol{\Sigma}^{(1)})$$
$$\boldsymbol{Y} \mid R = 0 \;\; \sim \;\; N(\boldsymbol{\mu}^{(0)}, \boldsymbol{\Sigma}^{(0)})$$
$$R \;\; \sim \;\; \mathrm{Ber}(\phi),$$

where $\boldsymbol{\mu}^{(r)} = (\mu_1^{(r)}, \mu_2^{(r)})^{\mathrm{T}}$ and $\boldsymbol{\Sigma}^{(r)}$ has elements $\boldsymbol{\sigma}^{(r)} = (\sigma_{11}^{(r)}, \sigma_{12}^{(r)}, \sigma_{22}^{(r)})$. For pattern r, define $\beta_0^{(r)}$, $\beta_1^{(r)}$, and $\sigma_{2|1}^{(r)}$ to be, respectively, the intercept, slope, and residual variance for the regression of Y_2 on Y_1. Under this reparameterization, the full-data model parameters are

$$\boldsymbol{\alpha} = \left\{ \mu_1^{(r)}, \sigma_{11}^{(r)}, \beta_0^{(r)}, \beta_1^{(r)}, \sigma_{2|1}^{(r)} : r = 0, 1 \right\}.$$

The nonidentified parameters are seen to be those indexing the conditional distribution of Y_2 given Y_1 among those with $R = 0$. Hence the distribution $p(\boldsymbol{y}_{\text{obs}}, \boldsymbol{y}_{\text{mis}} \mid r, \boldsymbol{\alpha})$ can be parameterized as

$$p(\boldsymbol{y}_{\text{obs}}, \boldsymbol{y}_{\text{mis}} \mid r, \boldsymbol{\alpha}) = p(\boldsymbol{y}_{\text{mis}} \mid \boldsymbol{y}_{\text{obs}}, r, \boldsymbol{\alpha}_{\text{E}}) \, p(\boldsymbol{y}_{\text{obs}} \mid r, \boldsymbol{\alpha}_{\text{O}}). \qquad (5.21)$$

To simplify notation, define $p_r(\boldsymbol{y}) = p(\boldsymbol{y} \mid r)$. Then the two factors of (5.21) can be written as

$$\begin{aligned}
p_r(\boldsymbol{y}_{\text{mis}} \mid \boldsymbol{y}_{\text{obs}}, \boldsymbol{\alpha}_{\text{E}}) &= \left\{ p_0(y_2 \mid y_1, \boldsymbol{\beta}^{(0)}, \sigma_{2|1}^{(0)}) \right\}^{1-r} \\
p_r(\boldsymbol{y}_{\text{obs}} \mid \boldsymbol{\alpha}_{\text{O}}) &= \left\{ p_0(y_1 \mid \mu_1^{(0)}, \sigma_{11}^{(0)}) \right\}^{1-r} \\
&\quad \times \left\{ p_1(y_2 \mid y_1, \boldsymbol{\beta}^{(1)}, \sigma_{2|1}^{(1)}) \, p_1(y_1 \mid \mu_1^{(1)}, \sigma_{11}^{(1)}) \right\}^{r}.
\end{aligned}$$

Hence $\boldsymbol{\alpha}_{\text{E}}$ can be partitioned as $(\boldsymbol{\alpha}_{\text{E:I}}, \boldsymbol{\alpha}_{\text{E:NI}})$, with

$$\begin{aligned}
\boldsymbol{\alpha}_{\text{E:I}} &= \emptyset \\
\boldsymbol{\alpha}_{\text{E:NI}} &= (\boldsymbol{\beta}^{(0)}, \sigma_{2|1}^{(0)}).
\end{aligned}$$

Furthermore,

$$\boldsymbol{\alpha}_{\text{O}} = (\boldsymbol{\mu}^{(1)}, \boldsymbol{\beta}^{(1)}, \sigma_{11}^{(1)}, \mu_0^{(0)}, \sigma_{11}^{(0)}).$$

It is a simple matter to show that the observed data likelihood contribution for individual i does not depend on $\boldsymbol{\alpha}_{\text{E:NI}}$,

$$\begin{aligned}
L(\boldsymbol{\alpha}_{\text{O}}, \phi \mid \boldsymbol{y}_{i,\text{obs}}, r_i) &\propto \left\{ (1 - \phi) \, p_0(y_{i1} \mid \mu_1^{(0)}, \sigma_{11}^{(0)}) \right\}^{1-r_i} \\
&\quad \times \left\{ \phi \, p_1(y_{i2} \mid y_{i1}, \boldsymbol{\beta}^{(1)}, \sigma_{2|1}^{(1)}) \, p_1(y_{i1} \mid \mu_1^{(1)}, \sigma_{11}^{(1)}) \right\}^{r_i}.
\end{aligned}$$

In this model, it is possible to show that setting $\boldsymbol{\beta}^{(0)} = \boldsymbol{\beta}^{(1)}$ and $\sigma_{2|1}^{(0)} = \sigma_{2|1}^{(1)}$ yields MAR; hence, in general, a function that maps identified parameters and sensitivity parameters to the space of nonidentified parameters can be used to quantify departures from MAR.

For example, if we constrain $\boldsymbol{\alpha}_{\text{E:NI}}$ such that $\beta_1^{(0)} = \beta_1^{(1)}$ and $\sigma_{2|1}^{(0)} = \sigma_{2|1}^{(1)}$, but let

$$\beta_0^{(0)} = \beta_0^{(1)} + \Delta,$$

then assigning a point mass prior at $\Delta = 0$ implies MAR. Fixing Δ at a nonzero value implies MNAR. Formal priors on Δ also can be used. More detail on this approach, including a general framework for model parameterization and prior specification, is given in Chapter 9. Illustrations using the Growth Hormone trial and the OASIS study are provided in Chapter 10. □

By contrast with the selection model for bivariate response in Example 5.10, this observed-data likelihood is completely free of parameters indexing the conditional distribution of $\boldsymbol{Y}_{\text{mis}}$ given $\boldsymbol{Y}_{\text{obs}}$. For the purposes of posterior inference, information about this distribution will derive at least in part from parameters that are informed solely by prior distributions.

To summarize, mixture models have the advantage that in many cases, one can find full-data parameters indexing the distribution of missing responses that are not identified by observed data. In one sense, this makes inferences about the full-data more transparent in that the source of information about the posterior is clear. A potential downside of the models relates to practical implementation: in the previous example, we confine attention to a bivariate distribution with no covariates, and even in this simple case there are three unidentified parameters. As the dimension of Y increases, so will the dimension of the unidentified parameter. In our case studies in Chapter 10, we address this concern by illustrating sensible ways to reduce the dimension of nonidentified parameters indexing the extrapolation distribution.

5.9.3 Shared parameter models

A third approach to specifying the full data distribution is to use an explicitly multilevel formulation, frequently called a *shared parameter model*, where random effects b are modeled jointly with Y and R. In some cases the random effects can be used to explain the association or to capture multiple sources of variation. Key papers tracing the development of these models include Wu and Carroll (1988); Follmann and Wu (1995); Pulkstenis et al. (1998), and Henderson et al. (2000). De Gruttola and Tu (1994), Wulfsohn and Tsiatis (1997), and Faucett and Thomas (1996) use similar formulations but with the objective of modeling a survival process as a function of stochastic longitudinal covariates.

The general form of the full-data model using a shared parameter approach is

$$p(\boldsymbol{y}, \boldsymbol{r} \mid \boldsymbol{x}, \boldsymbol{\omega}) = \int p(\boldsymbol{y}, \boldsymbol{r}, \boldsymbol{b} \mid \boldsymbol{x}, \boldsymbol{\omega}) \, d\boldsymbol{b}. \tag{5.22}$$

Specific shared parameter models are formulated by making assumptions about the joint distribution under the integral sign. Notice in (5.22) that the full-data parameter $\boldsymbol{\omega}$ also includes parameters indexing the distribution of the random effects.

Advantages to the SPM model include simplified specification for the response and missingness components. When Y is measured with error, SPMs provide a useful way for missingness to depend on the error-free version of Y, represented in terms of the random effects (this is a type of 'random-coefficient-dependent missingness'; see Wu and Carroll, 1988, and Little, 1995). Another potential advantage is that through the use of random effects, SPMs can be used to handle high-dimensional or multilevel response data. A disadvantage is that except in simple settings, the underlying missing data mechanism can be difficult to understand and may not even have a closed form: the representation of $p(r \mid y, \omega)$ requires integration over the random effects.

Example 5.12. *Random coefficients selection model.*
Wu and Carroll (1988) are concerned with estimating the population slope of longitudinal lung function measurements using data from a study where the dropout proportion was appreciable. They assume the lung function measures follow a linear random effects model

$$Y_i \mid x_i, b_i \sim N(x_i \beta + w_i b_i, \ \Sigma_i(\phi)),$$

where $w_i \subseteq x_i$ are the random effects covariates, with rows $w_{ij} = (1, t_{ij})$; hence each individual has a random slope and intercept. The random effects $b_i = (b_{i1}, b_{i2})^{\mathrm{T}}$ are assumed to follow a bivariate normal distribution

$$b_i \sim N(0, \Omega).$$

The hazard of dropout is Bernoulli, with

$$R_{ij} \mid R_{i,j-1} = 1, b_i \quad \sim \quad \mathrm{Ber}(\pi_{ij}),$$

where hazard of dropout depends on the random effects governing Y_i via

$$g(\pi_{ij}) = \psi_0 + \psi_1 b_{i1} + \psi_2 b_{i2}.$$

The model is seen as a special case of (5.22) where the joint distribution under the integral is factored as

$$
\begin{aligned}
p(y, r, b \mid x, \omega) &= p(r \mid y, b, x, \psi)\, p(y \mid b, x, \beta, \phi)\, p(b \mid x, \Omega) \\
&= p(r \mid b, \psi)\, p(y \mid b, x, \beta, \phi)\, p(b \mid \Omega),
\end{aligned}
$$

with $\omega = (\beta, \phi, \Omega, \psi)$. The first equality is just a factorization of the joint distribution, and the second equality follows from the key assumption that dropout R is independent of both Y_{obs} and Y_{mis}, conditionally on random effects. However, integrating over the random effects induces dependence between R and Y_{mis} conditionally on Y_{obs}; hence the model characterizes an MNAR mechanism. $\qquad\qquad\qquad\qquad\qquad\qquad\qquad\qquad\qquad\qquad\quad\Box$

The 'conditional linear model' (Wu and Bailey, 1989; Hogan and Laird, 1997a) also can be viewed as a version of an SPM, though its more standard representation is in terms of a mixture model. The conditional linear model specializes (5.22) to

$$p(y, r, b \mid x) = p(y \mid r, b, x)\, p(b \mid r, x)\, p(r \mid x).$$

As an example, Hogan and Laird (1997a) assume that mixture components $p(y \mid r, b, x)$, $r \in \mathscr{R}$ in the first factor follow random effects models where the regression coefficients β depend on r, and the random effects have a distribution that does not depend on r; i.e., $p(b \mid r, x) = p(b \mid x)$. Hogan et al. (2004a) generalize the model to allow continuous dropout time, using a mixture of varying coefficient models. For the case where $p(\boldsymbol{y} \mid \boldsymbol{r}, \boldsymbol{b}, \boldsymbol{x})$ and $p(\boldsymbol{b} \mid \boldsymbol{r}, \boldsymbol{x})$ follow normal distributions and $E(\boldsymbol{Y} \mid \boldsymbol{r}, \boldsymbol{b}, \boldsymbol{x})$ is linear in $\boldsymbol{b}$, integrating the random effects yields a standard mixture-of-normals model for the full-data distribution (as in Example 5.11).

5.10 Summary

In this chapter, we have provided a formal framework for conceptualizing inference from incomplete data, based on specifying a full-data model and drawing inference from an observed-data likelihood (and associated priors). In order to make clear the limits of drawing inference from incomplete data, we provided formal definitions for missing data mechanisms and applied them to settings with dropout. We then showed how to apply the notions of model specifications and missing data assumptions in terms of selection models, mixture models, and shared parameter models.

In the next chapter, we turn our focus to drawing inference about full-data models for longitudinal responses under the ignorability assumption. For longitudinal data, the key issue lies in specifying the dependence structure, because the extrapolation model under ignorability, $p(\boldsymbol{y}_{\mathrm{mis}} \mid \boldsymbol{y}_{\mathrm{obs}})$, depends critically on the dependence model.

5.11 Further reading

Models that accommodate dropout and intermittent missingness

Several papers have discussed approaches to handle dropout and intermittent missingness simultaneously. Albert (2000) and Albert et al. (2002) constructed models with a multinomial random variable specified at each time indicating still in study, intermittent missing, or dropout. Lin, McCulloch, and Rosenheck (2004) deal with intermittent missingness by grouping the missing data patterns using a latent pattern mixture approach. Troxel et al. (1998) use a pseudo-likelihood approach that models the missingness indicator at time t conditional on only the potentially missing response at time t.

Multiple cause dropout and dropout due to death

For a general discussion of multiple cause dropout, see Rotnitzky et al. (2001).
 There have been several important papers recently that address the issues involved with dropouts due to death. Frangakis and Rubin (2002) illustrate the use of principal stratification to estimate causal effects of treatment in the presence of death. Kurland and Heagerty (2005) address dropout due to death in observational studies by directly modeling the mean conditional on surviving past a specified time.

CHAPTER 6

Inference about Full-Data Parameters under Ignorability

6.1 Overview

In this chapter, we discuss inference about the full-data distribution under the ignorability assumption. Under the full-data model factorization

$$p(\boldsymbol{y}, \boldsymbol{r}, \mid \boldsymbol{x}, \boldsymbol{\omega}) \quad = \quad p(\boldsymbol{y} \mid \boldsymbol{x}, \boldsymbol{\theta}(\boldsymbol{\omega})) \, p(\boldsymbol{r} \mid \boldsymbol{y}, \boldsymbol{x}, \boldsymbol{\psi}(\boldsymbol{\omega})), \tag{6.1}$$

missing at random implies

$$p(\boldsymbol{r} \mid \boldsymbol{y}, \boldsymbol{x}, \boldsymbol{\psi}(\boldsymbol{\omega})) = p(\boldsymbol{r} \mid \boldsymbol{y}_{\text{obs}}, \boldsymbol{x}, \boldsymbol{\psi}(\boldsymbol{\omega})). \tag{6.2}$$

Ignorability arises under the additional restriction of *a priori* independence between the parameters of the full-data response model $\boldsymbol{\theta}$ and the parameters of the missing data mechanism $\boldsymbol{\psi}$. Under ignorability, we only need to specify the full-data response model, not the missing data mechanism. In the following we suppress the dependence on $\boldsymbol{x}$ to maintain clarity.

Recall that the full-data posterior of $\boldsymbol{\theta}$ is $p(\boldsymbol{\theta} \mid \boldsymbol{y}) = p(\boldsymbol{\theta} \mid \boldsymbol{y}_{\text{obs}}, \boldsymbol{y}_{\text{mis}})$. Under ignorability, inference is based on the observed-data posterior,

$$
\begin{aligned}
p(\boldsymbol{\theta} \mid \boldsymbol{y}_{\text{obs}}) &= \int p(\boldsymbol{\theta} \mid \boldsymbol{y}) p(\boldsymbol{y}_{\text{mis}} \mid \boldsymbol{y}_{\text{obs}}) \, d\boldsymbol{y}_{\text{mis}} \\
&\propto \int p(\boldsymbol{y}_{\text{obs}}, \boldsymbol{y}_{\text{mis}} \mid \boldsymbol{\theta}) p(\boldsymbol{\theta}) d\boldsymbol{y}_{\text{mis}}.
\end{aligned}
\tag{6.3}
$$

This integral, and hence $p(\boldsymbol{\theta} \mid \boldsymbol{y}_{\text{obs}})$, depends on the assumed full data response model $p(\boldsymbol{y} \mid \boldsymbol{\theta})$, where $\boldsymbol{y} = (\boldsymbol{y}_{\text{obs}}, \boldsymbol{y}_{\text{mis}})$. Thus, valid inference depends on its correct specification. We must keep in mind that neither the full-data response model nor the ignorability assumption itself is verifiable from the observed data. In longitudinal data, one of the most important aspects in specifying a full-data response model is the association (dependence) structure.

The main focus of this chapter is methods for posterior inference. Sampling from the observed data posterior can often be simplified by supplementing the Gibbs sampler with a data augmentation step. In particular, we can sample from

1. $p(\boldsymbol{y}_{\text{mis}} \mid \boldsymbol{y}_{\text{obs}}, \boldsymbol{\theta})$ using a data augmentation step, and

2. $p(\boldsymbol{\theta} \mid \boldsymbol{y})$ using Gibbs sampling.

The second step of this algorithm, which samples the full-data response model parameters, is the same as if there were no missing data (Section 3.4). This often simplifies computations considerably. We provide more details on this in Section 6.2.

The remainder of the chapter is organized as follows. First, issues of model specification are discussed, emphasizing the importance of correctly specifying the dependence structure in $p(\boldsymbol{y} \mid \boldsymbol{\theta})$. Second, the use of data augmentation to simplify posterior sampling will be described in the setting of missing data and its implementation in specific models shown. Third, we will illustrate modeling of dependence by giving examples of convenient parsimonious structures and computationally tractable ways to allow the dependence to be a function of model covariates. Fourth, we will illustrate the use of joint modeling of several processes as a way to incorporate information on auxiliary stochastic time-varying covariates (recall Section 5.4.4 from Chapter 5 and Definition 5.11). The chapter closes with some suggested approaches for assessing model fit and model checking. Throughout the chapter, we often refer the full-data response model and posterior as the full-data model and posterior without any loss of clarity.

6.2 General issues in model specification

Inference under ignorability relies on correct specification of the full-data response model $p(\boldsymbol{y} \mid \boldsymbol{\theta})$. It is convenient to view the the full-data response model in terms of assumptions about the mean, variability and (serial) dependence, and distribution. The following development focuses on the consequences of mis-specification of the dependence.

6.2.1 Mis-specification of dependence

In longitudinal settings, when primary interest is in the mean, the dependence is often viewed as a nuisance, with correct specification only leading to gains in efficiency. However, when dealing with incomplete data, correctly specifying the dependence takes on added importance.

To illustrate the consequences of mis-specification of dependence under ignorability, we return to the bivariate normal model (Example 5.7) where Y_{i1} is observed and Y_{i2} is possibly missing. Recall, $\boldsymbol{Y}_i \sim N(\boldsymbol{\mu}, \boldsymbol{\Sigma})$, with the j, k element of $\boldsymbol{\Sigma}$ denoted as σ_{jk}. To fill in missing values of Y_{i2} within the sampling algorithm, we sample from the (conditional) normal distribution $p(y_{i2} \mid y_{i1}, \boldsymbol{\mu}, \boldsymbol{\Sigma})$, where $\boldsymbol{\theta} = (\boldsymbol{\mu}, \boldsymbol{\Sigma})$. It has mean $\mu_2 + \phi_{21}(y_{i1} - \mu_1)$ and is a function of the covariance parameter through $\phi_{21} = \sigma_{21}/\sigma_{11}$. By contrast, with complete data, the data augmentation step is not necessary and the posterior of μ_2 is *not* a function of ϕ_{21}.

The simplest mis-specification would be to assume $\phi_{21} = 0$ when, in truth,

$\phi_{21} \neq 0$. A less obvious, but important mis-specification would be to assume ϕ_{21} is constant when, in truth, ϕ_{21} is a function of model covariates via $\phi_{21} = h(\boldsymbol{x})$ for some function h.

Of course, when $J > 2$, there are more dependence parameters and covariance model simplification is often necessary. For example, in a multivariate normal model, when $\boldsymbol{\Sigma}$ is large, a common strategy is to assume a simpler, parsimonious form for $\boldsymbol{\Sigma}$; we might (incorrectly) assume an AR-1 form for $\boldsymbol{\Sigma}$. Or we might introduce structure via random effects, but assume the wrong model for the random effects covariance matrix (Daniels and Zhao, 2003).

Several of these cases are explored in detail below. We start by examining the consistency of the posterior distribution of the mean when the dependence is mis-specified.

Example 6.1. *Posterior distribution of the mean for a bivariate normal under monotone missingness that is ignorable* (continuation of Example 5.7). Consider a bivariate response from a randomized controlled trial with binary treatment and response vector $\boldsymbol{Y} = (Y_1, Y_2)^{\mathrm{T}}$. Y_1 is always observed, but Y_2 may be missing. The missingness indicator R takes value 0 if Y_2 is missing. Conditional on a single scalar covariate x, we assume the full-data response model is a bivariate normal distribution

$$\boldsymbol{Y} \mid x, \boldsymbol{\mu}, \boldsymbol{\Sigma} \quad \sim \quad N(\boldsymbol{\mu}(x), \boldsymbol{\Sigma}(x)),$$

with $\boldsymbol{\Sigma}(x) = \{\sigma_{jk}(x)\}$ and $\phi_{21}(x) = \sigma_{21}(x)/\sigma_{11}(x)$. For $j = 1, 2$, define

$$
\begin{aligned}
\mu_j(x) &= E(Y_j \mid x) \\
\mu_j^{\mathrm{C}}(x) &= E(Y_j \mid x, R = 1) \\
\mu_j^{\mathrm{D}}(x) &= E(Y_j \mid x, R = 0).
\end{aligned}
$$

It can be shown that under ignorability,

$$
\begin{aligned}
\mu_2^{\mathrm{C}}(x) - \phi_{21}(x)\,\mu_1^{\mathrm{C}}(x) &= \mu_2^{\mathrm{D}}(x) - \phi_{21}(x)\,\mu_1^{\mathrm{D}}(x) \\
&= \mu_2(x) - \phi_{21}(x)\,\mu_1(x). \quad (6.4)
\end{aligned}
$$

We now examine the posterior distribution of $\mu_2(x)$ under mis-specification of the dependence $\phi_{21}(x)$; in particular, we incorrectly assume $\phi_{21}(x) = \phi_{21}$. Define $\widehat{\boldsymbol{\beta}} = (\widehat{\beta}_0, \widehat{\phi}_{21})^{\mathrm{T}}$ to be the ordinary least squares estimates for regressing Y_2 on Y_1 for those with $R = 1$ and $X = x$ where $\phi_{21}(x)$ is the slope. Assume there are n_x individuals with $X = x$.

Consider the *conditional* posterior mean of $\mu_2(x)$ given $\{\boldsymbol{y}_{\mathrm{obs}}, \mu_1(x)\}$; we do not need to condition on $\mu_1(x)$ here, but it will make the arguments more clear. We examine the behavior of the posterior mean of $\mu_2(x)$ as $n_x \to \infty$ (and the dropout proportion remains constant); in particular, we will determine whether the posterior distribution of $\mu_2(x)$ is consistent (recall Definition 3.2).

Under ignorability and a correctly specified covariance, condition (6.4) can be used to show that

$$
\begin{aligned}
E\{\mu_2(x) \mid \boldsymbol{y}_{\text{obs}}, \mu_1(x)\} \quad &= \quad \widehat{\beta}_0 + \widehat{\phi}_{21}\mu_1(x) \\
&\to \quad \mu_2^{\text{C}}(x) - \phi_{21}(x)\mu_1^{\text{C}}(x) + \phi_{21}(x)\mu_1(x) \\
&= \quad \mu_2(x).
\end{aligned} \tag{6.5}
$$

However, under the mis-specification $\phi_{21}(x) = 0$,

$$
\begin{aligned}
E\{\mu_2(x) \mid \boldsymbol{y}_{\text{obs}}, \mu_1(x)\} \quad &= \quad \bar{y}_2^{\text{C}}(x) \\
&\to \quad \mu_2^{\text{C}}(x) \\
&= \quad \mu_2(x) - \phi_{21}(x)\{\mu_1(x) - \mu_1^{\text{C}}(x)\},
\end{aligned}
$$

where

$$
\bar{y}_2^{\text{C}}(x) \quad = \quad \frac{1}{m_x} \sum_{\{i: x_i = x, r_i = 1\}} y_{i2}
$$

and m_x is the number of subjects with $\{x_i = x, r_i = 1\}$. So, under mis-specification of the dependence structure, the posterior distribution of $\mu_2(x)$ will be inconsistent with bias

$$
-\phi_{21}(x)\{\mu_1(x) - \mu_1^{\text{C}}(x)\}.
$$

Clearly, the bias increases with the magnitude of the dependence parameter. The bias is also a function of the difference in the means between the completers and dropouts at time 1.

Using similar arguments, bias in treatment effect at time 2, $\mu_2(1) - \mu_2(0)$, is equal to

$$
-\phi_{21}(1)\{\mu_1(1) - \mu_1^{\text{C}}(1)\} + \phi_{21}(0)\{\mu_1(0) - \mu_1^{\text{C}}(0)\}.
$$

$\square$

This example illustrates that even in the simplest case, a bivariate normal with ignorable monotone missingness, mis-specification of the dependence leads to an inconsistent posterior for the mean parameters. Similar biases occur when the mis-specification is $\phi_{21}(x) = \phi_{21} \neq 0$. These results readily generalize to higher-dimensional multivariate normals and other models discussed in Chapter 2.

6.2.2 Orthogonal parameters

A more general way to understand the importance of correctly modeling the dependence is by exploring the difference in the form of the information matrix for the mean and dependence parameters for complete data vs. incomplete data (under ignorability). Before going into the details, we first introduce the concept of orthogonal parameters. Let $\boldsymbol{\beta}$ represent the mean parameters, $\boldsymbol{\alpha}$ the dependence parameters, and let $\ell = \log L$.

Definition 6.1. *Orthogonal parameters.*
For parameters $\boldsymbol{\beta}$ and $\boldsymbol{\alpha}$, let $\boldsymbol{\beta}_0$ and $\boldsymbol{\alpha}_0$ be their true values, respectively. Then $\boldsymbol{\beta}$ and $\boldsymbol{\alpha}$ are orthogonal (Cox and Reid, 1987) if, for any component β_j of $\boldsymbol{\beta}$ and any component α_k of $\boldsymbol{\alpha}$,

$$\boldsymbol{I}_{\beta_j,\alpha_k}(\boldsymbol{\beta}_0,\boldsymbol{\alpha}_0) = -E\left\{\frac{\partial^2 \ell(\boldsymbol{\beta},\boldsymbol{\alpha})}{\partial\beta_j\partial\alpha_k}\right\}\bigg|_{\boldsymbol{\beta}_0,\boldsymbol{\alpha}_0} = 0. \tag{6.6}$$

□

The block diagonality of the information matrix implies that the maximum likelihood estimates of the orthogonal parameters are asymptotically independent (Cox and Reid, 1987).

In the Bayesian paradigm, in addition to the full-data likelihood, we also have a prior. Typically, the prior will be $O_p(1)$, whereas the likelihood is $O_p(n)$. We now state a similar definition to Definition 6.1 for the Bayesian setting.

Definition 6.2. *Orthogonal parameters in Bayesian inference.*
Consider parameters $\boldsymbol{\beta}$ and $\boldsymbol{\alpha}$ with prior $p(\boldsymbol{\beta},\boldsymbol{\alpha})$, and let $\boldsymbol{\beta}_0$ and $\boldsymbol{\alpha}_0$ be their true values. Then $\boldsymbol{\beta}$ and $\boldsymbol{\alpha}$ are orthogonal if, for any component of β_j of $\boldsymbol{\beta}$ and any component α_k of $\boldsymbol{\alpha}$,

$$\boldsymbol{I}^{\star}_{\beta_j,\alpha_k}(\boldsymbol{\beta}_0,\boldsymbol{\alpha}_0) = -E\left[\frac{\partial^2\{\ell(\boldsymbol{\beta},\boldsymbol{\alpha}) + \log(p(\boldsymbol{\beta},\boldsymbol{\alpha}))\}}{\partial\beta_j\partial\alpha_k}\right]\bigg|_{\boldsymbol{\beta}_0,\boldsymbol{\alpha}_0} = 0. \tag{6.7}$$

□

For identified parameters, if condition (6.6) holds, then condition (6.7) holds exactly under independent priors on $\boldsymbol{\beta}$ and $\boldsymbol{\alpha}$. If $p(\boldsymbol{\beta},\boldsymbol{\alpha}) \neq p(\boldsymbol{\beta})p(\boldsymbol{\alpha})$, but satisfies $p(\boldsymbol{\beta},\boldsymbol{\alpha}) = O_p(1)$, then condition (6.7) holds asymptotically.

When condition (6.7) holds, the joint posterior distribution of $(\boldsymbol{\beta},\boldsymbol{\alpha})$ is equal to the product of the marginal posteriors of $\boldsymbol{\beta}$ and $\boldsymbol{\alpha}$ (asymptotically). For orthogonality involving non- or weakly identified parameters, the *a priori* independence of $\boldsymbol{\beta}$ and $\boldsymbol{\alpha}$ is necessary for orthogonality.

Orthogonality of the mean and dependence parameters is only a necessary condition for the posterior distribution of the mean parameters to be consistent if the dependence is mis-specified. A sufficient condition for the posterior of the mean parameters to be consistent is given next. This condition is stronger than orthogonality conditions given in Definitions 6.1 and 6.2.

Theorem 6.1. *Consistency of the posterior for mean parameters.*
Let $\boldsymbol{\beta}_0$ be the true value for the mean parameters and $\boldsymbol{\alpha}^{\star}$ be any value for the dependence parameters. If

$$\boldsymbol{I}_{\beta,\alpha}(\boldsymbol{\beta}_0,\boldsymbol{\alpha}^{\star}) = -E\left\{\frac{\partial^2 \ell(\boldsymbol{\beta},\boldsymbol{\alpha})}{\partial\boldsymbol{\beta}\partial\boldsymbol{\alpha}}\right\}\bigg|_{\boldsymbol{\beta}_0,\boldsymbol{\alpha}^{\star}} = \boldsymbol{0}, \tag{6.8}$$

then the mle's of the mean parameters are consistent even if the dependence is mis-specified (Firth, 1987). The analogous condition for the consistency of the posterior for the mean parameters in Bayesian inference replaces $\ell(\boldsymbol{\beta}, \boldsymbol{\alpha})$ in (6.8) with $\ell(\boldsymbol{\beta}, \boldsymbol{\alpha}) + \log p(\boldsymbol{\beta}, \boldsymbol{\alpha})$.   □

We illustrate using a multivariate normal model.

Example 6.2. *Information matrix based on observed data log likelihood under ignorability with a multivariate normal model.*
Assume $\boldsymbol{Y}_i$ follows a multivariate normal distribution with mean $\boldsymbol{X}_i\boldsymbol{\beta}$ and covariance matrix $\boldsymbol{\Sigma}(\boldsymbol{\alpha})$. Let $\boldsymbol{\theta} = (\boldsymbol{\beta}, \boldsymbol{\alpha})$ and assume that $p(\boldsymbol{\beta}, \boldsymbol{\alpha}) = p(\boldsymbol{\beta})p(\boldsymbol{\alpha})$ (a common assumption). For the case of complete data, it is easy to show that the off-diagonal block of the information matrix, $\boldsymbol{I}_{\boldsymbol{\beta},\boldsymbol{\alpha}}$, is equal to zero for all values of $\boldsymbol{\alpha}$, thereby satisfying condition (6.8). For Bayesian inference, the posterior for $\boldsymbol{\beta}$ will be consistent even under mis-specification of $\boldsymbol{\Sigma}(\boldsymbol{\alpha})$. However, under ignorability, the submatrix of the information matrix, now based on the observed data log likelihood ℓ_{obs} (or observed data posterior) and given by

$$\boldsymbol{I}^{\text{obs}}_{\boldsymbol{\beta},\boldsymbol{\alpha}}(\boldsymbol{\beta}, \boldsymbol{\alpha}) = -E\left\{\frac{\partial^2 \ell_{\text{obs}}(\boldsymbol{\beta}, \boldsymbol{\alpha})}{\partial\boldsymbol{\beta}\partial\boldsymbol{\alpha}^{\text{T}}}\right\},$$

is no longer equal to zero even at the true value for $\boldsymbol{\Sigma}(\boldsymbol{\alpha})$ (Little and Rubin, 2002). Hence the weaker parameter orthogonality condition given in Definition 6.2 does not even hold. *As a result, in order for the posterior distribution of the mean parameters to be consistent, the dependence structure must be correctly specified.*

This lack of orthogonality can be seen in the setting of a bivariate normal linear regression, by making a simple analogy to univariate simple linear regression. This will also provide some additional intuition into how inferences change under missingness.

Suppose $E(\boldsymbol{Y}) = \boldsymbol{\mu}$, $R_i = 1$ for $i = 1, \ldots, n_1$ (y_{i2} observed), and $R_i = 0$ for $i = n_1+1, \ldots, n$ (y_{i2} missing). As in Chapter 5, we factor the joint distribution of $p(y_1, y_2)$ as $p(y_1)p(y_2 \mid y_1)$. For *complete* data, the conditional distribution of Y_2 given Y_1 (ignoring priors for the time being) as a function of μ_2 and $\phi_{21} = \sigma_{12}/\sigma_{11}$ is proportional to

$$\exp\left[-\sum_{i=1}^{n}\{y_{i2} - \mu_2 - \phi_{21}(y_{i1} - \mu_1)\}^2/2\sigma_{2|1}\right], \qquad (6.9)$$

where $\sigma_{2|1} = \sigma_{22} - \sigma_{21}\sigma_{11}^{-1}\sigma_{12}$. Note that ϕ_{21} and μ_2 do not appear in $p(y_1)$.

The orthogonality of μ_2 and ϕ_{21} is apparent by recognizing (6.9) as the same form as the log likelihood for a simple linear regression having a centered covariate $y_{i1} - \mu_1$ with intercept μ_2 and slope ϕ_{21}. It can be shown from this form that the element of the (expected) information matrix corresponding to μ_2 and ϕ_{21} is zero for all values of ϕ_{21}. However, with missing data (under

MAR), the analogue of (6.9) is

$$\exp\left[-\sum_{i=1}^{n_1}\{y_{i2} - \mu_2^C - \phi_{21}(y_{i1} - \mu_1^C)\}^2/2\sigma_{2|1}\right].$$

The sum is now only over the terms that correspond to $R_i = 1$.

Recall the mean of the completers at time j is $\mu_j^C = E(Y_{ij} \mid R = 1)$. Again using the analogy to simple linear regression, μ_2^C and ϕ_{21} are orthogonal, but μ_2 and ϕ_{21} are not. This is clear from the following, which holds under ignorability:

$$
\begin{aligned}
\mu_2 &= \mu_2^C\pi + \mu_2^D(1-\pi)\\
&= \mu_2^C - \phi_{21}(1-\pi)(\mu_1^C - \mu_1^D),
\end{aligned}
$$

where $\pi = P(R_i = 1)$. Thus, μ_2 is a function of ϕ_{21}. With no missing data, $\pi = 1$ (so $\mu_2 = \mu_2^C$). Under MCAR, $(\mu_1^C - \mu_1^D) = 0$ and the second term involving ϕ_{21} disappears. $\qquad\square$

Examples 6.1 and 6.2 demonstrate the importance of correctly specifying Σ even when primary interest is in μ. Thus, if we model the covariance matrix parsimoniously, we must be sure to consider whether Σ depends on covariates. Of course, such modeling decisions are only verifiable from the data *under the ignorability assumption*.

Similar results hold for directly specified models for binary data. It can be shown that the information matrix for the mean parameters β and the dependence parameters α in an MTM(1) (Example 2.5) satisfies condition (6.8) under no missing data or MCAR (Heagerty, 2002) , but not under MAR.

There are also situations where misspecification of dependence with complete data can lead to biased estimates of the mean parameters. For example, in marginalized transition model of order p (where $p \geq 2$), β and α are not orthogonal (Heagerty, 2002). Hence, the dependence structure must be correctly specified. Similar specification issues with complete data are also seen in conditionally specified models for binary data.

Of course, it is not possible to verify whether the dependence structure is correct when data are incomplete. As a practical matter, when ignorability is being assumed, it is recommended that the dependence model be selected based on the model that is most suitable for the observed data.

6.3 Posterior sampling using data augmentation

Data augmentation is an important tool for full data inference in the presence of missing data; it is related to the EM algorithm (Dempster, Laird, and Rubin, 1977) and its variations (van Dyk and Meng, 2001). As we illustrated in Chapter 5 the general strategy is to specify a model and priors for the full data and then to base posterior inference on the induced observed data posterior, $p(\boldsymbol{\theta} \mid \boldsymbol{y}_{\mathrm{obs}})$. However, the full-data posterior $p(\boldsymbol{\theta} \mid \boldsymbol{y})$ is often easier to sample than the observed-data posterior $p(\boldsymbol{\theta} \mid \boldsymbol{y}_{\mathrm{obs}})$. Specifically, full conditional distributions of the full-data posterior used in Gibbs sampling typically have simpler forms than the full conditionals derived using the observed-data posterior. This motivates augmenting the observed-data posterior with the missing data $\boldsymbol{y}_{\mathrm{mis}}$. We point out that if the model is specified directly for the observed data, i.e, specify the observed data response model instead of the full-data response model, and priors are put directly on the parameters of the observed-data response model (see Example 5.7), then data augmentation is not needed.

For data augmentation, at each iteration k of the sampling algorithm, we sample $(\boldsymbol{y}_{\mathrm{mis}}^{(k)}, \boldsymbol{\theta}^{(k)})$ via

1. $\boldsymbol{y}_{\mathrm{mis}}^{(k)} \sim p(\boldsymbol{y}_{\mathrm{mis}} \mid \boldsymbol{y}_{\mathrm{obs}}, \boldsymbol{\theta}^{(k-1)})$
2. $\boldsymbol{\theta}^{(k)} \sim p(\boldsymbol{\theta} \mid \boldsymbol{y}_{\mathrm{obs}}, \boldsymbol{y}_{\mathrm{mis}}^{(k)})$.

Thus, we can sample $\boldsymbol{\theta}$ using the tools we described in Chapter 3 for the full-data posterior, as if we had complete data. Implicitly, via Monte Carlo integration within the MCMC algorithm, we obtain a sample from the observed-data posterior $p(\boldsymbol{\theta} \mid \boldsymbol{y}_{\mathrm{obs}})$ given in (6.3).

Because data augmentation depends on sampling from $p(\boldsymbol{y}_{\mathrm{mis}} \mid \boldsymbol{y}_{\mathrm{obs}}, \boldsymbol{\theta})$, the augmentation depends heavily on the within-subject dependence structure. We illustrate by giving some examples of the data augmentation step for several models from Chapter 2 under ignorable dropout.

Example 6.3. *Data augmentation under ignorability with a multivariate normal model* (continuation of Example 2.3).
Without loss of generality, define $\boldsymbol{Y}_{\mathrm{obs},i} = (Y_{i1}, \ldots, Y_{iJ^\star})^{\mathrm{T}}$ and $\boldsymbol{Y}_{\mathrm{mis},i} = (Y_{i,J^\star+1}, \ldots, Y_{iJ})^{\mathrm{T}}$ with corresponding partitions of $\boldsymbol{X}_i$ and $\boldsymbol{\Sigma}$ given by

$$\boldsymbol{x}_i = \begin{pmatrix} \boldsymbol{x}_{\mathrm{obs},i} \\ \boldsymbol{x}_{\mathrm{mis},i} \end{pmatrix}$$

$$\boldsymbol{\Sigma} = \begin{pmatrix} \boldsymbol{\Sigma}_{\mathrm{obs}} & \boldsymbol{\Sigma}_{\mathrm{obs,mis}} \\ \boldsymbol{\Sigma}_{\mathrm{obs,mis}}^{\mathrm{T}} & \boldsymbol{\Sigma}_{\mathrm{mis}} \end{pmatrix}.$$

Within the sampling algorithm, the distribution of $p(\boldsymbol{y}_{\mathrm{mis},i} \mid \boldsymbol{y}_{\mathrm{obs},i}, \boldsymbol{\theta})$ takes the form

$$\boldsymbol{Y}_{\mathrm{mis},i} \mid \boldsymbol{Y}_{\mathrm{obs},i}, \boldsymbol{\theta} \sim N(\boldsymbol{\mu}^\star, \boldsymbol{\Sigma}^\star),$$

where

$$\boldsymbol{\mu}^\star = \boldsymbol{x}_{\mathrm{mis},i}\boldsymbol{\beta} + \boldsymbol{B}(\boldsymbol{\Sigma})\left(\boldsymbol{y}_{\mathrm{obs},i} - \boldsymbol{x}_{\mathrm{obs},i}\boldsymbol{\beta}\right)$$
$$\boldsymbol{\Sigma}^\star = \boldsymbol{\Sigma}_{\mathrm{mis}} - \boldsymbol{\Sigma}_{\mathrm{obs,mis}}^{\mathrm{T}} \boldsymbol{\Sigma}_{\mathrm{obs}}^{-1} \boldsymbol{\Sigma}_{\mathrm{obs,mis}},$$

and $\boldsymbol{B}(\boldsymbol{\Sigma}) = \boldsymbol{\Sigma}_{\mathrm{obs,mis}}^{\mathrm{T}}\boldsymbol{\Sigma}_{\mathrm{obs}}^{-1}$. The dependence of $\boldsymbol{Y}_{\mathrm{mis},i}$ on $\boldsymbol{Y}_{\mathrm{obs},i}$ is governed by $\boldsymbol{B}(\boldsymbol{\Sigma})$, the matrix of autoregressive coefficients from regressing $\boldsymbol{Y}_{\mathrm{mis},i}$ on $\boldsymbol{Y}_{\mathrm{obs},i}$. Clearly, $\boldsymbol{B}(\boldsymbol{\Sigma})$ is a function of the full-data covariance matrix $\boldsymbol{\Sigma}$.   □

Example 6.4. *Data augmentation under ignorability with random effects logistic regression* (continuation of Example 2.2).
Again, we define $\boldsymbol{Y}_{\mathrm{mis},i}$ and $\boldsymbol{Y}_{\mathrm{obs},i}$ as in Example 6.3. The distribution $p(\boldsymbol{y}_{\mathrm{mis},i} \mid \boldsymbol{y}_{\mathrm{obs},i}, \boldsymbol{b}_i, \boldsymbol{\theta})$ is a product of independent Bernoullis with probabilities

$$P(Y_{\mathrm{mis},ij} = 1 \mid \boldsymbol{y}_{\mathrm{obs},i}, \boldsymbol{b}_i, \boldsymbol{\theta}) = \frac{\exp(\boldsymbol{x}_{ij}\boldsymbol{\beta} + \boldsymbol{w}_{ij}\boldsymbol{b}_i)}{1 + \exp(\boldsymbol{x}_{ij}\boldsymbol{\beta} + \boldsymbol{w}_{ij}\boldsymbol{b}_i)}, \; j \geq J^\star + 1. \quad (6.10)$$

The dependence of these imputed values on the random effects covariance matrix $\boldsymbol{\Omega}$ is evidenced by the presence of $\boldsymbol{b}_i$ in (6.10). By integrating out $\boldsymbol{b}_i$ (as in Example 2.2), we have

$$\begin{aligned} p(\boldsymbol{y}_{\mathrm{mis},i} \mid \boldsymbol{y}_{\mathrm{obs},i}, \boldsymbol{\theta}) &= \int p(\boldsymbol{y}_{\mathrm{mis},i} \mid \boldsymbol{y}_{\mathrm{obs},i}, \boldsymbol{b}_i, \boldsymbol{\theta})\, p(\boldsymbol{b}_i \mid \boldsymbol{\theta}, \boldsymbol{y}_{\mathrm{obs},i})\, d\boldsymbol{b}_i \\ &= \int p(\boldsymbol{y}_{\mathrm{mis},i} \mid \boldsymbol{b}_i, \boldsymbol{\beta})\, p(\boldsymbol{b}_i \mid \boldsymbol{\beta}, \boldsymbol{\Omega}, \boldsymbol{y}_{\mathrm{obs},i})\, d\boldsymbol{b}_i. \quad (6.11) \end{aligned}$$

The integral (6.11) is not available in closed form; however, recalling that the population-averaged distribution can be approximated by

$$P(Y_{\mathrm{mis},ij} = 1 \mid \boldsymbol{y}_{\mathrm{obs},i}, \boldsymbol{\theta}) \approx \frac{\exp(\boldsymbol{x}_{ij}\boldsymbol{\beta}^\star)}{1 + \exp(\boldsymbol{x}_{ij}\boldsymbol{\beta}^\star)},$$

where $\boldsymbol{\beta}^\star \approx \boldsymbol{\beta} K(\boldsymbol{\Omega})$ and $K(\boldsymbol{\Omega})$ is a constant that depends on $\boldsymbol{\Omega}$, the dependence on the random effects covariance matrix is clear.   □

Data augmentation and multiple imputation

Certain types of multiple imputation (Rubin, 1987) can be viewed as approximations to data augmentation. Bayesianly proper multiple imputation (Schafer, 1997) provides an approximation to the fully Bayesian data augmentation procedure in (6.3), which is based on the full-data response model; for nonignorable missingness (Chapter 8), it would be based on the entire full-data model. This approximation is computed by sampling just a few values, say M, from $p(\boldsymbol{y}_{\mathrm{mis}} \mid \boldsymbol{y}_{\mathrm{obs}})$ (as opposed to full Monte Carlo integration). The M sets of $\boldsymbol{y}_{\mathrm{mis}}$ are then used to create M full datasets that are analyzed using full-data response log likelihoods, $\ell(\boldsymbol{\theta} \mid \boldsymbol{y})$ or full-data response model posteriors $p(\boldsymbol{\theta} \mid \boldsymbol{y})$. Inferences are then appropriately adjusted for the uncertainty in the missing values (Schafer, 1997).

Bayesianly 'improper' multiple imputation would sample M values from some distribution, say $p^\star(\boldsymbol{y}_{\mathrm{mis}} \mid \boldsymbol{y}_{\mathrm{obs}})$, where

$$p^\star(\boldsymbol{y}_{\mathrm{mis}} \mid \boldsymbol{y}_{\mathrm{obs}}) \neq p(\boldsymbol{y}_{\mathrm{mis}} \mid \boldsymbol{y}_{\mathrm{obs}}).$$

This might be implemented when the imputation model is specified and fit separately from the full-data response model (Rubin, 1987) or when auxiliary covariates V are being used under an MAR assumption.

6.4 Covariance structures for univariate longitudinal processes

In Examples 6.1 and 6.2, we showed the importance of covariance specification in incomplete data. We now describe a number of specific approaches to accomplish this. For multivariate normal models where the dimension of Y_i is large relative to the sample size, it is common to assume a parsimonious structure for Σ to avoid having to estimate a large number of parameters. We discuss two classes of models that are computationally convenient to do this. For the first class, we directly specify the covariance structure. For the second, we specify the covariance structure indirectly via random effects.

6.4.1 Serial correlation models

A natural parameterization on which to introduce structure for Σ in multivariate normal models is via the parameters in the modified Cholesky decomposition (Pourahmadi, 1999). The parameters of this decomposition correspond to the means and variances of the conditional distributions $p(y_j \mid y_1, \ldots, y_{j-1})$: $j = 1, \ldots, J$,

$$E(Y_j \mid y_1, \ldots, y_{j-1}) = \mu_j + \sum_{k=1}^{j-1} \phi_{jk}(y_k - \mu_k), \qquad (6.12)$$

$$\mathrm{var}(Y_j \mid y_1, \ldots, y_{j-1}) = \sigma_j^2. \qquad (6.13)$$

The autoregressive coefficients in (6.12),

$$\{\phi_{jk} : k = 1, \ldots, j-1; j = 2, \ldots, J\},$$

are called generalized autoregressive parameters (GARP) and characterize the dependence structure. The variance parameters in (6.13),

$$\{\sigma_j^2 : j = 1, \ldots, J\},$$

are called the innovation variances (IV). A major advantage of these parameters is that the GARP are unconstrained regression coefficients and the logs of the innovation variances are also unconstrained, unlike the variance and covariances $\{\sigma_{jk}\}$ of Σ.

The GARP/IV parameters are also natural for characterizing missingness due to dropout and for characterizing identifying restrictions given their connections to the conditional distributions $p(y_j \mid y_1, \ldots, y_{j-1})$. This was briefly discussed in Example 5.11 and will be discussed in detail in Section 8.4.2 in Chapter 8.

Before discussing a particular class of models based on the GARP and IV parameters, we review some approaches to explore the feasibility of different parsimonious structures based on these parameters.

Exploratory analysis of GARP/IV parameters

Exploratory model selection can be conducted by examining an unstructured estimate of Σ and by examining regressograms (Pourahmadi, 1999), which plot the GARP and IV parameters vs. both lag and time. We illustrate both these approaches on the schizophrenia data (Section 1.2). To simplify this demonstration, we ignore the fact that the lag between the last two measurements was 2 weeks, not 1 week.

Table 6.1 *Schizophrenia trial: GARP parameters from fitting a multivariate normal model. The elements in the matrix are* $\phi_{j,j+k}$.

	Week (j)				
Lag (k)	1	2	3	4	5
1	.81	.89	.85	.68	.80
2	−.07	.03	.30	.14	
3	.17	−.10	.06		
4	−.02	.03			
5	−.04				

Table 6.1 shows the estimated GARP parameters from fitting a multivariate normal model to the schizophrenia data. The lag-1 parameters (first row) are the largest and appear to characterize most of the dependence. We can also view the GARP graphically using regressograms. Figure 6.1 shows a regressogram of the lag-1 GARP as a function of week (corresponding to the first row of Table 6.1); there is little structure to exploit here other than potentially assuming the lag-1 GARP are constant over week $\phi_{j,j+1} = \phi_1$.

Figure 6.2 is another regressogram showing the GARP as a function of lag (for example, the first column in the figure has a dot for each of the five lag-1 GARP given in Table 6.1). The GARP for lags greater than 2 seem small, suggesting these parameters can be fixed at zero. An alternative approach, as suggested in Pourahmadi (1999) would be to model the GARP using a polynomial in lag. Based on Figure 6.2, a quadratic might be adequate for this data and would reduce the number of GARP parameters from fifteen to three. Note that this model implicitly assumes that for a given lag, the GARP are constant.

Figure 6.3 shows the log of the IV as a function of weeks. Clearly, the

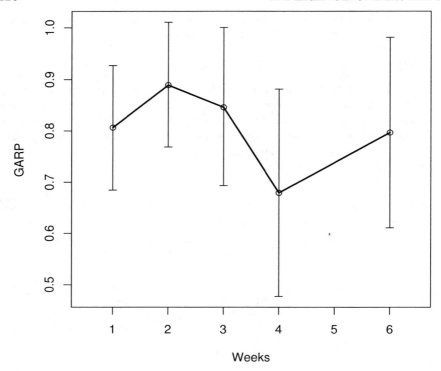

Figure 6.1 *Schizophrenia trial: posterior means of lag-1 GARP with 95% credible intervals as a function of weeks.*

innovation variances are decreasing over time and a simple linear trend in weeks would likely be adequate to model the IV, reducing the number of IV parameters from six to two.

Structured GARP/IV models for Σ

Structured GARP/IV models (Pourahmadi and Daniels, 2002) specify linear and log-linear models for the GARP and IV parameters, respectively, via

$$\phi_{jk} = \boldsymbol{v}_{jk}\boldsymbol{\gamma}, \quad k = 1, \dots, j-1; \ j = 2, \dots, J$$
$$\log(\sigma_j^2) = \boldsymbol{d}_j\boldsymbol{\lambda}, \quad j = 1, \dots, J. \tag{6.14}$$

The design vector $\boldsymbol{v}_{jk}$ can be specified as a smooth function in lag $(j-k)$ (cf. Figure 6.2) and/or a smooth function in j for a fixed lag $j-k$ (cf. Figure 6.1). The design vector $\boldsymbol{d}_j$ can be specified as a smooth function of j (cf. Figure 6.3). These smooth functions are typically chosen as low-order polynomials

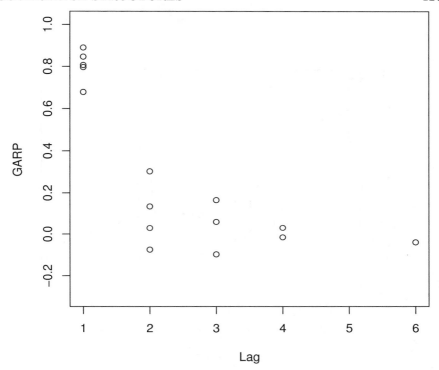

Figure 6.2 *Schizophrenia trial: posterior means of GARP as a function of lag.*

(or splines). Special cases of these models include setting $\phi_{jk} = \phi^\star_{|j-k|}$ for all j and k, i.e., constant within lag (stationary) GARP. A first-order structure would set $\phi^\star_{|j-k|} = 0$ for $|j - k| > 1$.

Based on our exploratory analysis of the schizophrenia data, we might specify a first order lag structure for the GARP. As such, the design vector for the GARP would be

$$\boldsymbol{v}_{jk} = \begin{cases} 1 & |j - k| = 1 \\ 0 & \text{otherwise} \end{cases}$$

with γ representing the lag-1 regression coefficient. We might specify a linear model for the log IV, setting $\boldsymbol{d}_j = (1, j)^\mathrm{T}$.

A practical advantage of structured GARP/IV models is that they allow simple computations since the GARP regression parameters $\boldsymbol{\gamma}$ have full conditional distributions that are normal when the prior on $\boldsymbol{\gamma}$ is normal (see Pourahmadi and Daniels, 2002). In general, fitting these models in WinBUGS

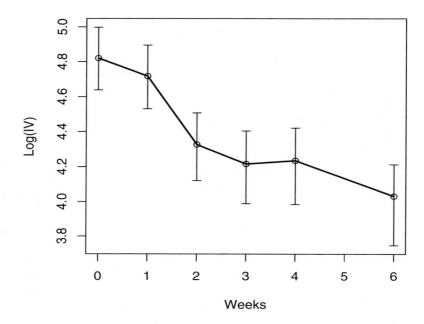

Figure 6.3 *Schizophrenia trial: posterior means of log of the IV as a function of weeks.*

can be difficult (slow mixing) because WinBUGS does not recognize that the full conditional distributions of the mean regression coefficients *and* the GARP regression parameters are both multivariate normal (see Pourahmadi and Daniels, 2002). However, for some simple cases, these can be fit efficiently in WinBUGS. In Chapter 7, we explore models based on these parameters further (including the computations) for the Growth Hormone trial (Section 1.3).

6.4.2 Covariance matrices induced by random effects

Random effects are another way to parsimoniously model the dependence structure. Whereas the GARP/IV models parameterize the covariance matrix directly, random effects models induce structure on the covariance matrix indirectly.

Recall the normal random effects model (Example 2.1),

$$\begin{aligned} \boldsymbol{Y}_i \mid \boldsymbol{x}_i, \boldsymbol{b}_i &\sim N(\boldsymbol{\mu}_i^b, \boldsymbol{\Sigma}^b) \\ \boldsymbol{b}_i \mid \boldsymbol{\Omega} &\sim N(0, \boldsymbol{\Omega}), \end{aligned}$$

with

$$\boldsymbol{\mu}_i^b = \boldsymbol{x}_i \boldsymbol{\beta} + \boldsymbol{w}_i \boldsymbol{b}_i.$$

Here, we set $\boldsymbol{\Sigma}^b = \sigma^2 I$. The marginal covariance structure for $\boldsymbol{Y}_i$, after integrating out the random effects, is

$$\boldsymbol{\Sigma} = \sigma^2 \boldsymbol{I} + \boldsymbol{w}_i \boldsymbol{\Omega} \boldsymbol{w}_i^{\mathrm{T}}$$

with j, k element

$$\sigma_{jk} = \sigma^2 I(j = k) + \boldsymbol{w}_{ij} \boldsymbol{\Omega} \boldsymbol{w}_{ik}^{\mathrm{T}},$$

where $\boldsymbol{w}_{ij}$ is the jth row of $\boldsymbol{w}_i$ and $\dim(\boldsymbol{\Omega}) = q$. The vector of covariance parameters has dimension $q(q + 1)/2 + 1$ whereas the vector of parameters of an unstructured $\boldsymbol{\Sigma}$ has dimension $J(J + 1)/2$. Typically, $q \ll J$.

For the schizophrenia clinical trial analysis in Chapter 7, we specify w_{ij} using an orthogonal quadratic polynomial ($q = 3$). As such, we reduce the number of covariance parameters from 21 to 7.

The random effects structure is an alternative to the GARP/IV structure and can also be used to introduce structured dependence in generalized linear mixed models (cf. Example 2.2). It is possible to combine structured GARP/IV models with random effects by decomposing $\boldsymbol{\Sigma}$ as $\boldsymbol{\Sigma} = \boldsymbol{\Sigma}^b + \boldsymbol{w}_i \boldsymbol{\Omega} \boldsymbol{w}_i^{\mathrm{T}}$ and modeling $\boldsymbol{\Sigma}^b$ using a parsimonious GARP/IV model (Pourahmadi and Daniels, 2002).

6.4.3 Covariance functions for misaligned data

As discussed in Example 2.4, a structured covariance is typically required to estimate the covariance function for misaligned temporal data. We review some examples next.

For temporally misaligned longitudinal data, the covariance structure is typically summarized via a covariance function,

$$\mathrm{cov}(Y_{ij}, Y_{ik} \mid \boldsymbol{x}_i; \boldsymbol{\phi}) = C(t_{ij}, t_{ik}; \boldsymbol{x}_i, \boldsymbol{\phi}).$$

To draw inference about this function, simplifying assumptions (about the process) or structure (on the function itself) are necessary because it is often the case that no (or few) replications are available for observations at certain times or pairs of times. In the remaining development, we drop the dependence on $\boldsymbol{x}$ in the covariance function for clarity.

A common assumption is (weak) stationarity, where the covariance function $C(t_{ij}, t_{ik}; \boldsymbol{\phi})$ is only a function of the difference between times, i.e.,

$C(t_{ij}, t_{ik}; \boldsymbol{\phi}) = g(|t_{ij} - t_{ik}|; \boldsymbol{\phi})$. An example is the exponential covariance function (cf. Example 2.4) given by

$$C(t_{ij}, t_{ik}; \sigma^2, \phi) = \sigma^2 \exp\{-\phi|t_{ij} - t_{ik}|\}.$$

The corresponding correlation function is $\exp(-\phi|t_{ij} - t_{ik}|)$, which for $\phi > 0$ decays exponentially with the lag between times. We use this covariance function to model the residual autocorrelation of CD4 trajectories in the HERS data (described in Section 1.5) in Section 7.6.

An example of a nonstationary covariance function is the one based on an integrated Ornstein-Uhlenbeck process (Taylor, Cumberland, and Sy, 1994), which takes the form

$$C(t_{ij}, t_{ik}; \sigma^2, \alpha) = \frac{\sigma^2}{2\alpha^3} \{2\alpha \min(t_{ij}, t_{ik}) + \exp(-\alpha t_{ij})$$
$$+ \exp(-\alpha t_{ik}) - 1 - \exp(-\alpha|t_{ij} - t_{ik}|)\}.$$

This is a structured *nonstationary* covariance function because the covariance depends on the lag $|t_{ij} - t_{ik}|$ between the observation times, *and* the times themselves. These covariance functions have been used for longitudinal CD4 counts (Taylor, Cumberland, and Sy, 1994).

6.5 Covariate-dependent covariance structures

The covariance structure can also depend on covariates. Recent work by Heagerty and Kurland (2001) and Kurland and Heagerty (2004) has shown that even for complete data, there can be bias in the mean regression coefficients in generalized linear mixed models if the random effects variance depends on a between-subject covariate and this dependence is not modeled. The problems for incomplete data have already been documented.

6.5.1 Covariance/correlation matrices

A complication in allowing components of a covariance matrix to depend on covariates is to ensure the resulting covariance matrices are positive definite. Several parameterizations have been proposed that provide a new set of parameters that are unconstrained, giving a natural parameterization on which to introduce covariates. As we did to introduce structure, we use the GARP/IV parameters of the modified Cholesky decomposition as a way to introduce covariates. Other approaches are mentioned in Further Reading.

Modeling Σ as a function of covariates using the GARP/IV parameters

Recall that the GARP, $\{\phi_{ijk}\}$, are unconstrained, as are the log of the innovation variances $\{\log(\sigma^2_{ij})\}$. Covariates can therefore be introduced as

$$\phi_{ijk} = v_{ijk}\gamma, \quad k = 1,\ldots,j-1, \quad j = 2,\ldots,J,$$
$$\log(\sigma^2_{ij}) = d_{ij}\lambda, \quad j = 1,\ldots,J,$$

where v_{ijk} and d_{ij} are design matrices for the GARP and log innovation variances, respectively. These design vectors contain covariates of interest. The form of these models is the same as the structured GARP/IV models, but now the design vectors are also indexed by i and include covariates.

We again illustrate this approach using the data from the schizophrenia clinical trial (described in Section 1.2). The main covariate of interest in this data was treatment, so we examine the GARP and IV parameters by treatment. Figure 6.4 shows the posterior means of the log innovation variances for each treatment. Within each treatment, the innovation variances do not show much structure as a function of time. However, there appear to be some large differences across treatments. For example, the innovation variances for the high dose at weeks 4 and 6 are considerably higher than for the other three treatments. These plots suggest that the innovation variances at weeks 4 and 6 could be modeled as a function of treatment.

Figure 6.5 shows the lag-1 GARP for each treatment, again plotted as a function of time. Here, the lag-1 GARP at week 6 for the medium dose is much smaller than for the other three treatments (with its credible interval not overlapping with the standard dose treatment) and we could allow this GARP to differ by treatment.

This exploratory analysis suggests the covariance matrix for the schizophrenia data does depend on treatment, and the GARP/IV parameterization provides a parsimonious way to model this as the individual parameters can depend (or not) on (a subset of) treatment groups and the resulting covariance matrices will be guaranteed to be positive definite.

These GARP/IV models will be explored more formally for the Growth Hormone data in Chapter 7. For a detailed application of these models, see Pourahmadi and Daniels (2002).

The modified Cholesky parameterization has also been used to introduce covariates into the random effects covariance matrix in the normal random effects model (Example 2.1) (Daniels and Zhao, 2003) and could also be used for the random effects covariance matrix in generalized linear mixed models (Example 2.2); however, there should be some implicit or explicit ordering of the random effects for this parameterization to be fully justified because the parameterization is *not* invariant to the ordering of the components of b_i. If the components of b_i were the coefficients of orthogonal polynomials or regression splines, there is an obvious ordering. For a detailed example of introducing

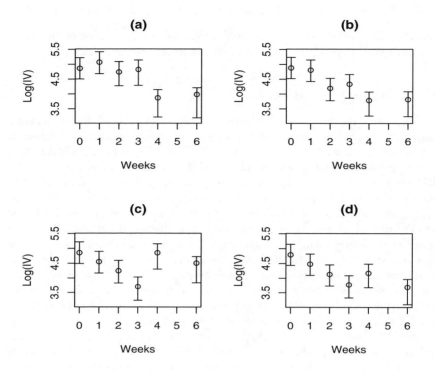

Figure 6.4 *Schizophrenia trial: posterior means of log of the innovation variances (as a function of week) with* 95% *credible intervals for each treatment: (a) low dose, (b) medium dose, (c) high dose, (d) standard treatment.*

covariates into the random effects matrix, we refer the reader to Daniels and Zhao (2003). In Section 7.3, we allow the random effects covariance matrix in the schizophrenia trial to differ by treatment.

There are other less computationally friendly ways to introduce covariates into a covariance/correlation matrix. In the setting of the multivariate normal model, there has been work on modeling the logarithm of the marginal variances while keeping the correlation matrix constant across covariates (Manly and Rayner, 1987; Barnard, McCulloch, and Meng, 2000; Pourahmadi, Daniels, and Park, 2007).

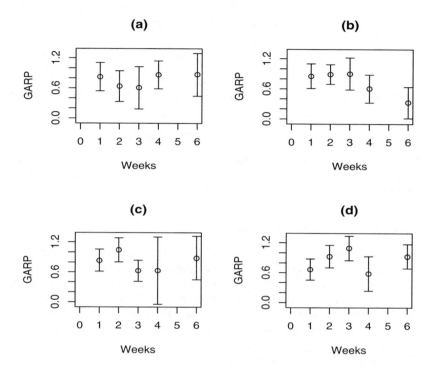

Figure 6.5 *Schizophrenia trial: posterior means of lag-1 GARP (as a function of week) with 95% credible intervals for each treatment: (a) low dose, (b) medium dose, (c) high dose, (d) standard treatment.*

Modeling the correlation matrix as a function of covariates in the multivariate probit model

In the setting of multivariate probit models, we can model the individual correlations as a function of covariates. The typical approach is to individually transform the correlations to $\mathbb{R}$ using Fisher's z-transform

$$z(\rho_{ijk}) = \tfrac{1}{2} \log \frac{(1 - \rho_{ijk})}{(1 + \rho_{ijk})} = \boldsymbol{v}_{ijk}\boldsymbol{\gamma}$$

(Czado, 2000). This transformation of the individual correlations does not guarantee the resulting correlation matrices are positive definite. Hence, the vector of regression coefficients for the correlations $\boldsymbol{\gamma}$ will be constrained; within a Gibbs sampling algorithm, values of $\boldsymbol{\gamma}$ sampled corresponding to a non-positive definite correlation matrix will have prior probability zero and will be rejected within a Metropolis-Hastings algorithm.

Modeling the covariance function in terms of covariates

For temporally misaligned data, similar approaches can be applied to covariance/correlation functions. Consider the exponential covariance function introduced in Example 2.4,

$$C(t_{ij}, t_{ik}; \sigma^2, \phi) = \sigma^2 \exp(-\phi|t_{ij} - t_{ik}|).$$

A natural way to introduce covariates here is through $\log(\sigma^2)$ and $\log(\phi)$ (assuming $\phi > 0$) as these transformations provide a set of unconstrained parameters to maintain positive definiteness of $C(\cdot, \cdot; \sigma^2, \phi)$.

6.5.2 Dependence in longitudinal binary models

For longitudinal binary data models based on underlying normal latent variables (Example 2.6) or random effects (Example 2.2), covariates can be introduced into their corresponding correlation/covariance matrix as discussed in the previous section. For (marginalized) transition models, covariates can be introduced through the (unconstrained) Markov dependence parameters.

Recall a marginalized transition model of order one (MTM(1)) has (Markov) dependence parameters γ_{ij}, $j = 2, \ldots, J$. The γ_{ij} are (unconstrained) log odds ratios that can be modeled as a function of covariates via

$$\gamma_{ij} = \boldsymbol{v}_{ij}\boldsymbol{\alpha},$$

with few additional computations compared to models without covariates. For higher-order MTMs, the dependence parameters corresponding to lags larger than one also are unconstrained and can be similarly modeled.

These models are fit to data from the CTQ I trial in Section 7.4.

6.6 Joint models for multivariate processes

Multiple longitudinal processes are typically modeled separately, despite potential gains in efficiency from modeling them jointly, unless there is interest specifically in the relationship between the multiple processes (Liu and Daniels, 2007). In the case of incomplete data, there are additional reasons to build and base inference on joint models. Joint modeling allows the use of all available data, on both the process of interest and the other processes, to 'impute' the missing values on the process of interest. This can be especially advantageous when (1) the processes are not all observed at the same times and observed responses for other processes are 'closer' temporally than observed responses from the same process and (2) the correlation between the processes is strong. By modeling the processes separately, we do not use the information from the other processes to fill in the missing values.

Our primary interest here in developing joint models will be to incorporate auxiliary variables under the auxiliary variable MAR (A-MAR) assumption (Definition 5.11), where missingness in Y is MAR after conditioning on the observed responses y_{obs}, model covariates X, *and* auxiliary covariates V. Let V_{obs} be the observed values of the auxiliary covariate process. Then A-MAR corresponds to the following form of the missing data mechanism,

$$p(r \mid y, x, v_{\mathrm{obs}}) = p(r \mid y_{\mathrm{obs}}, x, v_{\mathrm{obs}}).$$

In Bayesian inference, to allow for A-MAR mechanisms, we need to specify the joint distribution $p(y, v \mid x)$ and then integrate out v

$$p(y \mid x) = \int p(y, v \mid x) dv \qquad (6.15)$$

to obtain the full data response model of interest; from a practical perspective, joint models for which this integral can be obtained in closed form would be preferred. When the auxiliary covariate(s) are time-varying, joint longitudinal models can be constructed for the auxiliary process V and the process of interest Y.

For the CTQ II example (Section 1.4), the A-MAR assumption implies that probability of missingness of cessation status at time t is independent of the unobserved cessation status at time t conditional on the previous weeks' smoking responses *and* the previous weeks' weight change responses (as opposed to conditional on just the previous weeks' smoking responses). In a data augmentation step, we draw Y_{mis} from $p(y_{\mathrm{mis}} \mid y_{\mathrm{obs}}, v_{\mathrm{obs}})$, not $p(y_{\mathrm{mis}} \mid y_{\mathrm{obs}})$. Hence, the A-MAR assumption is weaker than MAR *to the extent that we can correctly specify the distribution* $p(y, v \mid x)$. In the following, we review and introduce some models for multivariate processes that are convenient when one of the processes is of primary interest and the other process is an auxiliary covariate process (Carey and Rosner, 2001; Gueorgeiva and Agresti, 2001; O'Brien and Dunson, 2004; Liu and Daniels, 2007). In particular, the models presented here lend themselves to the setting of auxiliary time-varying covariates in that the full data response model of interest can be obtained in closed form (cf. (6.15)). We fit one of these models to the CTQ II smoking cessation data in Section 7.5.

In the following examples, we denote Y_{ij} as the primary response of interest and V_{ij} as the auxiliary covariate process at time t_{ij}.

6.6.1 Continuous response and continuous auxiliary covariate

As a starting point, we assume both processes are (potentially) measured at the same set of observation times ($t_{ij} = t_j$). For continuous data, the multivariate normal distribution is an obvious choice for the joint distribution of the longitudinal processes. We model the joint distribution of the two processes, $(\boldsymbol{Y}_i, \boldsymbol{V}_i)$, as

$$\left(\begin{array}{c} \boldsymbol{Y}_i \\ \boldsymbol{V}_i \end{array}\right)\Bigg|\, \boldsymbol{x}_i \sim N(\boldsymbol{x}_i\boldsymbol{\beta}, \boldsymbol{\Sigma}),$$

where

$$\boldsymbol{x}_i \;=\; \left(\begin{array}{c} \boldsymbol{x}_{iy} \\ \boldsymbol{x}_{iv} \end{array}\right)$$

$$\boldsymbol{\Sigma} \;=\; \left(\begin{array}{cc} \boldsymbol{\Sigma}_{yy} & \boldsymbol{\Sigma}_{yv} \\ \boldsymbol{\Sigma}_{yv}^{\mathrm{T}} & \boldsymbol{\Sigma}_{vv} \end{array}\right).$$

The extension of this model to more than one auxiliary covariate process is obvious. The dimension of $\boldsymbol{\Sigma}$ can be quite large, so structure is often imposed on $\boldsymbol{\Sigma}$ using latent variables (Roy and Lin, 2000) or by inducing structure directly on $\boldsymbol{\Sigma}$ (Carey and Rosner, 2001). To illustrate, consider models for CD4 where the auxiliary covariate is viral load.

Consider the following (serial correlation) structure for $\boldsymbol{\Sigma}$ (Carey and Rosner, 2001),

$$\mathrm{cov}(Y_{ij}, Y_{ij'}) \;=\; \sigma_y^2 \gamma_y^{|t_j - t_{j'}|^{\theta_y}}$$

$$\mathrm{cov}(V_{ij}, V_{ij'}) \;=\; \sigma_v^2 \gamma_v^{|t_j - t_{j'}|^{\theta_v}}$$

$$\mathrm{cov}(Y_{ij}, V_{ij'}) \;=\; \sigma_y \sigma_v \gamma_{yv}^{|t_j - t_{j'}+1|^{\theta_{yv}}}.$$

The first two covariances correspond to within-process covariances for the primary response (CD4) and the auxiliary covariate (viral load), and the third corresponds to the between-process covariance. For each, it is assumed that the correlation decreases for responses farther apart in time by restricting γ_y, γ_v, and γ_{yv} to be in $[0, 1)$. The parameter γ_{yv} determines the relationship between CD4 and viral load; if $\gamma_{yv} = 0$, then they are independent processes.

Missing data is imputed in the data augmentation step by constructing the appropriate conditional distributions from this multivariate normal model with a structured covariance matrix. Clearly, this form of $\boldsymbol{\Sigma}$ implies that observed responses closer in time to the missing responses will carry more weight in the imputation and the magnitude of γ_{yv} will determine how much information is used from the auxiliary covariates process to fill in values for the primary process.

Multivariate normal models are natural for handling time-varying continuous (auxiliary) covariates as the conditional distribution of one process given the other can be written down in closed form (as a linear regression) as well as the marginal distribution of the primary process of interest using standard multivariate normal results.

The covariance structures described here also can be used for misaligned measurement times by inducing a structured covariance function within a Gaussian process.

6.6.2 Binary response and binary auxiliary covariate

Models for a binary response and binary auxiliary covariate can be specified similarly to those described in Section 6.6.1. Let $Y_{ij} = I\{Z_{ij}^Y > 0\}$ and $V_{ij} = I\{Z_{ij}^V > 0\}$ where $Z_i^Y = (Z_{i1}^Y, \ldots, Z_{iJ}^Y)^{\mathrm{T}}$ and $Z_i^V = (Z_{i1}^V, \ldots, Z_{iJ}^V)^{\mathrm{T}}$ are latent variables modeled as

$$\left(\begin{array}{c} Z_i^Y \\ Z_i^V \end{array} \right) \Bigg| \, x_i \sim N(x_i \beta, \Sigma)$$

with

$$x_i = \left(\begin{array}{c} x_{iy} \\ x_{iv} \end{array} \right)$$

$$\Sigma = \left(\begin{array}{cc} \Sigma_{yy} & \Sigma_{yv} \\ \Sigma_{yv}^{\mathrm{T}} & \Sigma_{vv} \end{array} \right).$$

The covariance matrix Σ is a correlation matrix for identifiability (cf. the multivariate probit model in Example 2.6). Models using latent normal formulations are convenient for specification of dependence and computations.

Recent work has also considered allowing the latent variables to follow a multivariate t-distribution (O'Brien and Dunson, 2004). One reason for using a t-distribution is that when the scale matrix and degrees of freedom are appropriately specified, the multivariate t-distribution approximates a multivariate logistic model; that is, the marginal distribution of Z_{ij}^Y follows (approximately) a logistic distribution, which in turn implies $P(Y_{ijk} = 1 \mid x_i)$ follows a logistic regression model. Computations are as easy as the probit model given that the multivariate t-distribution can be re-expressed as a gamma mixture of multivariate normals (see Example 3.16).

Similar to the model described in Section 6.6.1, this class of models is well-suited for handling time-varying auxiliary covariates as the conditional and marginal distributions of the process of interest take a simple form due to the underlying normal (or t-) latent structure.

We finish our discussion of joint modeling in the next section by introducing a model for a binary primary process of interest and a continuous auxiliary covariate (or vice versa). This is motivated by the CTQ II smoking cessation data (Section 1.4) where the primary process of interest was smoking cessation (binary) and the auxiliary covariate was weight change (continuous).

6.6.3 Binary response and continuous auxiliary covariate

A straightforward way to induce correlation between a continuous and a binary longitudinal process is to specify a joint multivariate normal distribution for the continuous responses and the latent variables underlying the binary process, with the restriction that the marginal covariance matrix corresponding to the binary responses is a correlation matrix. This model has appeared in various forms in the literature (Catalano and Ryan, 1992; Gueorgeiva and Agresti, 2001; Liu and Daniels, 2007).

As in Section 6.6.2, $Y_{ij} = I\{Z_{ij} > 0\}$ where Z is the vector of latent variables underlying the vector of longitudinal binary responses Y, and (Z, V) are jointly multivariate normal as follows:

$$\left(\begin{array}{c} Z_i \\ V_i \end{array} \right) \bigg| \, x_i \sim N(x_i \beta, \Sigma),$$

where, as above,

$$x_i \;\; = \;\; \left(\begin{array}{c} x_{iy} \\ x_{iv} \end{array} \right)$$

$$\Sigma \;\; = \;\; \left(\begin{array}{cc} \Sigma_{yy} & \Sigma_{yv} \\ \Sigma_{yv}^{T} & \Sigma_{vv} \end{array} \right),$$

with Σ_{yy} a correlation matrix. Notice that the marginal distribution of Z_i is multivariate normal and the marginal distribution of Y_i follows a multivariate probit model. The submatrix Σ_{yv} characterizes the dependence between the continuous and binary longitudinal processes. If $\Sigma_{yv} = 0$, then the two longitudinal processes are independent.

This model is used to address an auxiliary covariate (process) in the CTQ II smoking cessation data in Section 7.5.

6.7 Model selection and model fit under ignorability

We now propose some modifications of the techniques for model comparison and for assessing model fit, first introduced in Chapter 3, that can be extended to incomplete data settings (under ignorability). Before we discuss these modifications, we remind the reader that we can only assess the fit of the full-data model to the *observed data*; the adequacy of the model for the missing data cannot be ascertained. Different full-data models that provide the same fit to the observed data *should* be indistinguishable when using sensible model selection criteria even though they may make very different assumptions about the missing data.

Because the focus of this chapter has been on ignorable missingness, we assess the fit of the model using only the full-data response model $p(\boldsymbol{y} \mid \boldsymbol{\theta})$, and not the full data model $p(\boldsymbol{y}, \boldsymbol{r} \mid \boldsymbol{\omega})$ because it was unnecessary to specify $p(\boldsymbol{r} \mid \boldsymbol{y}_{\text{obs}}, \boldsymbol{\psi})$. For nonignorable mechanisms, we do need to use $p(\boldsymbol{y}, \boldsymbol{r} \mid \boldsymbol{\omega})$; this will be presented in detail in Chapter 8.

6.7.1 Deviance information criterion (DIC)

With complete data, the DIC is constructed using the likelihood based on the full-data response model. The complication in using the DIC with incomplete data under ignorability is that we do not observe the full data. The literature is sparse on recommendations for constructing the DIC from incomplete data. We consider two constructions based on those used in Pourahmadi and Daniels (2002), Ilk and Daniels (2007) and Celeux et al. (2007). The recommendations in Celeux et al., however, may not generalize well to our setting as they deal with missing data that are latent (e.g., in the context of mixture and random effects models). Latent data are of course fundamentally different than missing response data. First, we can never observe the latent data, but we could have observed the missing response data. Second, the observed vector of missingness indicators has no analog in random effects and mixture models.

DIC based on the observed data likelihood

A first, and perhaps most obvious construction based on the development in Chapter 5 and Section 6.1 would be to construct the DIC based on the observed data likelihood $L(\boldsymbol{\theta} \mid \boldsymbol{y}_{\text{obs}})$,

$$\mathrm{DIC}_O = \mathrm{DIC}_{\text{obs}}(\boldsymbol{y}_{\text{obs}}) = -4E\{\ell(\boldsymbol{\theta} \mid \boldsymbol{y}_{\text{obs}})\} + 2\ell(\overline{\boldsymbol{\theta}} \mid \boldsymbol{y}_{\text{obs}}),$$

where $\overline{\boldsymbol{\theta}} = E(\boldsymbol{\theta} \mid \boldsymbol{y}_{\text{obs}})$. This approach has been applied by Pourahmadi and Daniels (2002) and Ilk and Daniels (2007) and typically can be computed directly in WinBUGS.

DIC based on the full-data likelihood

An alternative approach would be to construct the DIC based on the full-data response model likelihood,

$$\mathrm{DIC}_{\text{full}}(\boldsymbol{y}_{\text{obs}}, \boldsymbol{y}_{\text{mis}}) \;=\; -4E_{\boldsymbol{\theta}}\{\ell(\boldsymbol{\theta} \mid \boldsymbol{y}_{\text{obs}}, \boldsymbol{y}_{\text{mis}})\} + 2\ell\{\overline{\boldsymbol{\theta}}(\boldsymbol{y}_{\text{mis}}) \mid \boldsymbol{y}_{\text{obs}}, \boldsymbol{y}_{\text{mis}}\},$$

where $E_{\boldsymbol{\theta}}(\cdot)$ is the expectation with respect to $p(\boldsymbol{\theta} \mid \boldsymbol{y})$ and $\overline{\boldsymbol{\theta}}(\boldsymbol{y}_{\text{mis}}) = E(\boldsymbol{\theta} \mid \boldsymbol{y}_{\text{obs}}, \boldsymbol{y}_{\text{mis}})$. Since $\boldsymbol{y}_{\text{mis}}$ is not observed, we take the expectation of $\mathrm{DIC}_{\text{full}}(\boldsymbol{y}_{\text{obs}}, \boldsymbol{y}_{\text{mis}})$ with respect to the posterior predictive distribution $p(\boldsymbol{y}_{\text{mis}} \mid$

$\boldsymbol{y}_{\text{obs}}$) to obtain

$$
\begin{aligned}
\text{DIC}_F &= \text{DIC}_{\text{full}}(\boldsymbol{y}_{\text{obs}}) \\
&= E_{\boldsymbol{y}_{\text{mis}}} \{ \text{DIC}_{\text{full}}(\boldsymbol{y}_{\text{obs}}, \boldsymbol{y}_{\text{mis}}) \} \\
&= -4 E_{\boldsymbol{y}_{\text{mis}}} \left[E_{\boldsymbol{\theta}} \{ \ell(\boldsymbol{\theta} \mid \boldsymbol{y}_{\text{obs}}, \boldsymbol{y}_{\text{mis}}) \} \right] \\
&\quad + 2 E_{\boldsymbol{y}_{\text{mis}}} \left[\ell(\overline{\boldsymbol{\theta}}(\boldsymbol{y}_{\text{mis}}) \mid \boldsymbol{y}_{\text{obs}}, \boldsymbol{y}_{\text{mis}}) \right], \quad (6.16)
\end{aligned}
$$

where $E_{\boldsymbol{y}_{\text{mis}}}(\cdot)$ is the expectation with respect to $p(\boldsymbol{y}_{\text{mis}} \mid \boldsymbol{y}_{\text{obs}})$. Note that we took a similar expectation with respect to $p(\boldsymbol{y}_{\text{mis}} \mid \boldsymbol{y}_{\text{obs}})$ to derive the observed data posterior from the full data posterior in (6.3).

DIC_F cannot be computed directly in WinBUGS. To compute (6.16), we use the samples from the data augmented Gibbs sampling algorithm; denote this sample as $\{ \boldsymbol{y}_{\text{mis}}^{(m)}, \boldsymbol{\theta}^{(m)} : m = 1, \dots, M \}$. The first term in (6.16) is computed by averaging $\ell(\boldsymbol{\theta} \mid \boldsymbol{y}_{\text{obs}}, \boldsymbol{y}_{\text{mis}})$ over the sample $\{ \boldsymbol{y}_{\text{mis}}^{(m)}, \boldsymbol{\theta}^{(m)} : m = 1, \dots, M \}$. For the second term, we need to compute

$$
\overline{\boldsymbol{\theta}}(\boldsymbol{y}_{\text{mis}}^{(m)}) = E(\boldsymbol{\theta} \mid \boldsymbol{y}_{\text{obs}}, \boldsymbol{y}_{\text{mis}}^{(m)})
$$

at each iteration m, and then average the entire term, $\ell \{ \overline{\boldsymbol{\theta}}(\boldsymbol{y}_{\text{mis}}) \mid \boldsymbol{y}_{\text{obs}}, \boldsymbol{y}_{\text{mis}} \}$, over the sample $\{ \boldsymbol{y}_{\text{mis}}^{(m)} : m = 1, \dots, M \}$.

A computational problem with this approach is that $\overline{\boldsymbol{\theta}}(\boldsymbol{y}_{\text{mis}}^{(m)})$ will typically not be available in closed form. A straightforward, but computationally impractical approach would be to run a new Gibbs sampler for every value of $\boldsymbol{y}_{\text{mis}}$ sampled and compute $\overline{\boldsymbol{\theta}}(\boldsymbol{y}_{\text{mis}})$ from these draws.

We recommend two alternative approaches that are more practical computationally. First, recall we have a sample $\{ \boldsymbol{y}_{\text{mis}}^{(m)}, \boldsymbol{\theta}^{(m)} : m = 1, \dots, M \}$ from $p(\boldsymbol{y}_{\text{mis}}, \boldsymbol{\theta} \mid \boldsymbol{y}_{\text{obs}})$. The first approach is to reweight this sample to estimate $E(\boldsymbol{\theta} \mid \boldsymbol{y}_{\text{obs}}, \boldsymbol{y}_{\text{mis}}^{(m)})$ for all values $\boldsymbol{y}_{\text{mis}}^{(m)}$ in the sample. In particular, we can use the weighted average*

$$
\overline{\boldsymbol{\theta}}(\boldsymbol{y}_{\text{mis}}^{(m)}) = E(\boldsymbol{\theta} \mid \boldsymbol{y}_{\text{obs}}, \boldsymbol{y}_{\text{mis}}^{(m)}) \approx \frac{\sum_{a=1}^{M} w_a^{(m)} \boldsymbol{\theta}^{(a)}}{\sum_{a=1}^{M} w_a^{(m)}}, \quad (6.17)
$$

with weights $w_a^{(m)}$ given by

$$
w_a^{(m)} = \frac{p(\boldsymbol{\theta}^{(a)}, \boldsymbol{y}_{\text{mis}}^{(m)}, \boldsymbol{y}_{\text{obs}}) / p(\boldsymbol{y}_{\text{mis}}^{(m)} \mid \boldsymbol{y}_{\text{obs}})}{p(\boldsymbol{\theta}^{(a)}, \boldsymbol{y}_{\text{mis}}^{(a)}, \boldsymbol{y}_{\text{obs}})},
$$

and $p(\boldsymbol{\theta}^{(l)}, \boldsymbol{y}_{\text{mis}}^{(l)}, \boldsymbol{y}_{\text{obs}}) = p(\boldsymbol{y}_{\text{obs}}, \boldsymbol{y}_{\text{mis}}^{(l)} \mid \boldsymbol{\theta}) p(\boldsymbol{\theta})$. The term $p(\boldsymbol{y}_{\text{mis}}^{(m)} \mid \boldsymbol{y}_{\text{obs}})$ can be

---

* The approximate equalities in this section, signified by $\approx$, are due only to Monte Carlo error.

estimated by

$$p(\boldsymbol{y}_{\text{mis}}^{(m)} \mid \boldsymbol{y}_{\text{obs}}) \approx \frac{1}{M} \sum_{j=1}^{M} p(\boldsymbol{y}_{\text{mis}}^{(m)} \mid \boldsymbol{y}_{\text{obs}}, \boldsymbol{\theta}^{(j)}),$$

where $\{\boldsymbol{\theta}^{(j)} : j = 1, \ldots, M\}$ is the sample from $p(\boldsymbol{\theta} \mid \boldsymbol{y}_{\text{obs}})$. This approach does not require re-running the sampling algorithm and we expect these weights to be fairly stable.

A second computational approach would be to run *one* additional Gibbs sampler for a fixed value of $\boldsymbol{y}_{\text{mis}}$, say $\overline{\boldsymbol{y}}_{\text{mis}} = E(\boldsymbol{y}_{\text{mis}} \mid \boldsymbol{y}_{\text{obs}})$; denote this sample as $\{\boldsymbol{\theta}^{(l)} : l = 1, \ldots, L\}$. We can then use the simpler weights

$$w_l^{(m)} = \frac{p(\boldsymbol{\theta}^{(l)}, \boldsymbol{y}_{mis}^{(m)}, \boldsymbol{y}_{\text{obs}})}{p(\boldsymbol{\theta}^{(l)}, \overline{\boldsymbol{y}}_{\text{mis}}, \boldsymbol{y}_{\text{obs}})},$$

in (6.17), which are available in closed form. The sum is taken over the second sample, $\{\boldsymbol{\theta}^{(l)} : l = 1, \ldots, L\}$

Parts of DIC_F can be computed in WinBUGS while other parts need to be computed in other software (like R).

Summary

Both DIC constructions need further exploration and comparison to determine their behavior in the setting of ignorable missingness. DIC_O removes the missing data by averaging over the predictive distribution conditional on $\boldsymbol{\theta}$,

$$p(\boldsymbol{y}_{\text{obs}} \mid \boldsymbol{\theta}) = \int p(\boldsymbol{y}_{\text{obs}}, \boldsymbol{y}_{\text{mis}} \mid \boldsymbol{\theta}) d\boldsymbol{y}_{\text{mis}},$$

to obtain the observed data likelihood with which the DIC is then constructed. Using the observed-data log likelihood in DIC_O is very similar to the AIC, a frequentist model selection criterion.

On the other hand, DIC_F removes the missing data by averaging the DIC based on the full-data model, $\text{DIC}_{\text{full}}(\boldsymbol{y}_{\text{obs}}, \boldsymbol{y}_{\text{mis}})$ with respect to the posterior predictive distribution of $\boldsymbol{y}_{\text{mis}}$, $p(\boldsymbol{y}_{\text{mis}} \mid \boldsymbol{y}_{\text{obs}})$. So, essentially, this form of the DIC is a weighted average of the DIC based on the complete data with weights equal to how likely these 'completed' datasets are under the model and the observed data. That the full-data response model and its parameters are typically of primary interest provides some support for this approach. For the data examples in Chapter 7, we use DIC_O.

6.7.2 Posterior predictive checks

In Section 3.5.3, we discussed posterior predictive checks based on complete data. Assessing model fit for incomplete data using posterior predictive checks can be based on statistics computed from replications of the observed data or replications of completed datasets. A recent paper by Gelman et al. (2005) advocates doing checks using completed (full-) data. The idea is to use the value of y_{mis} at each iteration of the data-augmented Gibbs sampler to create an 'observed complete' dataset to then compare to a 'replicated complete' dataset. We provide more details below. In using this approach, we must keep in mind that the fit of the model is *still* only assessed to the observed data since the missing data components of the observed datasets are filled in using data augmentation (conditional on the model).

Basing the checks on completed datasets offers several advantages. In principle, as discussed throughout this chapter, interest most often is on the full-data response model $p(y \mid \boldsymbol{\theta})$. Therefore, diagnostics on this model (as opposed to the implied observed data model) will typically be of primary interest. As such, the choices of test statistics discussed in the complete data setting in Chapter 3 would be appropriate here. Second, model building under an assumption of ignorability does not require explicit specification of the missing data mechanism, $p(r \mid y_{\mathrm{obs}}, x, \psi)$. By using diagnostics based on completed datasets, the replicated realizations of the full data vector y do not need to differentiate the missing and observed components as specified by the response indicator vector r; hence there is no need to specify or check $p(r \mid y_{\mathrm{obs}}, x, \psi)$. For these reasons, we conduct posterior predictive checks on the completed datasets.

To assess the fit of certain features of the model to the observed data, appropriate test statistics need to be constructed. These test statistics have the full data as their argument. For each MCMC sample, the test statistics evaluated at the 'observed' data will have as their argument the observed data y_{obs} combined with the current realization of the posterior predictive distribution of the missing data sampled in the data augmentation step. For example, at iteration l, the full data would be $(y_{\mathrm{obs}}, y_{\mathrm{mis}}^{(l)})$, where within the data augmented Gibbs sampler, we sample $y_{\mathrm{mis}}^{(l)}$ from $p(y_{\mathrm{mis}}^{(l)} \mid y_{\mathrm{obs}}, \boldsymbol{\theta}^{(l)})$. The test statistics for the replicated data $y_{rep}^{(l)}$ will be sampled from the full-data response model $p(y_{\mathrm{rep}}^{(l)} \mid \boldsymbol{\theta}^{(l)})$ given the current value of the parameter at iteration l, $\boldsymbol{\theta}^{(l)}$ (cf. posterior predictive distribution in (3.15)). This approach has been implemented recently in several papers, including Ilk and Daniels (2007). Choice of test statistics would be similar to those suggested in Chapter 3.

To be more specific, for some data summary $\boldsymbol{T}(\cdot; \boldsymbol{\theta})$ of interest, we compute $T(y_{\mathrm{obs}}, y_{\mathrm{mis}}^{(l)}; \boldsymbol{\theta})$ and $T(y_{\mathrm{rep}}^{(l)}; \boldsymbol{\theta})$ at iteration l and compare the two realizations at each iteration. The corresponding posterior predictive probability is defined as

$$p = \iiint H(y_{\mathrm{obs}}, y_{\mathrm{mis}}, y_{\mathrm{rep}}, c; \boldsymbol{\theta}) \, p(y_{\mathrm{rep}}, y_{\mathrm{mis}}, \boldsymbol{\theta} \mid y_{\mathrm{obs}}) \, d\boldsymbol{\theta} \, dy_{\mathrm{mis}} \, dy_{\mathrm{rep}},$$

$$(6.18)$$

where

$$p(\boldsymbol{y}_{\mathrm{rep}}, \boldsymbol{y}_{\mathrm{mis}}, \boldsymbol{\theta} \mid \boldsymbol{y}_{\mathrm{obs}}) = p(\boldsymbol{y}_{\mathrm{mis}} \mid \boldsymbol{y}_{\mathrm{obs}}, \boldsymbol{\theta})p(\boldsymbol{y}_{\mathrm{rep}} \mid \boldsymbol{\theta})p(\boldsymbol{\theta} \mid \boldsymbol{y}_{\mathrm{obs}})$$

and

$$H(\boldsymbol{y}_{\mathrm{obs}}, \boldsymbol{y}_{\mathrm{mis}}, \boldsymbol{y}_{\mathrm{rep}}, c; \boldsymbol{\theta}) = I\left\{h(T(\boldsymbol{y}_{\mathrm{obs}}, \boldsymbol{y}_{\mathrm{mis}}; \boldsymbol{\theta}), T(\boldsymbol{y}_{rep}; \boldsymbol{\theta})) > c\right\}.$$

In Sections 7.2 and 7.5, we conduct checks for the growth hormone data and CTQ II data, respectively.

For nonignorable missing data mechanisms, $p(\boldsymbol{r} \mid \boldsymbol{y}, \boldsymbol{\omega})$ is specified and is therefore part of the full-data model. We discuss posterior predictive checks for these models in Chapter 8.

6.8 Further reading

Model specification: dependence

Daniels and Pourahmadi (2002) propose prior distributions to shrink toward structured GARP/IV models. Structured antedependence models (SAD) are an alternative class of models that can be used to introduce structure into a covariance matrix (Núñez-Antón and Zimmerman, 2000) and are similar to structured GARP/IV models. The log matrix parameterization is an alternative unconstrained reparameterization of a covariance matrix that can be used to introduce covariates (Leonard and Hsu, 1992; Chiu, Leonard, and Tsui, 1996). For correlation matrices, Daniels and Pourahmadi (2007) have been exploring alternative parameterizations for both structure and covariates.

Another line of research on parsimoniously modeling covariance matrices lets the data itself find a parsimonious structure. Wong et al. (2003) model dependence parsimoniously (within the multivariate normal models in Example 2.3) by using Bayesian model averaging with priors that can 'zero-out' partial correlations. Liu and Daniels (2007) use a similar approach in the context of the joint model in Section 6.6.3. Liechty et al. (2004) develop Bayesian models to parsimoniously estimate a correlation matrix through the construction of appropriate priors. Chen and Dunson (2003) re-parametrize the random effects covariance matrix in order to 'zero-out' random effects variances. This parametrization is related to the modified Cholesky decomposition (Pourahmadi, 2007).

For misaligned measurement times in binary data, Su and Hogan (2007) have developed continuous-time MTMs for longitudinal binary data that allow the dependence to depend on covariates.

Joint models

For the covariance structure for multivariate continuous longitudinal data with misaligned times, additional structures can be found in Sy et al. (1997), Sammel et al. (1999), Henderson et al. (2000), and Ferrer and McArdle (2003).

Directly specified models for binary data that do not use an underlying multivariate normal or multivariate t latent structure, but provide marginal logistic regressions for each component of $\boldsymbol{Y}$, Y_{ijk}, have been proposed. Fitzmaurice and Laird (1993) introduce a very general model for multivariate (longitudinal) data based on the canonical log-linear parameterization. However, given the parameterization of this model, it is difficult to parsimoniously exploit the longitudinal correlation using serial correlation structures. Ilk and Daniels (2007) constructed a model specifically for multivariate longitudinal data, building on earlier work by Heagerty (1999, 2002) mixing both transition (Markov) and random effects structures; computations are complex though an efficient MCMC algorithm is proposed and implemented.

Indirectly specified conditional models (via random effects or latent classes) are another approach for modeling (multivariate) longitudinal binary data. Relevant work includes Bandeen-Roche et al. (1997), Ribaudo and Thompson (2002), Dunson (2003), and Miglioretti (2003).

There are many other approaches for general mixed longitudinal data that have been proposed in the literature. Models for a set of mixed longitudinal outcomes (more than just one continuous and binary longitudinal process) have been developed without using an underlying normal latent structure, at the cost of more complex computations. Some of these models use latent variables or latent classes to connect the processes, e.g., see Sammel et al. (1997), Dunson (2003), and Miglioretti (2003). Instead of using latent variables, Lambert et al. (2002) developed models for mixed outcomes using copulas (Nelson, 1999). Other authors have used the general location model for mixtures of categorical and continuous outcomes (Fitzmaurice and Laird, 1997; Liu and Rubin, 1998), but not specifically in the setting of multivariate longitudinal data.

We did not discuss joint modeling of longitudinal and time to event data. However, there is an extensive literature on this topic, both in the context of surrogate markers and for missing data where dropout is modeled as a time to event process. See DeGruttola and Tu (1994), Faucett and Thomas (1996), Wulfsohn and Tsiatis (1997), Wang and Taylor (2001), and Xu and Zeger (2001), among others. For an early review of these methods, see Hogan and Laird (1997b), and more recently, Tsiatis and Davidian (2004).

CHAPTER 7

Case Studies: Ignorable Missingness

7.1 Overview

In this chapter, we re-analyze several examples from Chapter 4 using all available data, under an ignorability assumption for the missing data. We also analyze two additional examples. The first uses the CTQ II smoking cessation data (Section 1.4) to illustrate joint modeling and implementation of the auxiliary variable MAR assumption. The second uses the CD4 data from HERS (Section 1.5) to illustrate modeling irregularly spaced longitudinal data with missingness. In all these analyses we focus on the importance of correctly modeling/specifying the dependence under MAR.

7.2 Structured covariance matrices: Growth Hormone study

7.2.1 Models

The notation and model used here are the same as in Section 4.2. The distribution of quadriceps strength (QS), $\boldsymbol{Y}_i = (Y_{i1}, Y_{i2}, Y_{i3})^{\mathrm{T}}$, for treatment group $Z_i = k$ is

$$\boldsymbol{Y}_i \mid Z_i = k \sim N(\boldsymbol{\mu}_k, \boldsymbol{\Sigma}_k), \tag{7.1}$$

with $\boldsymbol{\mu}_k = (\mu_{1k}, \mu_{2k}, \mu_{3k})^{\mathrm{T}}$ and $\boldsymbol{\Sigma}_k = \boldsymbol{\Sigma}(\boldsymbol{\eta}_k)$, with $\boldsymbol{\eta}_k$ containing the six nonredundant parameters in the covariance matrix.

A convenient parameterization of the covariance matrix is based on the GARP and IV parameters. To fit these models in WinBUGS, we re-parameterize the multivariate normal model as

$$
\begin{aligned}
Y_{i1} \mid Z_i = k &\sim N(\mu_{1k}, \sigma_{1k}^2) \\
Y_{i2} \mid y_{i1}, Z_i = k &\sim N(\beta_{0k} + \phi_{21,k} y_{i1}, \sigma_{2k}^2) \\
Y_{i3} \mid y_{i1}, y_{i2}, Z_i = k &\sim N(\beta_{1k} + \phi_{31,k} y_{i1} + \phi_{32,k} y_{i2}, \sigma_{3k}^2),
\end{aligned}
$$

where, for $k = 1, \ldots, 4$,

$$\boldsymbol{\phi}_k = (\phi_{21,k}, \phi_{31,k}, \phi_{32,k})$$

is the set of GARP for each treatment and $\boldsymbol{\sigma}_k^2 = (\sigma_{1k}^2, \sigma_{2k}^2, \sigma_{3k}^2)$ are the IV for each treatment. We set $\boldsymbol{\eta}_k = (\boldsymbol{\phi}_k, \boldsymbol{\sigma}_k^2)$. Using this re-parameterized model, the marginal means $\boldsymbol{\mu}_k$ can be computed recursively as

$$\mu_{jk} = \beta_{j-2,k} + \sum_{l=1}^{j-1} \phi_{jl,k}\mu_{jk}, \ j = 2, 3.$$

7.2.2 Priors

Conditionally conjugate priors as described in Daniels and Pourahmadi (2002) were used for the GARP,

$$\boldsymbol{\phi}_k \ \sim \ N(0, 10^6 \boldsymbol{I}), \ k = 1, \dots, 4.$$

We used diffuse normal priors on the intercepts (β_{0k}, β_{1k}) and truncated uniforms on the square root of the IV parameters,

$$\beta_{jk} \ \sim \ N(0, 10^6 \boldsymbol{I}_3), \quad j = 0, 1; \ k = 1, \dots, 4,$$
$$\sigma_{jk} \ \sim \ U(0, 100), \quad j = 1, \dots, 3; \ k = 1, \dots, 4.$$

7.2.3 MCMC details

For all the models, we ran three chains, each with $55{,}000$ iterations with a burn-in of 5000 iterations. The chains mixed very well with minimal autocorrelation.

Table 7.1 *Growth hormone trial: posterior mean of the GARP/IV parameters of the covariance matrices by treatment. The entries on the main diagonal are the innovation variances (IV) and below the main diagonal are the generalized autoregressive parameters (GARP).*

	EG			G		
697			552		
.98	563		.90	176	
.45	.65	241	.26	.61	91

	EP			P		
741			622		
.89	199		.74	203	
.21	.59	82	−.01	.78	154

7.2.4 Model selection and fit

We first fit a model with $\mathbf{\Sigma}_k$ distinct and unstructured for each treatment (k), which we will call covariance model (1). The posterior means of the GARP/IV parameters $\boldsymbol{\eta}_k$ for covariance model (1) are given in Table 7.1. Examination of this table suggests most of the GARP/IV parameters do not vary substantially across the treatments. As a result, we fit two more parsimonious models: covariance model (2) $\{\boldsymbol{\eta}_k = \boldsymbol{\eta} : k = 1, \ldots, 4\}$ (common $\mathbf{\Sigma}$ across treatments) and covariance model (3),

$$
\begin{aligned}
(\sigma_{1k}^2, \phi_{31,k}, \phi_{32,k}) &= (\sigma_1^2, \phi_{31}, \phi_{32}) & k &= 1, 2, 3, 4 \\
(\sigma_{2k}^2, \sigma_{3k}^2) &= (\sigma_2^2, \sigma_3^2) & k &= 2, 3, 4 \\
\phi_{21,k} &= \phi_{21} & k &= 1, 2, 3,
\end{aligned}
$$

which allowed individual GARP/IV parameters to vary across (subsets) of the treatments. Covariance model (3) was chosen based on our examination of the estimates from covariance model (1) in Table 7.1. We use DIC$_O$ with

$$
\boldsymbol{\theta} = (\boldsymbol{\mu}_1, \boldsymbol{\beta}_0, \boldsymbol{\beta}_1, \phi_{21}, \phi_{31}, \phi_{32}, \{1/\sigma_{jk}^2 : j = 1, \ldots, 3; \ k = 1, \ldots, 4\})
$$

to compare the fit of the models. These results, which appear in Table 7.2, support covariance model (3).

We also conducted two posterior predictive checks based on the residuals to assess the fit of covariance model (3). The posterior predictive probability based on the multivariate version of Pearson's χ^2 given in (3.19) was .65. The one based on largest deviation in the empirical cdf's, given in (3.22), was .40. Both measures suggest the model fits well.

Table 7.2 *Growth hormone trial: DIC$_O$ for the three covariance models.*

Model	DIC	Dev($\bar{\theta}$)	p_D
(1)	2674	2599	37
(2)	2667	2631	18
(3)	2652	2609	21

7.2.5 Results and comparison with completers-only analysis

Posterior means and 95% credible intervals for the mean parameters in covariance model (3) are given in Table 7.3; the posterior means and credible intervals were similar under the three covariance models (not shown). In comparing these results to the complete case results in Table 4.3, we point out

Table 7.3 *Growth hormone trial: posterior means and 95% credible intervals of the mean parameters for each treatment assuming covariance model (3).*

		Month	
Treatment	0	6	12
EG	69 (62, 77)	82 (71, 94)	81 (70, 92)
G	68 (61, 76)	66 (58, 74)	65 (58, 73)
EP	66 (58, 74)	81 (73, 90)	73 (65, 80)
P	65 (58, 73)	62 (55, 70)	63 (56, 70)

Table 7.4 *Growth hormone trial: posterior probabilities that each of the pairwise differences at month 12 is greater than zero under covariance model (3) and the independence model.*

	Treatment Group Contrast					
Model	EG-G	EG-EP	EG-P	G-EP	G-P	EP-P
(3)	.99	.90	1.00	.08	.68	.97
Indep.	1.00	.98	1.00	.07	.52	.94

a few interesting differences. For example, the credible intervals are narrower under ignorability; this is expected as we are now using *all* the data. The difference in the posterior mean of the 12-month mean in treatment EG illustrates the potentially strong bias from the completers-only analysis; the posterior mean was 88 (74, 103) in the completers-only analysis vs. 81 (70, 92) in the MAR analysis). This is related to the fact that at baseline, there were large differences between the mean of completers and the mean of all the subjects (cf. Tables 4.3 and 7.3).

Posterior probabilities that each of the pairwise differences of the month 12 across the treatments are greater than zero are given in Table 7.4. The posterior probabilities under ignorability showed differences from the completers-only analysis (cf. the results in the caption of Figure 4.1). For example, the posterior probabilities for the difference between treatment EG and EP changed from .97 for the completers-only analysis to .90 for the MAR analysis. There were also important differences between the three covariance models. For example, in comparing covariance models (1) and (2) under ignorability, the posterior probability for EG vs. EP was .90 and .80, respectively.

To further illustrate the importance of modeling dependence under MAR, we fit an independence model, $\Sigma(\eta_k) = \sigma_k^2 I_3$. The DIC for this model was 2956 and the effective number of parameters was 16. Based on the DIC, this model fit considerably worse than the dependence models (cf. Table 7.2). The largest difference from the dependence models was seen in the month 12 mean for treatment EG. Under the independence model, this mean was 7 to 9 units higher than under the dependence models. This large difference was due to the data augmentation step under the independence model not using the observed values at month 0 and month 6 to 'fill-in' the month 12 values for the dropouts. This difference also led to more extreme (but incorrect) posterior probabilities for the differences between month 12 means for treatment EG and the other three treatments (see Table 7.4). It is also interesting to note that the posterior distribution of the month 12 means under the independence model are essentially equivalent to the posterior distribution under models that allow dependence based on the completers only data.

7.2.6 Conclusions

The covariance model that allowed the individual GARP/IV parameters to vary by treatment, covariance model (3), was the preferred model for this analysis (as measured by the DIC) and also seemed to provide a good fit itself as measured by the posterior predictive checks. Mean QS at month 12 was significantly higher on EG vs. the other three treatments, with posterior probabilities ranging from .90 to 1.00. EP was significantly higher than P with a difference of 10 (ft.-lbs. of torque) and a posterior probability of .97. An analysis of this data considering MNAR and sensitivity analysis can be found in Chapter 10.

7.3 Normal random effects model: Schizophrenia trial

7.3.1 Models and priors

The notation, models, and priors used here are the same as in Section 4.3. Recall the goal of this 6-week trial is to compare the mean change in BPRS scores (measure of schizophrenia severity) from baseline to week 6 between the four treatment groups. For the full data $Y_i = (Y_{i1}, \ldots, Y_{i6})^T$, we assume

$$Y_i \mid b_i, Z_i = k \quad \sim \quad N(x_i(\beta_k + b_i), \sigma^2 I)$$
$$b_i \mid \Omega(\phi_k), Z_i = k \quad \sim \quad N(0, \Omega(\phi_k)),$$

where the jth row of x_i, x_{ij} is an orthogonal quadratic polynomial and $Z_i \in \{1, 2, 3, 4\}$ is treatment group.

7.3.2 MCMC details

For all the models, we ran four chains, each with $10,000$ iterations with a burn-in of 100 iterations. The autocorrelation was negligible by lag 30.

Table 7.5 *Schizophrenia trial: DIC_O for the two random effects models under MAR.*

Model	DIC	Dev($\bar{\boldsymbol{\theta}}$)	p_D
MAR(1)	6407	6369	19
MAR(2)	6434	6365	34

7.3.3 Model selection

We fit two separate covariance models. For model (1), we assumed the random effects covariance matrix was constant across the four treatments $\{\boldsymbol{\phi}_k = \boldsymbol{\phi} : k = 1, \ldots, 4\}$. For model (2), we allowed this matrix to be different across treatments.

We compared the fit using DIC_O (based on the observed-data likelihood) with the random effects, $\boldsymbol{b}_i$ integrated out. Based on the DIC (see Table 7.5), the more parsimonious model with a common random effects covariance matrix provided the best fit.

7.3.4 Results and comparison with completers-only analysis

Figure 7.1 plots the posterior means of the trajectories for each treatment over the 6 weeks of the trial. Mean BPRS for the medium and high dose individuals started to go back up by week 6, unlike with the completers-only analysis (cf. Figure 4.2). This is because those dropping out were more likely to have been doing poorly under their respective treatments, as evidenced in Figure 1.1.

Tables 7.6 and 7.7 are of primary interest for inference as they contain summary information from the posterior on the change from baseline to 6 weeks for all four treatments and the differences in the change from baseline among the treatments, respectively. All the changes from baseline were apparently non-zero (credible intervals exclude zero) except for the low dose treatment. The low dose arm under the best-fitting MAR model exhibited the smallest improvement with a posterior mean of -5 and a 95% credible interval of $(-10, 1)$.

Table 7.7 shows the differences between treatments in the change from baseline with credible intervals. In the completers only analysis, the credible intervals for all these differences covered zero. However, under the MAR analyses with the common random effects covariance matrix, the credible intervals for the differences between the low dose arm and the other three treatments all had credible intervals that excluded zero.

Table 7.6 *Schizophrenia trial: posterior mean changes from baseline to week 6 for all four treatments. Results for random effects model with Ω constant across treatments (model 1) and different across treatments (model 2) under MAR using all available data, and for model (1) using data on completers only.*

	Treatment Group			
Model	Low	Medium	High	Standard
MAR (1)	−5	−14	−12	−12
	(−10, 1)	(−19, −10)	(−16, −7)	(−16, −7)
MAR (2)	−4	−15	−12	−13
	(−10, 4)	(−19, −10)	(−17, −6)	(−17, −8)
Comp (1)	−12	−16	−16	−17
	(−17, −7)	(−20, −11)	(−21, −12)	(−21, −13)

7.3.5 Conclusions

There were non-zero differences in the change from baseline between the treatments, unlike in the completers-only analysis. In particular, subjects on the low dose arm did significantly worse than the two other dose arms of the new drug and the standard dose of halperidol. In addition, the point estimates and corresponding estimates of uncertainty were very different between the MAR and completers-only analysis as expected. There were also small differences between the two covariance models under MAR, e.g., for the low vs. high dose comparison, the credible interval for model (2) covered zero, but not the interval for model (1). We based inference on the random effects model that provided the best fit to the observed data as measured by DIC_O.

7.4 Marginalized transition model: CTQ I trial

We revisit the CTQ I data (Section 1.4) using all available data under an ignorability assumption. Recall that the outcomes in this study are binary weekly quit status for weeks 1 to 12, and the objective is to compare time-

Table 7.7 *Schizophrenia trial: pairwise differences of the changes among the four treatments under random effects models (1) and (2) under MAR and random effects model (1) for completers only.*

	Treatment Group Contrast					
Model	L vs. M	L vs. H	L vs. S	M vs. H	M vs. S	H vs. S
MAR (1)	10	7	7	−2	−2	1
	(2 ,16)	(0, 14)	(1, 14)	(−9, 4)	(−9, 4)	(−6, 7)
MAR (2)	11	8	9	−3	−2	1
	(3, 19)	(−1, 17)	(1, 17)	(−10, 4)	(−8, 4)	(−6, 8)
Comp (1)	4	4	5	1	1	1
	(−3, 11)	(−3, 11)	(−2, 12)	(−6 ,7)	(−5, 8)	(−6, 7)

averaged treatment effect from weeks 5 through 12 (where week 5 was the quit week). Let $Y_{ij} : j = 1, \ldots, 12$ denote quit status of woman i at week j. Treatment is denoted by $X_i = 1$ (exercise) and $X_i = 0$ (wellness).

7.4.1 Models

We focus on marginalized transition models here and consider two models for the marginal mean structure and two models for the serial dependence. Serial correlation is modeled only for the weeks following the quit date (i.e., weeks 5 through 12). Define the corresponding indicator as $Z_j = I\{j \geq 5\}$.

For the marginal mean structure, we consider the following two formulations. Mean model (1) assumes an unstructured mean from week 5 to 12, with a constant treatment effect β_1:

$$\text{logit}(\mu_{ij}) = (1 - Z_j)\alpha + Z_j(\beta_{0j} + \beta_1 X_i).$$

Mean model (2) assumes a constant mean with a constant treatment effect:

$$\text{logit}(\mu_{ij}) = (1 - Z_j)\alpha + Z_j(\beta_0 + \beta_1 X_i).$$

The serial correlation follows one of two models. Serial correlation model (1) allows the serial dependence to differ between week 5 and weeks 6–12 with both parameters depending on treatment; i.e.,

$$\text{logit}(\phi_{ij}) = \Delta_{ij} + \gamma_{i1} I\{j = 5\} Y_{i,j-1} + \gamma_{i2} I\{j > 5\} Y_{i,j-1},$$

where $\gamma_{ik} = \phi_{0k} + \phi_{1k} x_i$, $k = 1, 2$. Serial correlation model (2) assumes the

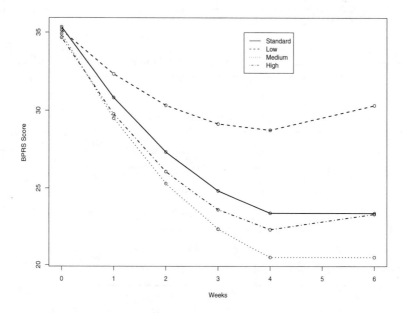

Figure 7.1 *Schizophrenia trial: posterior means of the trajectories for each of the four treatments for the random effects model (1) under MAR.*

serial dependence is constant for weeks 5–12 and differs by treatment,

$$\text{logit}(\phi_{ij}) = \Delta_{ij} + \gamma_i Y_{i,j-1},$$

where $\gamma_i = \phi_0 + \phi_1 x_i$. For comparison, we also fit the two marginal means models under independence ($\phi = 0$). The priors for (α, ϕ, β) are specified as diffuse normal priors as in the completers-only analysis (Section 4.4).

7.4.2 MCMC details

We ran three chains each with 55,000 iterations with a burn-in of 5000. There was minimal autocorrelation in the chains.

7.4.3 Model selection

We use DIC_O to compare the fit of all the models. The results appear in Table 7.8. The serial dependence models fit considerably better than the two independence models with the DIC decreasing by 50%. The best fitting model has the combination of the simpler mean model, mean model (2) (constant mean after week 4), and the more complex dependence model, serial model (1) (serial dependence differs between week 5 and weeks 6–12). We base inference on this model.

Table 7.8 *CTQ I: model selection summaries for the four MTM models.*

Mean	Model Dependence	DIC_O	$Dev(\overline{\boldsymbol{\theta}})$	p_D
(1)	(1)	1288	1260	14
(1)	(2)	1296	1272	12
(2)	(1)	1285	1271	7
(2)	(2)	1293	1284	5
(1)	indep	2404	2384	10
(2)	indep	2394	2388	3

7.4.4 Results

The posterior means and credible intervals for the treatment effect parameters β_1 appear in Table 7.9. The posterior mean for the best-fitting model was .27 with a 95% credible interval of $(-.18, .73)$.

Posterior estimates of the dependence parameters are given in Table 7.10. The dependence parameter for weeks 6–12 was almost three times as large as the dependence parameter for week 5 (compare ϕ_{01} vs. ϕ_{02}). Treatment did not appear to affect the dependence parameters; both ϕ_{11} and ϕ_{12} had credible intervals that covered zero.

Inference under the poorly fitting independence models (cf. Table 7.8) had treatment effects 50% larger with credible intervals that exclude zero. Clearly, ignoring the dependence here results in an incorrect determination of a treatment effect.

7.4.5 Conclusions

The effect of exercise on smoking cessation was to lower the quit rate, but the effect (over wellness) was not significant in our analysis under an ignorability assumption. DIC_O indicated the best-fitting model had a constant mean after the quit date and a non-constant serial dependence structure after the quit

Table 7.9 *CTQ I: posterior mean and 95% credible interval for treatment effect β_1 across models.*

| | Model | | |
Mean	Dependence	Posterior Mean	95% C.I.
(1)	(1)	.31	$(-.14, .76)$
(1)	(2)	.34	$(-.10, .78)$
(2)	(1)	.27	$(-.18, .73)$
(2)	(2)	.31	$(-.14, .74)$
(1)	indep	.48	$(.28, .68)$
(2)	indep	.47	$(.27, .68)$

Table 7.10 *CTQ I: posterior means and credible intervals for the dependence parameters ϕ from the best-fitting model.*

Parameter	Posterior Mean	95% C.I.
ϕ_{01}	1.7	$(.29, 3.4)$
ϕ_{11}	1.8	$(-.85, 5.3)$
ϕ_{02}	4.7	$(4.1, 5.2)$
ϕ_{12}	.69	$(-.13, 1.5)$

date. The importance of modeling the dependence was clear from comparing the results to models fit under independence.

7.5 Joint modeling with auxiliary variables: CTQ II trial

The objective of the following analysis is to illustrate joint modeling as an approach to accommodate the auxiliary variable MAR assumption with a stochastic time-varying auxiliary covariate. We illustrate this using the CTQ II data (described in Section 1.4) where the primary response of interest is smoking status and the auxiliary process is weight change. Weight change is generally associated with smoking cessation and we use it here to implement the auxiliary MAR (A-MAR) assumption (and to help fill in the missing cessation outcomes). The A-MAR assumption here says that the probability of dropout at week j is independent of quit status at week j only after conditioning on quit status at weeks 1 through $j-1$ *and* weight change at weeks 1 through $j-1$.

We focus on outcomes after the quit date (week 3), so there are 6 weeks of data (weeks 3–8). Dropout was high on both treatment arms, over 40%. Let $\boldsymbol{Y}_i = (Y_{i1}, \ldots, Y_{i6})^{\mathrm{T}}$ be the vector of cessation outcomes and $\boldsymbol{V}_i = (V_{i1}, \ldots, V_{i6})^{\mathrm{T}}$ be the vector of weight change outcomes.

7.5.1 Models

We use the joint model for binary and continuous longitudinal data introduced in Example 6.6.3. Let $\boldsymbol{Z}_i = (Z_{i1}, \ldots, Z_{i6})^{\mathrm{T}}$ be the vector of latent variables underlying the longitudinal vector of binary quit status responses, $\boldsymbol{Y}_i$, with components defined such that $Y_{ij} = I\{Z_{ij} > 0\}$. The model is given by

$$\left. \begin{pmatrix} \boldsymbol{Z}_i \\ \boldsymbol{V}_i \end{pmatrix} \right| \boldsymbol{x}_i \quad \sim \quad N(\boldsymbol{x}_i\boldsymbol{\beta}, \boldsymbol{\Sigma})$$

where

$$\boldsymbol{X}_i \quad = \quad \begin{pmatrix} \boldsymbol{X}_{iy} \\ \boldsymbol{X}_{iv} \end{pmatrix},$$

$$\boldsymbol{\Sigma} \quad = \quad \begin{pmatrix} \boldsymbol{\Sigma}_{yy} & \boldsymbol{\Sigma}_{yv} \\ \boldsymbol{\Sigma}_{yv}^{\mathrm{T}} & \boldsymbol{\Sigma}_{vv} \end{pmatrix},$$

and $\boldsymbol{\Sigma}_{yy}$ is a correlation matrix. The design matrix $\boldsymbol{X}_y$ is specified as a separate quadratic polynomial in time for each treatment and $\boldsymbol{X}_v$ is an unstructured mean over time for each treatment.

Following the approach in Liu and Daniels (2007), we re-parameterize $\boldsymbol{\Sigma}$ in terms of $(\boldsymbol{\Sigma}_{yy}, \boldsymbol{B}_{yv}, \boldsymbol{\Sigma}_{vv}^{\star})$, where

$$\boldsymbol{B}_{yv} \quad = \quad \boldsymbol{\Sigma}_{yv}^{\mathrm{T}} \boldsymbol{\Sigma}_{yy}^{-1}$$
$$\boldsymbol{\Sigma}_{vv}^{\star} \quad = \quad \boldsymbol{\Sigma}_{vv} - \boldsymbol{B}_{yv}\boldsymbol{\Sigma}_{yv}.$$

The matrices $\boldsymbol{B}_{yv}$ and $\boldsymbol{\Sigma}_{vv}^{\star}$ can be recognized as the regression coefficients and residual covariance matrix, respectively, from the regression of the auxiliary variables $\boldsymbol{V}$ on the latent response variables $\boldsymbol{Z}$; in fact,

$$\boldsymbol{V} \mid \boldsymbol{Z} \sim N(\boldsymbol{X}_v\boldsymbol{\beta} + \boldsymbol{B}_{yz}(\boldsymbol{Z} - \boldsymbol{X}_y\boldsymbol{\beta}), \boldsymbol{\Sigma}_{vv}^{\star}).$$

We fit models assuming dependence between the two processes, $\boldsymbol{B}_{yv} \neq \boldsymbol{0}$

(or equivalently, $\Sigma_{yv} \neq 0$), independence between the processes ($B_{yv} = 0$), and independence within and between processes (diagonal Σ).

7.5.2 Priors

The vector of regression coefficients β is given an improper uniform prior. Parameterizing Σ as ($\Sigma_{yy}, B_{yv}, \Sigma_{vv}^{\star}$), we placed a (proper) uniform prior on the correlation matrix Σ_{yy}, an (improper) uniform prior on the components of the coefficient matrix B_{zy}, and a flat prior on $\Sigma_{yy}^{\star}$. See Liu and Daniels (2007) for verification that this specification gives a proper posterior. However this prior specification is not allowed in WinBUGS.

7.5.3 Posterior sampling

The key to posterior sampling in this model is moving between two forms of the likelihood in deriving the full conditional distributions. Given the standard multivariate normal likelihood and the priors specified in the previous section, the full conditional distribution for β can easily be derived as a multivariate normal distribution. To derive the full conditional distributions of B_{yv} and $(\Sigma_{vv}^{\star})^{-1}$, which will be multivariate normal and Wishart, respectively, the multivariate normal likelihood can be factored into $p(v \mid z)p(z)$. For full details on the forms and derivations of the full conditional distributions for these models, see Liu and Daniels (2007).

7.5.4 Model selection and fit

To compare the fit of the models with different assumptions on the covariance structure, we use DIC_O. To compute the observed-data likelihood, which is not available in closed form (cf. Example 3.5), we use Monte Carlo integration and re-weighting as discussed in Chapter 3. The joint model that allows dependence between the smoking and weight gain process provides the best fit, with a DIC value of 2746; the model that assumed $B_{yv} = 0$ had a DIC of 2843 and the model with $\Sigma = I$ had a DIC of 3776.

We also conducted posterior predictive checks using the techniques outlined in Section 6.6. In particular, at each iteration of the sampling algorithm, we computed test statistics for each week (j) as the differences in quit rates and weight changes, respectively, for the replicated full data at each iteration and the completed dataset at each iteration using the observed data and the missing data sampled during the data augmentation step. For example, at week j, for the cessation outcomes, we compute

$$h\{T(\boldsymbol{y}_{\mathrm{obs},j}, \boldsymbol{y}_{\mathrm{mis},j}), T(\boldsymbol{y}_{\mathrm{rep},j})\} \;=\; \sum_{\{i:r_{ij}=1\}} \boldsymbol{y}_{\mathrm{obs},ij} + \sum_{\{i:r_{ij}=0\}} \boldsymbol{y}_{\mathrm{mis},ij}$$
$$- \sum_{i=1}^{n} \boldsymbol{y}_{\mathrm{rep},ij}. \tag{7.2}$$

For weight change, replace $\boldsymbol{y}$ with $\boldsymbol{v}$ in (7.2).

These statistics were chosen to examine consistent over- or underestimation of the mean for each process at each measurement time. The posterior predictive probabilities were defined as the probabilities that these differences were bigger than zero. All the probabilities were between .2 and .6, providing no evidence of lack of fit.

7.5.5 Results

Table 7.11 shows the posterior means of the quit rates under the full joint model ($\boldsymbol{B}_{yv} \neq \boldsymbol{0}$), the model with $\boldsymbol{B}_{yv} = \boldsymbol{0}$ (independence between weight change and smoking), and the one assuming $\boldsymbol{\Sigma} = \boldsymbol{I}$ (complete independence). As discussed earlier, we base inference on the model with $\boldsymbol{B}_{yv} \neq \boldsymbol{0}$.

The reason for the *small* differences between the model that assumes $\boldsymbol{B}_{yv} = \boldsymbol{0}$ and the full joint model is the fact that the dependence here is much stronger within each process than between the processes; for example, the correlations between the quit process over time are all greater than .5, while they are all less than .2 for the correlations between the quit and weight processes. Clearly, the A-MAR assumption here has little impact on inference over the MAR assumption (equivalent to model with $\boldsymbol{B}_{yv} = \boldsymbol{0}$).

In general, if the missingness is A-MAR, but

$$p(\boldsymbol{y}, \boldsymbol{v} \mid \boldsymbol{x}) \approx p(\boldsymbol{y} \mid \boldsymbol{x})p(\boldsymbol{v} \mid \boldsymbol{x}),$$

then joint modeling is probably not needed. See Liu and Daniels (2007) for a more detailed exploration of the dependence structure and a discussion addressing inference on the relationship between smoking and weight change between the two treatments, which we do not address here.

Table 7.12 contains the treatment difference at the final week with 95% credible intervals. The wellness arm had a higher quit rate than the exercise arm (with a 95% credible interval that had a lower bound of zero). Clearly, the independence model underestimates the treatment difference. This is due to the fact that in the dependence models, the values filled in for the missing responses using data augmentation used the previous quit responses to fill in smokers more often than quitters vs. the independence models, which did not use the previous responses.

Table 7.11 *CTQ II: posterior means and 95% credible intervals for weekly cessation rates on the exercise (E) and wellness W) treatment arms, computed for each of the three models. Observed-data sample means included for comparison.*

Treatment	Model			Observed
	A-MAR	MAR	Indep	
W	.39 (.34,.45)	.39 (.34, .45)	.38 (.31, .46)	.39
	.50 (.44,.55)	.50 (.44, .55)	.50 (.43, .56)	.49
	.55 (.49, .60)	.55 (.49, .60)	.56 (.48, .64)	.57
	.53 (.47, .59)	.53 (.47, .59)	.56 (.48, .64)	.56
	.53 (.47, .59)	.53 (.47, .59)	.53 (.45, .62)	.53
	.46 (.40, .52)	.46 (.40, .52)	.46 (.40, .52)	.46
E	.45 (.39, .51)	.45 (.39, .51)	.47 (.40,.54)	.44
	.37 (.32, .43)	.37 (.32, .43)	.38 (.30, .47)	.38
	.44 (.38, .49)	.44 (.38, .49)	.45 (.39, .51)	.44
	.42 (.37, .48)	.42 (.37, .48)	.47 (.40, .55)	.46
	.46 (.40, .52)	.46 (.40, .52)	.50 (.43, .57)	.49
	.38 (.32, .43)	.38 (.32, .43)	.43 (.36, .51)	.42

7.5.6 Conclusions

The analysis provided evidence of a higher quit rate on the wellness arm. Posterior predictive checks suggested the best model (as chosen by DIC_O) fit the observed data well. The inclusion of the auxiliary longitudinal process weight change had little effect on the conclusions in this example because it was not *strongly* correlated with smoking cessation.

7.6 Bayesian p-spline model: HERS CD4 data

In HIV epidemiology, the effect of HAART on disease dynamics remains an important question. Our analysis here examines CD4 trajectory relative to initiation of HAART in the HER study (described in Section 1.5). Because HAART use was incident in HERS, these data afford an opportunity to capture the

Table 7.12 *CTQ II: posterior means of difference in quit rates with 95% credible intervals at week 8 for each of the three models considered. Observed-data sample means included for comparison.*

Model	Posterior Mean	95% C.I.
A-MAR	.08	(.00, .16)
MAR	.08	(.00, .16)
Independence	.03	(−.06, .10)
Observed	.04	

population-level variation in CD4 before and after initiation of HAART. Here we present a comparative analysis based on p-splines.

7.6.1 Models

Define $\boldsymbol{Y}_i$ to be the J_i-dimensional vector of CD4 counts measured at times $(t_{i1}, \ldots, t_{i,J_i})$. We model the CD4 trajectories using p-splines (see Examples 2.7 and 3.10) with linear ($q = 1$) and quadratic ($q = 2$) bases using a multivariate normal model with an exponential covariance function

$$\boldsymbol{Y}_i \mid \boldsymbol{X}_i \sim N(\boldsymbol{X}_i \boldsymbol{\beta}, \boldsymbol{\Sigma}_i),$$

where the elements $\sigma_{ikl}(\boldsymbol{\phi})$ of $\boldsymbol{\Sigma}_i$ take the form

$$\sigma_{ikl} = \sigma^2 \exp(-\phi^{|t_{ik} - t_{il}|}). \tag{7.3}$$

The matrices $\boldsymbol{x}_i$ are specified using a truncated power basis with 18 equally spaced knots placed at the sample quantiles of the observation times. So, the jth row is given by

$$\boldsymbol{x}_{ij} = \{1, t_{ij}, t_{ij}^2, \ldots, t_{ij}^q, (t_{ij} - s_1)_+^q, (t_{ij} - s_2)_+^q, \ldots, (t_{ij} - s_{18})_+^q\}^{\mathrm{T}},$$

where $\{s_1, \ldots, s_{18}\}$ are the knots and $\boldsymbol{\beta}$ is partitioned as

$$\boldsymbol{\beta} = (\alpha_0, \alpha_1, \ldots, \alpha_q, a_1, \ldots, a_{18})^{\mathrm{T}} = (\boldsymbol{\alpha}^{\mathrm{T}}, \boldsymbol{a}^{\mathrm{T}})^{\mathrm{T}}.$$

In this study, not all the data were observed and the observed data were irregularly spaced. The analysis here implicitly assumes that the unobserved data are MAR.

7.6.2 Priors

Bounded uniform priors are specified for the residual standard deviation σ and the correlation parameter ϕ for the exponential covariance function in (7.3), with lower bounds of 0 and upper bounds of 100 and 10, respectively.

The components of the $(q+1)$-dimensional vector of coefficients $\boldsymbol{\alpha}$ are given 'just proper' independent normal priors with mean 0 and variance 10^4; the coefficients corresponding to the knots $\boldsymbol{a}$ are given normal priors with mean zero and variance τ^2. The standard deviation τ is then given a Unif$(0, 10)$ prior.

7.6.3 MCMC details

Three chains were run each for $15,000$ iterations with a burn-in of 5000. The autocorrelation in the chains was minimal.

7.6.4 Model selection

In addition to the model with an exponential covariance function for the residuals, we also fit a model that assumes independent residuals ($\phi = 0$) for comparison. To assess fit among the models, we computed the DIC with $\boldsymbol{\theta} = (\boldsymbol{\beta}, \phi, 1/\sigma^2, 1/\tau^2)$; see Table 7.13. The best-fitting model was the quadratic p-spline with the exponential covariance function. The greatly improved fit of both the linear and quadratic p-splines with residual autocorrelation over independence of the residuals is clear from the large differences in the DIC.

Table 7.13 *HERS CD4 data: DIC under $q = 1$, $q = 2$ (linear and quadratic spline bases), under independence and exponential covariance functions.*

Model	DIC	Dev$(\overline{\theta})$	p_D
Independence, $q = 1$	19417	19400	8
Independence, $q = 2$	19417	19402	7
Exponential, $q = 1$	18008	17985	11
Exponential, $q = 2$	17999	17980	9

7.6.5 Results

Posterior means and 95% credible intervals for the covariance parameters in the best-fitting model (as measured by the DIC) with $q = 2$ appear in

Table 7.14. The correlation parameter ϕ was significantly different from zero, with the posterior mean corresponding to a lag 1 (week) correlation of .52. Posterior means of the curves for $q = 1$ and $q = 2$ appear in Figures 7.2 and 7.3, respectively.

The posterior mean of the spline curves were relatively similar for $q = 1$ and $q = 2$ and captured more local variation than the independence models; time zero on the plot represents self-reported initiation of HAART. An increase can be seen near self-reported initiation (the effect of HAART), followed by a decrease later on, possibly attributable to noncompliance and development of resistant viral strains. The results also suggest that the self-reports of HAART initiation are subject to a delay relative to the actual initiation of HAART because the decrease in CD4 levels occurs prior to self-reported initiation time $(t = 0)$.

Table 7.14 *Posterior means and 95% credible intervals for exponential covariance function parameters and smoothing standard deviation under $q = 2$ for the HERS CD4 data.*

Parameter	Posterior Mean	95% C.I.
ϕ	.65	(.60, .70)
σ	3.9	(3.8, 4.0)
τ	1.9	(.81, 3.6)

7.7 Summary

The five examples in this chapter have demonstrated a variety of fundamental concepts for properly modeling incomplete data under an assumption of ignorability. The importance of correctly specifying the dependence structure was emphasized in all the examples and approaches for choosing the best-fitting model and assessing the fit of such models under ignorability using DIC_O and posterior predictive checks were demonstrated.

In addition, we illustrated the use of joint modeling to properly address auxiliary stochastic time-varying covariates when an assumption of auxiliary variable MAR (A-MAR) is thought plausible.

Under ignorability, model specification only involved the full data response model. In the next three chapters we discuss model-based approaches for nonignorable missing data that require specification of the full-data model, i.e., the full-data response model *and* the missing data mechanism (explicitly). In Chapter 10 we present three detailed case studies.

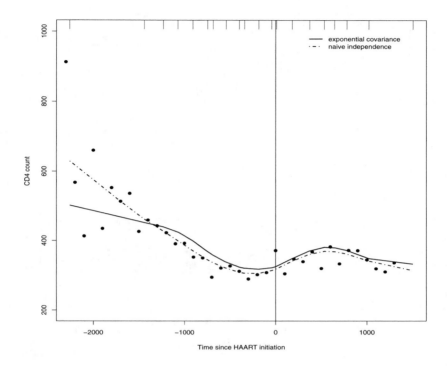

Figure 7.2 *HERS CD4 data: plot of observed data with fitted p-spline curve with q = 1 under independence and exponential covariance function. The observed data (dots) are averages over 100 day windows. The tick marks on the top of the plot denote percentiles.*

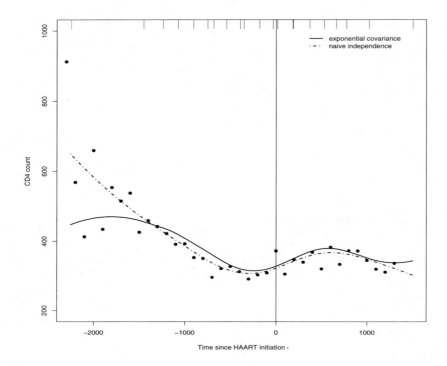

Figure 7.3 *HERS CD4 data: plot of observed data with fitted p-spline curve with q = 2 under independence and exponential covariance function. The observed data (dots) are averages over 100 day windows. The tick marks on the top of the plot denote percentiles.*

CHAPTER 8

Models for Handling Nonignorable Missingness

8.1 Overview

This chapter covers many of the central concepts and modeling approaches for dealing with nonignorable missingness and dropout. A prevailing theme is the factorization of the full-data model into the observed data distribution and the extrapolation distribution, where the latter characterizes assumptions about the conditional distribution of missing data given observed information (observed responses, missing data indicators, and covariates). The extrapolation factorization (as we refer to it) factors the full-data distribution into its identified and nonidentified components. We argue that models for nonignorable dropout should be parameterized in such a way that one or more parameters for the extrapolation distribution is completely nonidentified by data. These also should have transparent interpretation so as to facilitate sensitivity analysis or incorporation of prior information about the missing data distribution or missing data mechanism.

In Section 8.2, we describe the extrapolation distribution and state conditions for parameters to be *sensitivity parameters*. Section 8.3 reviews selection models, with particular emphasis on how the models are identified. We show that parametric selection models cannot be factored into identified and nonidentified parts, resulting in an absence of sensitivity parameters. Semiparametric selection models are introduced as a potentially useful alternative. Section 8.4 covers mixture models for both discrete and continuous responses, and for discrete and continuous dropout times. We show that in many cases, mixture models are easily factored using the extrapolation factorization, enabling interpretable sensitivity analyses and formulation of missing data assumptions. Computation of covariate effects for mixture models also is discussed. Section 8.5 provides an overview of shared parameters models. In Section 8.6, we provide a detailed discussion of possible approaches to model selection and model fit. This is still an open area of research, and our coverage highlights several of the key conceptual and technical issues. Model fit and model selection under nonignorable missingness differ from the ignorable case because with nonignorability, the full-data distribution includes a specification for $f(r \mid y)$ that must be checked against observed r. Suggestions for further reading are given in Section 8.7.

8.2 Extrapolation factorization and sensitivity parameters

In Chapter 5, we introduced a classification of dropout mechanisms that, with respect to posterior inference, can be classified as ignorable or nonignorable. In Chapter 6, we reviewed and discussed concepts and models for ignorable dropout, and illustrated using case studies in Chapter 7. In this chapter, we describe several different models for nonignorable dropout that are characterized via specification of the full data distribution $f(\boldsymbol{y}, \boldsymbol{r} \mid \boldsymbol{\omega})$.

Recall that the full data distribution can be factored into an *extrapolation model* and an *observed data model*,

$$p(\boldsymbol{y}, \boldsymbol{r} \mid \boldsymbol{\omega}) \;=\; p(\boldsymbol{y}_{\mathrm{mis}} \mid \boldsymbol{y}_{\mathrm{obs}}, \boldsymbol{r}, \boldsymbol{\omega}_{\mathrm{E}}) \, p(\boldsymbol{y}_{\mathrm{obs}}, \boldsymbol{r} \mid \boldsymbol{\omega}_{\mathrm{O}}).$$

Here, $\boldsymbol{\omega}_{\mathrm{E}}$ and $\boldsymbol{\omega}_{\mathrm{O}}$ denote parameters indexing the extrapolation and observed data models, respectively. They are potentially, but not necessarily, overlapping functions of $\boldsymbol{\omega}$. The observed data distribution $p(\boldsymbol{y}_{\mathrm{obs}}, r \mid \boldsymbol{\omega}_{\mathrm{O}})$, is identified and can (in principle) be estimated nonparametrically. The extrapolation distribution $p(\boldsymbol{y}_{\mathrm{mis}} \mid \boldsymbol{y}_{\mathrm{obs}}, \boldsymbol{r}, \boldsymbol{\omega}_{\mathrm{E}})$ cannot be identified without modeling assumptions or constraints on the parameter space.

In general we advocate using parameterizations in which one or more parameters indexing the extrapolation cannot be identified by observed data. To formalize this notion, we define a class of parameters for full-data models that can be used for sensitivity analysis or incorporation of informative prior information. Generally they are not identifiable from observed data, but when their values are fixed, the remainder of the full-data model is identified. Formally we call them *sensitivity parameters* and in general, we use the term to mean parameters that satisfy this definition.

Definition 8.1. *Sensitivity parameter.*
Let $p(\boldsymbol{y}, \boldsymbol{r} \mid \boldsymbol{\omega})$ denote a full-data model. Its extrapolation factorization is given by

$$p(\boldsymbol{y}, \boldsymbol{r} \mid \boldsymbol{\omega}) \;=\; p(\boldsymbol{y}_{\mathrm{mis}} \mid \boldsymbol{y}_{\mathrm{obs}}, \boldsymbol{r}, \boldsymbol{\omega}_{\mathrm{E}}) \, p(\boldsymbol{y}_{\mathrm{obs}}, \boldsymbol{r} \mid \boldsymbol{\omega}_{\mathrm{O}}).$$

If there exists a reparameterization $\boldsymbol{\xi}(\boldsymbol{\omega}) = (\boldsymbol{\xi}_{\mathrm{S}}, \boldsymbol{\xi}_{\mathrm{M}})$ such that:

1. $\boldsymbol{\xi}_{\mathrm{S}}$ is a nonconstant function of $\boldsymbol{\omega}_{\mathrm{E}}$,

2. the observed-data likelihood

$$L(\boldsymbol{\xi}_{\mathrm{S}}, \boldsymbol{\xi}_{\mathrm{M}} \mid \boldsymbol{y}_{\mathrm{obs}}, \boldsymbol{r})$$

   is constant as a function of $\boldsymbol{\xi}_{\mathrm{S}}$, and

3. at a fixed value of $\boldsymbol{\xi}_{\mathrm{S}}$, the observed data likelihood is a nonconstant function of $\boldsymbol{\xi}_{\mathrm{M}}$,

then $\boldsymbol{\xi}_{\mathrm{S}}$ is a *sensitivity parameter*. $\square$

The first condition requires that $\boldsymbol{\xi}_{\mathrm{S}}$ be a nonconstant function of the parameters of the extrapolation distribution. The second condition implies that the fit of the model to the observed data is not affected by the sensitivity parameters. In any particular full-data model, there may be several sets of parameters that satisfy the first two conditions. However, there will often be only one set that satisfies the third condition as well, which states when the sensitivity parameters are fixed, the full data model is identified. These conditions will be helpful in determining sensitivity parameters for semiparametric selection models in Section 8.3.7. For mixture models, finding sensitivity parameters is usually very easy (see Section 8.4.3).

The framework we use is very similar to the one based on nonparametric identified (NPI) full-data models proposed by Robins (1997) and discussed in some detail by Vansteelandt et al. (2006). A class $\mathcal{M}(\gamma)$ of full-data models is nonparametric identified if, for an observed data distribution $p(\boldsymbol{y}_{\mathrm{obs}}, \boldsymbol{r})$, there exists a unique full-data model $p(\boldsymbol{y}_{\mathrm{obs}}, \boldsymbol{y}_{\mathrm{mis}}, \boldsymbol{r} \mid \gamma) \in \mathcal{M}(\gamma)$ that marginalizes to $p(\boldsymbol{y}_{\mathrm{obs}}, \boldsymbol{r})$; i.e.

$$\int p(\boldsymbol{y}_{\mathrm{obs}}, \boldsymbol{y}_{\mathrm{mis}}, \boldsymbol{r} \mid \gamma) \, d\boldsymbol{y}_{\mathrm{mis}} = p(\boldsymbol{y}_{\mathrm{obs}}, \boldsymbol{r}).$$

The observed data do not contain information about γ, but knowing γ points to a specific $p(\boldsymbol{y}_{\mathrm{obs}}, \boldsymbol{y}_{\mathrm{mis}}, \boldsymbol{r} \mid \gamma)$ in the class $\mathcal{M}(\gamma)$ that coincides with $p(\boldsymbol{y}_{\mathrm{obs}}, \boldsymbol{r})$. Vansteelandt et al. (2006) call γ a sensitivity parameter if it indexes a NPI class of full-data models $\mathcal{M}(\gamma)$.

Definition 8.1 is designed for the likelihood framework, but essentially possesses the same attributes. In short, sensitivity parameters index the missing data extrapolation and cannot be identified by observed data. They provide the framework for assessing sensitivity of model-based inferences to assumptions about the missing data mechanism, or incorporating those assumptions formally in terms of informative prior distributions.

As it turns out, not all models are amenable to this type of parametrization. Parametric selection models in particular cannot generally be factored to permit sensitivity parameterizations. Mixture models tend to be easier to work with. For models described below, we provide examples of parameterizations that satisfy Definition 8.1, or indicate when they are difficult or even impossible to find.

8.3 Selection models

In most of this section, we assume a common set, $\{t_1, \ldots, t_J\}$, of observation times and without loss of generality $t_j = j$.

8.3.1 Background and history

Recall from Chapter 5 that the selection model factors the full-data distribution as

$$p(\boldsymbol{y}, \boldsymbol{r} \mid \boldsymbol{\omega}) = p(\boldsymbol{r} \mid \boldsymbol{y}, \boldsymbol{\psi}(\boldsymbol{\omega})) \, p(\boldsymbol{y} \mid \boldsymbol{\theta}(\boldsymbol{\omega})); \tag{8.1}$$

this factorization underlies the missing data taxonomy described in Section 5.4.

The typical strategy to model the two components is to specify a parametric model for each and to assume that ψ and θ are distinct. In the following, we will assume ψ and θ are distinct and *a priori* independent unless stated otherwise. Note that this does not correspond to the $(\omega_{\mathrm{E}}, \omega_{\mathrm{O}})$ partition in (8.1).

An early example of selection modeling for multivariate data can be found in Heckman (1979), where the joint distribution of a bivariate response Y with missing Y_2 was specified using a multivariate normal distribution (implicitly, a probit model for the binary indicators of missingness that is linear in $(Y_1, Y_2)^{\mathrm{T}}$). Diggle and Kenward (1994) expanded the Heckman model to the case of dropout in longitudinal studies using a logistic model for the hazard of dropout. Many subsequent articles adapted and extended their approach within a likelihood framework (see, e.g., Fitzmaurice, Molenberghs, and Lipsitz, 1995; Baker, 1995; Molenberghs, Kenward, and Lesaffre, 1997; Liu, Waternaux, and Petkova, 1999; Albert, 2000; Heagerty and Kurland 2004). A framework for semiparametric inference can be found in Robins et al. (1995) and Scharfstein et al. (1999).

We begin our review of parametric selection models by demonstrating via several examples the difficulty in finding sensitivity analysis parameterizations.

8.3.2 Absence of sensitivity parameters in the missing data mechanism

In many parametric selection models, all the parameters are identified. Identification is driven by parametric assumptions on both the full-data response model and the missing data mechanism (MDM).

To illustrate using a simple example, consider a cross-sectional setting where Y may or may not be missing. Suppose the histogram of the observed y's looks like that given in Figure 8.1 and that we specify the following model for the missing data mechanism:

$$\mathrm{logit}\{P(R = 1 \mid y)\} = \psi_0 + \psi_1 y,$$

where $R = 1$ corresponds to observing Y. With no further assumptions about the distribution of Y, we cannot identify ψ_1 because when $R = 0$, we do not observe Y. It can also be shown (Scharfstein et al., 2003) that ψ_1 is a sensitivity parameter.

However, suppose we further assume that the full-data response model is a normal distribution, $N(\mu, \sigma^2)$. By looking at the histogram of the *observed* y's in Figure 8.1, it is clear that for this histogram to be consistent with normally distributed *full-data* y's, we need to fill in the right tail; this implies $\psi_1 < 0$. On the other hand, if we assumed a parametric model for the full-data response model that was consistent with the histogram for the observed y's (e.g., a skew-normal), it would suggest that $\psi_1 = 0$. Thus, inference about

ψ_1 depends heavily on the distributional assumptions about $p(y)$, and ψ_1 is in fact *identified* by observed data. Hence it does not meet the criteria in Definition 8.1, and therefore is not a sensitivity parameter.

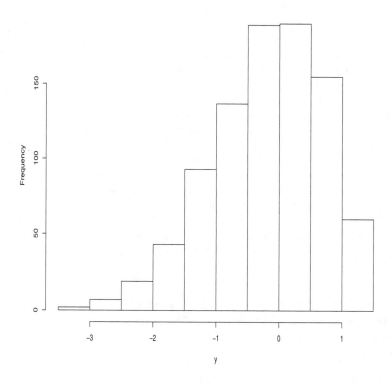

Figure 8.1 *Histogram of observed responses.*

This univariate example extends readily to the longitudinal setting. For example, consider a bivariate normal full-data response having missingness in Y_2, with missing data mechanism

$$\text{logit}\{P(R = 1 \mid y)\} = \psi_0 + \psi_1 y_1 + \psi_2 y_2, \qquad (8.2)$$

where $R = 1$ corresponds to observing Y_2. Under a bivariate normal distribution for the full-data response model, the same argument used above would be applicable in terms of the conditional distribution, $p(y_2 \mid y_1)$ and the identification of ψ_2.

Sensitivity of MDM parameters to assumptions on the full-data response model was studied in a simple but very informative empirical example by Kenward (1998). He considered a bivariate full-data response with Y_2 potentially missing and missing data mechanism given in (8.2), and assessed the sensitivity of inference about ψ to distributional assumptions about $p(y_2 \mid y_1)$. If this distribution was assumed to be normal, the estimates and standard errors of ψ implied MNAR. However, if this distribution was assumed to follow a t-distribution with only a few degrees of freedom, the estimates and standard errors of ψ implied MAR. So by just changing the tails of the distribution of the full-data response, inference concerning the missing data mechanism changed significantly. This and the previous hypothetical examples illustrate the prominent role of modeling assumptions in parametric selection models: widely differing conclusions can be drawn based on unverifiable modeling assumptions about the full-data response.

The preceding discussion has focused entirely on the full-data response model in identifying potential sensitivity parameters in the missing data mechanism. However, we have focused on the simple (and common) situation where the response y is entered into the MDM linearly. This is a strong assumption and is not always appropriate. For example, in the schizophrenia clinical trial (Section 1.2), it is conceivable that participants may be more likely to drop out when they are doing much better or much worse, suggesting that the missing data mechanism should be quadratic in y. Revisiting the previous example with cross-sectional data, assume $p(y_{\mathrm{obs}})$ follows the histogram in Figure 8.1, $p(y)$ is a normal distribution, and the MDM follows

$$\mathrm{logit}\{P(R = 1 \mid y)\} = \psi_0 + \psi_1 y + \psi_2 y^2. \qquad (8.3)$$

This scenario is consistent with an MNAR mechanism having $\psi_1 < 0$ and $\psi_2 = 0$, because the right tail needs to be filled in for the full-data responses to be normally distributed (the left tail is already consistent with a normal distribution). Alternatively, suppose that $p(y_{\mathrm{obs}})$ resembled the histogram in Figure 8.2. The quadratic MDM (8.3) is now consistent with $\psi_2 > 0$, which is MNAR. On the other hand, with the same observed data response distribution, an MDM that is linear in y ($\psi_2 = 0$) is consistent with $\psi_1 = 0$, which is MAR.

These simple examples demonstrate that for a fully specified parametric selection model, all parameters are identified. As a consequence, there are no obvious sensitivity parameters. By contrast, in an ideal sensitivity analysis, the distribution of Y_{mis} given Y_{obs} and R is governed by parameters that affect the full-data distribution but not the observed-data distribution, so that perturbations of these parameter values do not affect fit of the full-data model to observables. By this criterion, parametric selection models are not well-suited to sensitivity analysis or to incorporation of prior information about $p(y_{\mathrm{mis}} \mid y_{\mathrm{obs}}, r)$. A detailed discussion of this point follows in Sections 8.3.3 through 8.3.6; in Section 8.3.7 we introduce semiparametric selection mod-

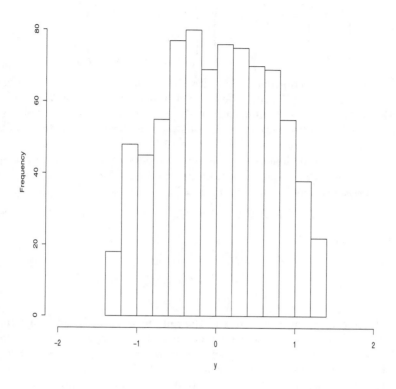

Figure 8.2 *Histogram of observed responses.*

els, which offer a more viable selection-model-based framework for sensitivity analysis.

8.3.3 Heckman selection model for a bivariate response

We start out our review of parametric selection models by providing details on the Heckman model (Heckman, 1979). Let $Y = (Y_1, Y_2)^T$ denote a bivariate outcome, and let R be an indicator of whether Y_2 is observed ($R = 1$ corresponds to Y_2 being observed). Heckman proposed to jointly model the outcome and missingness using a trivariate normal distribution as follows:

$$
\begin{pmatrix} Y_1 \\ Y_2 \\ Z \end{pmatrix} \sim N \left(\begin{pmatrix} \mu_1 \\ \mu_2 \\ \mu_z \end{pmatrix}, \begin{pmatrix} \sigma_{11} & & \\ \sigma_{21} & \sigma_{22} & \\ \sigma_{31} & \sigma_{32} & \sigma_{33} \end{pmatrix} \right). \tag{8.4}
$$

In this setup, Z is a continuous latent variable underlying the missingness indicator, where $R = I\{Z > 0\}$. This model implies $\boldsymbol{Y} \sim N(\boldsymbol{\mu}_Y, \boldsymbol{\Sigma}_Y)$, where $\boldsymbol{\Sigma}_Y$ is the upper left 2×2 submatrix of the covariance matrix in (8.4). The Heckman model requires some restrictions on the the σ_{jk}, which we detail below.

Let $\boldsymbol{\Gamma} = \boldsymbol{\Sigma}_Y^{-1}$, and denote the unique elements of $\boldsymbol{\Gamma}$ by $\{\gamma_{11}, \gamma_{12}, \gamma_{22}\}$. The selection model above implies that the missing data mechanism follows a probit regression that is linear in Y_1 and Y_2,

$$
P(R = 1 \mid \boldsymbol{y}) = P(Z > 0 \mid \boldsymbol{y}) = \Phi(\psi_0 + \psi_1 y_1 + \psi_2 y_2). \tag{8.5}
$$

Referring to (8.4),

$$
\begin{aligned}
\psi_0 &= \mu_z - \psi_1 \mu_1 - \psi_2 \mu_2 \\
\psi_1 &= \sigma_{31} \gamma_{11} + \sigma_{32} \gamma_{12} \\
\psi_2 &= \sigma_{31} \gamma_{12} + \sigma_{32} \gamma_{22}.
\end{aligned}
$$

Note that the regression parameters in the missing data mechanism are functions of the mean *and* covariance parameters of the full-data response model $p(\boldsymbol{y} \mid \boldsymbol{\theta})$, which is bivariate normal with mean $\boldsymbol{\mu}_Y = (\mu_1, \mu_2)^T$ and covariance matrix $\boldsymbol{\Gamma}^{-1}$. If the missing data mechanism was thought to be nonlinear in $\boldsymbol{y}$, the trivariate normal model given in (8.4) would not be appropriate as it induces the linear missing data mechanism in (8.5).

For identifiability, the model in (8.4) is typically parameterized such that the conditional variance of the latent variable Z is fixed, i.e.,

$$
\operatorname{var}(Z \mid y_1, y_2) = \sigma_{33} - \boldsymbol{\sigma}_3 \boldsymbol{\Gamma} \boldsymbol{\sigma}_3^T = 1,
$$

where $\boldsymbol{\sigma}_3 = (\sigma_{31}, \sigma_{32})$. When the conditional variance is set to 1, the missing data mechanism coefficients $\boldsymbol{\psi}$ in (8.5) correspond to a standard deviation change in the latent variable Z.

Individual contributions to the observed data likelihood take the form

$$
\left\{ \int_{-\infty}^{0} p(y_1, z \mid \boldsymbol{\theta}, \boldsymbol{\psi}) dz \right\}^{1-r} \left\{ \int_{0}^{\infty} p(y_1, y_2, z \mid \boldsymbol{\theta}, \boldsymbol{\psi}) dz \right\}^{r},
$$

where the first integrand is a bivariate normal derived from (8.4) and the second integrand is the trivariate normal given in (8.4).

The Heckman selection model can be extended in several ways. First, there are obvious extensions to settings where $J > 2$ by increasing the dimension of the model. Second, binary longitudinal responses can be accommodated by making the multivariate normal in (8.4) into a multivariate probit model. Third, the means $\boldsymbol{\mu}_Y$ can be modified to include covariates. Because the model relies on a normal specification, integrating over $\boldsymbol{Y}_{\text{mis}}$ is relatively straightforward. But the normality assumption also identifies all model parameters, including coefficients of incompletely observed Y's in the missing data mechanism. We continue our review of parametric selection models by introducing more general models for longer series of longitudinal responses.

8.3.4 Specification of the missing data mechanism for longitudinal data

Recall that selection models for longitudinal data require specification of two separate models: (1) a model $p(\boldsymbol{r} \mid \boldsymbol{y}, \boldsymbol{\psi})$ for the missing data mechanism; and (2) a model $p(\boldsymbol{y} \mid \boldsymbol{\theta})$ for the full-data response. General issues in specifying $p(\boldsymbol{y} \mid \boldsymbol{\theta})$ were discussed in detail in Chapter 6 and still are applicable here. The assumption of nonignorable dropout requires further specification of $p(\boldsymbol{r} \mid \boldsymbol{y}, \boldsymbol{\psi})$.

As discussed in Chapter 5, it is convenient to specify the MDM for longitudinal data with dropout using the hazard of dropout. As in Chapter 5, let $\overline{\boldsymbol{Y}}_j = \{Y_1, \ldots, Y_j\}$ denote the history of responses for an individual up to and including time t_j. For the case of J fixed measurement times, the hazard of dropout at $U = t_j$ is a function of the full data $\boldsymbol{Y}$ via

$$h(t_j \mid \overline{\boldsymbol{y}}_J, \boldsymbol{\psi}) = P(R_j = 0 \mid R_1 = \cdots = R_{j-1} = 1, \overline{\boldsymbol{y}}_J, \boldsymbol{\psi}). \qquad (8.6)$$

A transformation of $h(t_j \mid \overline{\boldsymbol{y}}_J, \boldsymbol{\psi})$ is often specified as additive in the components of $\overline{\boldsymbol{Y}}_J$. For example, we might assume

$$h(t_j \mid \overline{\boldsymbol{y}}_J, \boldsymbol{\psi}) = g^{-1}(\boldsymbol{\psi}_j^{\mathrm{T}} \boldsymbol{y}_i),$$

where g is an appropriate link function and $\boldsymbol{\psi}_j = (\psi_{j1}, \ldots, \psi_{jJ})^{\mathrm{T}}$. Under MNAR, the parameters

$$\{\psi_{jj}, \psi_{j,j+1}, \ldots, \psi_{jJ}\}$$

are potential sensitivity parameters.

The general model in (8.6), even under additivity in the components of $\overline{\boldsymbol{Y}}_j$, has a lot of potential sensitivity parameters. Different simplifications can be made to constrain the parameter space fully or partially identify the model; choices include restricting the dependence on elements of $\overline{\boldsymbol{Y}}_J$ to one or two informative variables (e.g., Y_j and $Y_j - Y_{j-1}$); assuming constant baseline hazard; or imposing restrictions on the missing data mechanism.

Related to the third choice, a parsimonious class of models for the missing data mechanism, referred to as *missing non-future dependence* (Kenward et al., 2003) is characterized by

$$h(t_j \mid \overline{\boldsymbol{y}}_J; \boldsymbol{\psi}) = h(t_j \mid \overline{\boldsymbol{y}}_j; \boldsymbol{\psi}), \qquad (8.7)$$

where the hazard of dropout at time j depends on past responses $\overline{Y}_{j-1}$, on the potentially missing response Y_j, but not on future responses. The non-future dependent missing data mechanisms also can be applied in mixture models (see Section 8.4.2). A simple version of the hazard of dropout under non-future dependent missingness is

$$h(t_j \mid \overline{\boldsymbol{y}}_j, \boldsymbol{\psi}) = g^{-1}(\psi_0 + \psi_1 y_{j-1} + \psi_2 y_j), \qquad (8.8)$$

where the hazard of dropout at t_j depends only on the current response y_j and the most recent past response y_{j-1}. In this specification, there is only one potential sensitivity parameter, ψ_2, which considerably simplifies sensitivity analyses in certain classes of semiparametric selection models (see Section 8.3.7 and Chapter 9). In addition, in this simplified form, a test for MNAR vs. MAR reduces to a test of $\psi_2 = 0$. However, this test is valid only under the assumption that the *full-data model* is specified correctly.

8.3.5 Parametric selection models for longitudinal data

In this section, we present some details on parametric selection models for continuous longitudinal responses and binary longitudinal responses, respectively.

Example 8.1. *A parametric selection model for continuous responses.*
Diggle and Kenward (1994) proposed a parametric selection model for a J-dimensional vector of longitudinal responses $\boldsymbol{Y}$ with monotone dropout. We describe this model in the context of the schizophrenia clinical trial (Section 1.2). In this trial, measurements of schizophrenia symptomatology (BPRS scores) were intended to be collected for 6 weeks but there were dropouts at each week after baseline. The full-data response model (for the six weekly BPRS measurements) might be specified as

$$\boldsymbol{Y}_i \mid \boldsymbol{x}_i \;\sim\; N(\boldsymbol{x}_i \boldsymbol{\beta}, \boldsymbol{\Sigma}),$$

where $\boldsymbol{\theta} = (\boldsymbol{\beta}, \boldsymbol{\Sigma})$. The hazard of dropout can be specified using a simple form of the hazard with a logistic link,

$$\text{logit}\{h(t_j \mid \overline{\boldsymbol{y}}_{ij}, \boldsymbol{\psi})\} \;=\; \psi_0 + \psi_1 y_{i,j-1} + \psi_2 y_{ij}.$$

Thus, dropout at time j is allowed to depend on the immediate previous response $y_{i,j-1}$ and the potentially unobserved current response y_{ij}. Equivalently, the model could be specified using y_{ij} and the difference $y_{ij} - y_{i,j-1}$.

The observed data likelihood for this model takes a complex form. In particular, with monotone missingness caused by dropout, the contribution to the observed data likelihood for an individual who drops out just prior to observing response k is

$$L(\boldsymbol{\theta}, \boldsymbol{\phi} \mid \boldsymbol{y}_{\mathrm{obs},i}, r_i) \quad \propto \quad \left[\prod_{j<k} \{1 - h(t_j \mid \overline{y}_{ij}, \boldsymbol{\psi})\} \, p(y_{ij} \mid \overline{y}_{i,j-1}, \boldsymbol{\theta}) \right]$$

$$\times \int h(t_k \mid \overline{y}_{i,k-1}, y_k, \boldsymbol{\psi}) \, p(y_k \mid \overline{y}_{i,k-1}, \boldsymbol{\theta}) \, dy_k. \quad (8.9)$$

The complexity of this likelihood is due to the integral with respect to the missing response y_{ik}. Bayesian MCMC approaches avoid direct evaluation of this integral through data augmentation. Details are provided in Section 8.3.8.

The inability in parametric selection models to partition the full-data model parameters vector into identified and nonidentified parameters is evident from (8.9), where the observed data likelihood is a function of all the model parameters, including the coefficient ψ_2 of the potentially missing response data; hence, the parameters of the extrapolation distribution cannot be disentangled from the parameters of the observed data distribution. □

There is a considerable literature on parametric selection models for longitudinal binary responses (Baker, 1995; Fitzmaurice, Molenberghs, and Lipsitz, 1995; Albert, 2000; and Kurland and Heagerty, 2004) which mostly differs in how the the full-data response model is specified. For example, Kurland and Heagerty (2004) proposed a parametric selection model for binary responses that assumes the same form for the missing data mechanism as in Example 8.1 and then replaces the multivariate normal likelihood for the full-data response model with a first-order marginalized transition model, MTM(1) (see Example 2.5). The observed-data log likelihood takes the same form as in (8.9), with the terms involving the full-data response model now being derived based on an MTM and integrals replaced by sums.

We fit parametric selection model to longitudinal binary responses in Chapter 10 using data from the OASIS study (Section 1.6) to examine the effect of two treatments on smoking cessation among alcoholic smokers.

8.3.6 Feasibility of sensitivity analysis for parametric selection models

In the setting of parametric selection models, several authors have suggested fixing parameters like ψ_2 in (8.2) at reasonable values (via expert opinion), and then maximizing the observed data log likelihood over the remaining parameters (see Little and Rubin, 1999; Kurland and Heagerty, 2004). This is similar in spirit to the sensitivity analysis approaches of Robins and colleagues (e.g., Rotnitzky et al., 1998; Scharfstein et al., 1999), but there is an important difference. In the work of Robins and colleagues, the observed data response model was specified *nonparametrically*. Thus, fixing the potential sensitivity parameters in the MDM did not impact the fit of the model to the observed responses. In a *parametric* selection model, different choices of ψ_2 yield different models for the observable data because ψ_2 does not satisfy condition 2 of Definition 8.1. In a frequentist analysis, there is a best fit to the observed data within a given parametric specification (at the mle of $\boldsymbol{\psi}$).

In the next section, we discuss a more flexible class of selection models. These (Bayesian) semiparametric selection models allow sensitivity analysis to be done without affecting fit to the observed data, and provide a sensitivity analysis framework similar to those in the work of Robins and colleagues.

8.3.7 Semiparametric selection models

Semiparametric selection models typically use a parametric model for the missing data mechanism and a semi- or nonparametric model for the observed-data response distribution (or the full-data response distribution). In a framework that is likelihood-based, it is sometimes the case that one can find parameters in the missing data mechanism that very nearly satisfy Definition 8.1 in that they are only weakly identified by observed data. As such, we can then specify an informative prior on these parameters in the missing data mechanism — e.g., ψ_2 in (8.8) — and not affect the fit of the model to the observed data. Fully parametric selection models typically do not allow this flexibility.

For a (univariate) continuous response Y, a semi-parametric selection model was proposed in Scharfstein et al. (1999) and extended to a fully Bayesian model by Scharfstein et al. (2003). Unfortunately, the multivariate nature in the longitudinal case makes direct extensions difficult due to the curse of dimensionality (see, e.g., Robins and Ritov, 1997). However, in the case of binary (categorical) longitudinal data, such extensions are more feasible.

Binary longitudinal responses

We begin with the simplest longitudinal setting of a bivariate binary response with missingness only in the second component (Y_2) and no covariates. This case provides a useful platform for understanding fully nonparametric model specification for selection models with longitudinal binary data and will provide a starting point for a semiparametric specification for the full-data model.

Let $\boldsymbol{Y} = (Y_1, Y_2)^{\mathrm{T}}$ denote the full-data response, with $R = 1$ if Y_2 is observed and $R = 0$ if it is missing. The entire full-data distribution $p(\boldsymbol{y}, r \mid \boldsymbol{\omega})$ can be enumerated using a multinomial distribution with probabilities

$$\omega_{y_1,y_2}^{(r)} \;=\; P(Y_1 = y_1, Y_2 = y_2, R = r),$$

shown in Table 8.1. The multinomial model has seven distinct parameters (noting that $\sum_{r,y_1,y_2} \omega_{y_1,y_2}^{(r)} = 1$). The parameters $\omega_{00}^{(1)}, \ldots, \omega_{11}^{(1)}$ corresponding to $p(y_1, y_2, r = 1)$ all are identifiable from the observed data. Among those with $R = 0$, we can identify only $P(Y_1 = 0, R = 0) = \omega_{00}^{(0)} + \omega_{01}^{(0)}$ and

$P(Y_1 = 1, R = 0) = \omega_{10}^{(0)} + \omega_{11}^{(0)}$. We denote the *identified* parameters as

$$\boldsymbol{\omega}_{\mathrm{I}} \;\; = \;\; (\omega_{00}^{(1)}, \omega_{01}^{(1)}, \omega_{10}^{(1)}, \omega_{11}^{(1)}, \omega_{0+}^{(0)}, \omega_{1+}^{(0)}), \tag{8.10}$$

where

$$\begin{aligned}
\omega_{0+}^{(0)} &= \omega_{00}^{(0)} + \omega_{01}^{(0)} \\
\omega_{1+}^{(0)} &= \omega_{10}^{(0)} + \omega_{11}^{(0)}.
\end{aligned}$$

Table 8.1 *Multinomial parameterization of full-data distribution for bivariate binary data with possibly missing* Y_2.

R	Y_1	Y_2	$p(y_1, y_2, r \mid \boldsymbol{\omega})$
0	0	0	$\omega_{00}^{(0)}$
0	0	1	$\omega_{01}^{(0)}$
0	1	0	$\omega_{10}^{(0)}$
0	1	1	$\omega_{11}^{(0)}$
1	0	0	$\omega_{00}^{(1)}$
1	0	1	$\omega_{01}^{(1)}$
1	1	0	$\omega_{10}^{(1)}$
1	1	1	$\omega_{11}^{(1)}$

In this setting, the selection model can be viewed as a reparameterization of the joint distribution in Table 8.1 using the factorization

$$p(y_1, y_2, r) \;\; = \;\; p(r \mid y_1, y_2)\, p(y_2 \mid y_1)\, p(y_1),$$

where

$$\begin{aligned}
Y_1 &\sim \mathrm{Ber}(\theta_1) \\
Y_2 \mid Y_1 &\sim \mathrm{Ber}(\theta_{2|1}) \\
R \mid Y_1, Y_2 &\sim \mathrm{Ber}(\pi),
\end{aligned}$$

and

$$\begin{aligned}
\mathrm{logit}(\theta_1) &= \alpha \\
\mathrm{logit}(\theta_{2|1}) &= \beta_0 + \beta_1 y_1 \\
\mathrm{logit}(\pi) &= \psi_0 + \psi_1 y_1 + \psi_2 y_2 + \psi_3 y_1 y_2.
\end{aligned}$$

This model has seven unique parameters, so it is saturated in time and missingness pattern. It is therefore nonparametric. Clearly, with the selection model parameterization, α is directly identified from the observed data. However, none of the other parameters can be identified without imposing untestable constraints.

To identify the remaining parameters without affecting the fit of the model to the observed data, we have two degrees of freedom with which to work. Equivalently, we have two sensitivity parameters. In the logit parameterization, the missing data mechanism is conveniently represented by the sensitivity parameters (ψ_2, ψ_3) that satisfy the three conditions in Definition 8.1. Here, ψ_2 and ψ_3 are zero under MAR and correspond to log odds ratios of missingness in Y_2 given Y_1.

Fixing ψ_2 and ψ_3 identifies all remaining model parameters. To see this, we first express the parameters in the missing data mechanism as functions of the original multinomial probabilities $\boldsymbol{\omega}$:

$$\psi_0 = \text{logit} \frac{\omega_{00}^{(1)}}{\omega_{00}^{(1)} + \omega_{00}^{(0)}} \tag{8.11}$$

$$\psi_0 + \psi_1 = \text{logit} \frac{\omega_{10}^{(1)}}{\omega_{10}^{(1)} + \omega_{10}^{(0)}} \tag{8.12}$$

$$\psi_0 + \psi_2 = \text{logit} \frac{\omega_{01}^{(1)}}{\omega_{01}^{(1)} + \omega_{01}^{(0)}} \tag{8.13}$$

$$\psi_0 + \psi_1 + \psi_2 + \psi_3 = \text{logit} \frac{\omega_{11}^{(1)}}{\omega_{11}^{(1)} + \omega_{11}^{(0)}}. \tag{8.14}$$

By subtracting (8.12) from (8.14), we obtain $\psi_2 + \psi_3$. After some algebra, it is possible to obtain a closed-form (but not simple) expression for $\omega_{10}^{(0)}/\omega_{11}^{(0)}$. By combining this with the identified sum $\omega_{10}^{(0)} + \omega_{11}^{(0)}$, we can obtain closed-form expressions for $\omega_{10}^{(0)}$ and $\omega_{11}^{(0)}$ individually. By taking a similar approach with (8.11) and (8.13), we can identify $\omega_{00}^{(0)}$ and $\omega_{01}^{(0)}$.

It turns out that if we specify an informative prior on (ψ_2, ψ_3) that is independent of the identified parameters $\boldsymbol{\omega}_I$ given in (8.10), the posterior for (ψ_2, ψ_3) will be equal to the prior

$$p(\psi_2, \psi_3 \mid \boldsymbol{y}_{\text{obs}}, \boldsymbol{r}) = p(\psi_2, \psi_3). \tag{8.15}$$

In Chapter 9 we actually recommend priors such that (ψ_2, ψ_3) are *a priori dependent* on $\boldsymbol{\omega}_I$. In this case, the equality in (8.15) will not hold exactly. We defer details on this to Chapter 9.

Unfortunately, when $J \gg 2$, a fully nonparametric selection model is less practical. For monotone dropout (and no dropouts at the first observation time), there are $(2^J - 1) + (J-1)2^J (= J2^J - 1)$ parameters, of which only $(2^J -$

$1) + \sum_{j=1}^{J-1} 2^j$ are identified; this leaves $\sum_{j=1}^{J-1}(2^J - 2^j)$ sensitivity parameters. For example, for $J = 4$, there are 34 sensitivity parameters — and this does not even include covariates! However, a semiparametric selection model can be a viable alternative. By semiparametric, we mean a parametric model for the missing data mechanism and a nonparametric model for the full-data response. We provide an example for a trivariate binary response next.

Example 8.2. *Semiparametric selection model for longitudinal binary data with $J = 3$.*
Let $\boldsymbol{Y} = (Y_1, Y_2, Y_3)^{\mathrm{T}}$ be the full-data response and $\boldsymbol{R} = (R_1, R_2, R_3)^{\mathrm{T}}$ be the observed data indicators (assuming monotone dropout), where $R_j = 0$ corresponds to Y_j being missing. We assume $R_1 \equiv 1$. The nonparametric selection model can be written as

$$
\begin{aligned}
Y_1 &\sim \mathrm{Ber}(\theta_1) \\
Y_2 \mid Y_1 &\sim \mathrm{Ber}(\theta_{2|1}) \\
Y_3 \mid Y_1, Y_2 &\sim \mathrm{Ber}(\theta_{3|21}) \\
R_2 \mid Y_1, Y_2, Y_3 &\sim \mathrm{Ber}(\pi_2) \\
R_3 \mid R_2 = 1, Y_1, Y_2, Y_3 &\sim \mathrm{Ber}(\pi_3)
\end{aligned}
$$

with

$$
\begin{aligned}
\mathrm{logit}(\theta_1) &= \alpha \\
\mathrm{logit}(\theta_{2|1}) &= \beta_0 + \beta_1 y_1 \\
\mathrm{logit}(\theta_{3|21}) &= \beta_2 + \beta_3 y_1 + \beta_4 y_2 + \beta_5 y_1 y_2 \\
\mathrm{logit}(\pi_2) &= \psi_0 + \psi_1 y_1 + \psi_2 y_2 + \psi_3 y_3 + \psi_4 y_1 y_2 \\
&\quad + \psi_5 y_1 y_3 + \psi_6 y_2 y_3 + \psi_7 y_1 y_2 y_3. \\
\mathrm{logit}(\pi_3) &= \psi_8 + \psi_9 y_1 + \psi_{10} y_2 + \psi_{11} y_3 + \psi_{12} y_1 y_2 \\
&\quad + \psi_{13} y_1 y_3 + \psi_{14} y_2 y_3 + \psi_{15} y_1 y_2 y_3.
\end{aligned}
$$

There are 23 parameters, including 10 sensitivity parameters. Clearly, some model simplification is needed. We can reduce the number of sensitivity parameters used in a sensitivity analysis by fixing some of these at zero. For example, we could consider the following simplified MDM,

$$
\begin{aligned}
\mathrm{logit}(\pi_2) &= \psi_0 + \psi_1 y_1 + \psi_2 y_2 \\
\mathrm{logit}(\pi_3) &= \psi_3 + \psi_4 y_1 + \psi_5 y_2 + \psi_6 y_1 y_2 + \psi_7 y_3.
\end{aligned}
$$

This model now has only two sensitivity parameters, ψ_2 and ψ_7. They are not identified by data, but fixing their values identifies the full-data model.

Even if we reduce the number of sensitivity parameters, there are still 13 parameters that need to be identified by the observed data. To reduce the number of identified parameters, we can consider a semiparametric specification with a nonparametric full-data response model and a simpler, parametric

missing data mechanism. For example, the hazard of dropout can be simplified
to allow dependence only on the most current and most recent outcome,

$$\text{logit}(\pi_2) = \psi_0 + \psi_1 y_1 + \psi_2 y_2$$
$$\text{logit}(\pi_3) = \psi_0 + \psi_1 y_2 + \psi_2 y_3.$$

The simplifying assumptions in this missing data mechanism are: (1) non-
future dependence; (2) the hazard of dropout is constant over time; (3) dropout
at time j only depends on the response at time j and the response at the previous time $j-1$; (4) the effect of the current and past response on dropout is the
same at each dropout time (i.e., the coefficients of y_j and y_{j-1} do not depend
on j). We have further reduced the number of sensitivity parameters via the
fourth simplifying assumption that equates the sensitivity parameter in the
regression for π_2 and for π_3. This model has one sensitivity parameter, ψ_2 and
only nine parametrs for the data to identify. This model is no longer nonparametric as we have assumed a (simpler) parametric form for the MDM. Given
that we now have a parametric model for the MDM, the sensitivity parameter may be weakly identified by data. Specification of such semiparametric
models, particularly in higher dimensions, and exploration of their properties,
including the degree of identification of the potential sensitivity parameters,
is an area of ongoing work.   □

We close out our discussion of selection models by providing some guidance
on posterior sampling.

8.3.8 Posterior sampling strategies

We provide some strategies for posterior sampling for selection models. First,
we point out that by using data augmentation to fill in $\boldsymbol{y}_{\text{mis}}$, posterior sampling
of $\boldsymbol{\psi}$ and $\boldsymbol{\theta}$ proceeds using complete data techniques for the missing data
mechanism and the full-data response model, respectively, i.e., sampling $p(\boldsymbol{\psi} \mid \boldsymbol{y}, \boldsymbol{r})$ and $p(\boldsymbol{\theta} \mid \boldsymbol{y})$ 'separately'. As a result, intractable integrals, such as those
found in (8.9), do *not* need to be evaluated directly. The main *new* computing
issue to discuss here — in the sense that it is specific to selection models —
is sampling the missing data, $\boldsymbol{y}_{\text{mis}}$ from $p(\boldsymbol{y}_{\text{mis}} \mid \boldsymbol{y}_{\text{obs}}, \boldsymbol{r}, \boldsymbol{\theta}, \boldsymbol{\psi})$.

For binary response data, individual components Y_{ij} in the distribution
$p(\boldsymbol{y}_{\text{mis}} \mid \boldsymbol{y}_{\text{obs}}, \boldsymbol{r}, \boldsymbol{\theta}, \boldsymbol{\psi})$ will be sampled from a Bernoulli distribution with probability

$$P(Y_{ij} = 1 \mid \boldsymbol{y}_{-ij}, \boldsymbol{r}, \boldsymbol{\theta}, \boldsymbol{\psi}) = \frac{p(\boldsymbol{r} \mid \boldsymbol{y}_{-ij}, y_{ij} = 1, \boldsymbol{\psi}) p(\boldsymbol{y}_{-ij}, y_{ij} = 1 \mid \boldsymbol{\theta})}{\sum_{y=0}^{1} p(\boldsymbol{r} \mid \boldsymbol{y}_{-ij}, y_{ij} = y, \boldsymbol{\psi}) p(\boldsymbol{y}_{-ij}, y_{ij} = y \mid \boldsymbol{\theta})},$$

where $\boldsymbol{y}_{-ij}$ corresponds to the vector of responses with y_{ij} removed. Considerable simplification of this expression for specific models is typical.

For continuous responses, a simple Metropolis-Hastings algorithm can be
used to sample from $p(\boldsymbol{y}_{\text{mis}} \mid \boldsymbol{y}_{\text{obs}}, \boldsymbol{r}, \boldsymbol{\theta}, \boldsymbol{\psi})$. At iteration k, sample a candidate

$\boldsymbol{y}_{\mathrm{mis}}^{(k)}$ from $p(\boldsymbol{y}_{\mathrm{mis}}^{(k)} \mid \boldsymbol{y}_{\mathrm{obs}}, \boldsymbol{\theta})$, i.e., the conditional distribution of $\boldsymbol{y}_{\mathrm{mis}}$ based on *only* the full-data response model. This conditional distribution typically takes a known form (see Chapter 6). Accept this candidate with probability

$$p^{\star} = \min \left\{ 1, \frac{p(\boldsymbol{r} \mid \boldsymbol{y}_{\mathrm{mis}}^{(k)}, \boldsymbol{y}_{\mathrm{obs}}, \boldsymbol{\psi})}{p(\boldsymbol{r} \mid \boldsymbol{y}_{\mathrm{mis}}^{(k-1)}, \boldsymbol{y}_{\mathrm{obs}}, \boldsymbol{\psi})} \right\}.$$

If this candidate distribution results in low acceptance, a Laplace approximation to $p(\boldsymbol{y}_{\mathrm{mis}} \mid \boldsymbol{y}_{\mathrm{obs}}, \boldsymbol{r}, \boldsymbol{\theta}, \boldsymbol{\psi})$ can be used instead.

For certain models with continuous responses, draws from the predictive distribution (or a suitably data augmented version of this distribution) can be taken directly (e.g., the Heckman selection model described earlier).

8.3.9 Summary of pros and cons of selection models

The main advantages of SMs are (1) the ability to directly specify both the full data response model and the missing data mechanism using familiar models; and (2) the correspondence between the specified missing data mechanism and the MAR-MNAR hierarchy. As a result, the primary parameters of interest are explicit in the model, and the nature of the dependence between missingness and response has a transparent representation.

The main disadvantage in parametric selection models — for most practical situations — is the inability to partition the full-data parameter vector $\boldsymbol{\omega}$ into identified and unidentified components as discussed in Section 8.2. In particular, the extrapolation distribution $p(\boldsymbol{y}_{\mathrm{mis}} \mid \boldsymbol{y}_{\mathrm{obs}}, \boldsymbol{r}, \boldsymbol{\omega}_{\mathrm{E}})$ is generally identified for parametric selection models, and the distribution of observables is indexed by parameters that govern the missing data mechanism. As a consequence, any strategy for sensitivity analysis based on refitting the model under different missing data assumptions will have the unfortunate side effect of changing the fit to observed data as well. Semiparametric selection models offer a more flexible alternative, and are revisited in Chapter 9.

8.4 Mixture models

8.4.1 Background, specification, and identification

The development of mixture models for handling missing data can be traced at least to Rubin (1977), who described methods for using informative priors in surveys to capture and use subjective information about nonresponse. Since then a number of key papers have expanded mixture models for handling informative dropout in longitudinal data. Little (1993, 1994) developed a general theory for finite mixtures of multivariate distributions in discrete-time settings. For longer follow-up times, mixtures of random effects models proved useful (Wu and Bailey, 1988, 1989; Mori et al., 1992; Hogan and Laird, 1997a).

Fitzmaurice and Laird (2000) developed moment-based approaches based on mixtures of generalized linear models. Roy (2003) and Roy and Daniels (2007) addressed the issue of having a large number of dropout categories by using mixtures over latent classes. Hogan et al. (2004a) developed approaches for continuous dropout times based on mixtures of varying coefficient models. Reviews of model-based approaches can be found in Little (1995), Hogan and Laird (1997b), Kenward and Molenberghs (1999), Fitzmaurice (2003), and Hogan et al. (2004b).

Another key thread of research concerns model identification. Here, Molenberghs and colleagues have developed an important body of work for the case of discrete-time dropout (cf. Molenberghs et al., 1998; Kenward et al., 2003), much of which is described in this section.

The mixture model approach factors the full-data model as

$$p(\boldsymbol{y}, \boldsymbol{r} \mid \boldsymbol{x}, \boldsymbol{\omega}) = p(\boldsymbol{y} \mid \boldsymbol{r}, \boldsymbol{x}, \boldsymbol{\omega}) \, p(\boldsymbol{r} \mid \boldsymbol{x}, \boldsymbol{\omega}). \tag{8.16}$$

The full-data response distribution is obtained by averaging (8.16) over the distribution of $\boldsymbol{r}$,

$$p(\boldsymbol{y} \mid \boldsymbol{x}, \boldsymbol{\omega}) \;\; = \;\; \sum_{r \in \mathscr{R}} p(\boldsymbol{y} \mid \boldsymbol{r}, \boldsymbol{x}, \boldsymbol{\omega}) \, p(\boldsymbol{r} \mid \boldsymbol{x}, \boldsymbol{\omega}),$$

where $\mathscr{R}$ is the sample space of $\boldsymbol{R}$ (see also Section 5.9.2). In this section, we describe some specific formulations that give a broad representation of the settings where the models can be used, and describe formal methods for identifying them. In general, when $\boldsymbol{R}$ is discrete, the component distributions comprise the set $\{p(\boldsymbol{y} \mid \boldsymbol{r}, \boldsymbol{x}) : \boldsymbol{r} \in \mathscr{R}\}$. When missingness is caused by dropout at some time U, then the component distributions $p(\boldsymbol{y} \mid u, \boldsymbol{x})$ may also be a discrete set of distributions, or may be specified in terms of a continuous u. To maintain focus on key ideas related to specification and identification, we defer discussion of covariates to Section 8.4.7.

Mixture models for either discrete or continuous dropout time are under-identified and require specific constraints for fitting to data. There are various approaches to identifying the models; in this section, we focus on strategies that divide the full-data distribution into identified and nonidentified components (see the extrapolation factorization (8.1)).

In the case of discrete-time dropout, the MAR assumption is used as a basis for identifying the full-data distribution and for building a larger class of models that accommodate MNAR. For continuous-time dropout (or discrete-time dropout with large number of support points), models can be identified by making assumptions about the mean of the full-data response as a function of time; e.g., assuming $E\{Y(t) \mid U = u\}$ is linear in t, with intercept and slope depending on U. Usually one assumes that given dropout at $U = u$, mean response prior to and after U — i.e., $E\{Y(t) \mid U = u, t < u\}$ and $E\{Y(t) \mid U = u, t \geq u\}$ — are either equivalent or related through some known function; for example,

$$E\{Y(t) \mid U = u, t > u\} = q(u,t)E\{Y(t) \mid U = u, t \leq u\}$$

for some known $q(u,t)$. In this example, $q(u,t)$ cannot be identified, but fixing its value, or placing a prior on it, identifies the full-data model.

8.4.2 Identification strategies for mixture models

In this section we describe several approaches to model identification, emphasizing the case where follow-up times are discrete.

Identification via MAR constraints

Although the usual goal of fitting a mixture model is to represent a MNAR mechanism, for the purposes of discussing model identification it helps to begin by showing how to impose the MAR condition in a mixture model having discrete measurement times and monotone dropout. In Section 5.5, we saw that for the case of monotone dropout, MAR can be represented in terms of the hazard of dropout. For pattern mixture models, Molenberghs et al. (1998) show that MAR is equivalent to the available case missing value (ACMV) constraint (Little, 1994).

To simplify notation in our discussion of discrete-time mixture models, we write the conditional distribution of $\boldsymbol{Y}$ given follow-up time $S = k$ as

$$p_k(\boldsymbol{y}) \quad = \quad p(\boldsymbol{y} \mid S = k).$$

Similarly, we write $p_{\geq k}(\boldsymbol{y}) = p(\boldsymbol{y} \mid S \geq k)$. In the case of discrete-time follow-up with monotone dropout, MAR has a specific characterization given by the following theorem.

Theorem 8.1. *MAR for discrete-time pattern mixture models under monotone dropout.*
Let $Y_1, \ldots, Y_J$ denote the full-data responses, with measurements scheduled at times $t_1, \ldots, t_J$. Without loss of generality, assume $t_j = j$. Let $S \in \{1, 2, \ldots, J\}$ denote follow-up time, with $S = J$ for those with complete follow-up. MAR holds if and only if, for each $j \geq 2$ and $k < j$,

$$p_k(y_j \mid y_1, \ldots, y_{j-1}) \quad = \quad p_{\geq j}(y_j \mid y_1, \ldots, y_{j-1}). \qquad (8.17)$$

The proof can be found in Molenberghs et al. (1998).   □

The theorem states that the conditional distribution of Y_j given past responses for those whose follow-up terminates at some time prior to j – i.e., $p_k(y_j \mid y_1, \ldots, y_{j-1})$ for some $k < j$ – is equivalent to the corresponding distribution for those who have observed data at or beyond j – i.e., $p_{\geq j}(y_j \mid y_1, \ldots, y_{j-1})$. Those still in follow-up at time j are an aggregation of those having $S \in \{j, j+1, j+2, \ldots, J\}$. Under monotone dropout, the RHS of (8.17) is identifiable from observed data, while the LHS is not.

Tables 8.2 and 8.3 illustrate Theorem 8.1 using a schematic for the case $J = 4$. Table 8.2 shows the identified observed data distributions in a general pattern mixture model. Nonidentified components of the full-data distribution are denoted by question marks ('?'). In Table 8.3, they are filled in using the MAR constraint given in Theorem 8.1. The format of this table is used to illustrate several models and identification strategies discussed in this section.

Table 8.2 *Identifiable components of pattern-mixture model with $J = 4$ under monotone dropout. Nonidentified distributions labeled using '?'.*

	$j = 2$	$j = 3$	$j = 4$
$S = 1$	?	?	?
$S = 2$	$p_2(y_2\|y_1)$	?	?
$S = 3$	$p_3(y_2\|y_1)$	$p_3(y_3\|y_1, y_2)$	?
$S = 4$	$p_4(y_2\|y_1)$	$p_4(y_3\|y_1, y_2)$	$p_4(y_4\|y_1, y_2, y_3)$

Table 8.3 *Schematic representation of Theorem 8.1, illustrating identification of pattern-mixture model with $J = 4$ and monotone dropout using MAR constraints. Distributions above the dividing line are identified via MAR, using an aggregation of distributions in the same column appearing below the dividing line.*

	$j = 2$	$j = 3$	$j = 4$
$S = 1$	$p_{\geq 2}(y_2\|y_1)$	$p_{\geq 3}(y_3\|y_1, y_2)$	$p_4(y_4\|y_1, y_2, y_3)$
$S = 2$	$p_2(y_2\|y_1)$	$p_{\geq 3}(y_3\|y_1, y_2)$	$p_4(y_4\|y_1, y_2, y_3)$
$S = 3$	$p_3(y_2\|y_1)$	$p_3(y_3\|y_1, y_2)$	$p_4(y_4\|y_1, y_2, y_3)$
$S = 4$	$p_4(y_2\|y_1)$	$p_4(y_3\|y_1, y_2)$	$p_4(y_4\|y_1, y_2, y_3)$

To impose the MAR condition, one approach is to assume a parametric model for observables, and then use the MAR assumption to constrain the distribution of the missing data. Although this chapter is primarily concerned with MNAR models, we advocate embedding an MAR model within a broader class of MNAR specifications, so as to make clear the assumptions or parameter constraints that differentiate MNAR from MAR. Theorem 8.1 motivates this approach in the pattern mixture context and is discussed further in Chapter 9. The examples later in this chapter illustrate the application of the theorem to specific mixture models.

Interior family constraints

Consider the case of discrete-time measurement at times $j = 1, \ldots, J$, with individual-specific follow-up time denoted by $S = \sum_j R_j$. We can identify the nonidentified distributions $p_k(y_j \mid y_1, \ldots, y_{k-1})$ for $j \geq 2$ and $k < j$ using the constraints

$$p_k(y_j \mid y_1, \ldots, y_{j-1}) = \sum_{s=j}^{J} \Delta_{jks} \, p_s(y_j \mid y_1, \ldots, y_{j-1}),$$

where, for any $k < j$, the Δ_{jks} are fixed weights satisfying $\sum_{s=j}^{J} \Delta_{jks} = 1$. Constraints with this form (linear combinations of the corresponding identified distributions) are called 'interior family constraints' (Kenward et al., 2003). They can be used to represent commonly used mixture model identification strategies such as 'complete case', 'nearest-neighbor', and 'available case' constraints, which we detail below. In the absence of these types of constraints, the Δ_{jks} parameters satisfy the conditions of Definition 8.1 and can be used as sensitivity parameters.

With complete case missing value (CCMV) constraints, the distributions

$$\{p_k(y_j \mid y_1, \ldots, y_{j-1}) : j \geq 2, \ k < j\} \tag{8.18}$$

for those who drop out are equated to the distribution of the completers, $p_J(y_j \mid y_1, \ldots, y_{j-1})$. For example, if $J = 4$, then the unindentified distribution $p_1(y_2 \mid y_1)$ is identified by equating it to $p_4(y_2 \mid y_1)$. This corresponds to

$$\Delta_{jks} = \begin{cases} 0 & s < 4 \\ 1 & s = 4 \end{cases}$$

for $k = 1, 2, 3$ and $j > k$.

Nearest-neighbor or adjacent-pattern constraints equate each of the unidentified distributions in (8.18) to the 'nearest-neighbor' — in terms of dropout pattern — having an identified distribution for $(Y_j \mid Y_1, \ldots, Y_{j-1})$. Continuing with the example above, $p_1(y_2 \mid y_1)$ is identified under nearest-neighbor constraints by equating it to $p_2(y_2 \mid y_1)$, because $S = 2$ is the 'nearest neighbor' to $S = 1$. Thus,

$$\Delta_{jks} = \begin{cases} 0 & s \neq j + 1 \\ 1 & s = j + 1 \end{cases}.$$

Rather than rely on nearest-neighbor or complete-case patterns, the available case missing value (ACMV) constraints use all available patterns where $p_k(y_j \mid y_1, \ldots, y_{j-1})$ is identified by data. Molenberghs et al. (1998) show that for discrete-time monotone dropout, MAR as stated in Theorem 8.1 is equivalent to ACMV.

Table 8.4 gives the general interior family constraint identification strategy for $J = 4$. A more detailed illustration of interior family constraints, including expressions for Δ_{jks} under MAR, is given in Example 8.4 and the discussion following the example.

Table 8.4 *Identification for discrete-time pattern mixture model with $J = 4$ and monotone dropout using interior family constraints. Distributions identified by observed data appear below the dividing line. Nonidentified distributions appear above the dividing line and are equated to weighted averages of identified distributions in the same column.*

	$j = 2$	$j = 3$	$j = 4$
$S = 1$	$\displaystyle\sum_{s=2}^{4}\Delta_{21s}\,p_s(y_2|y_1)$	$\displaystyle\sum_{s=3}^{4}\Delta_{31s}\,p_s(y_3|y_1,y_2)$	$p_4(y_4|y_1,y_2,y_3)$
$S = 2$	$p_2(y_2|y_1)$	$\displaystyle\sum_{s=3}^{4}\Delta_{32s}\,p_s(y_3|y_1,y_2)$	$p_4(y_4|y_1,y_2,y_3)$
$S = 3$	$p_3(y_2|y_1)$	$p_3(y_3|y_1,y_2)$	$p_4(y_4|y_1,y_2,y_3)$
$S = 4$	$p_4(y_2|y_1)$	$p_4(y_3|y_1,y_2)$	$p_4(y_4|y_1,y_2,y_3)$

Non-future dependence missing value restrictions

Non-future dependence missing value restrictions (Kenward et al., 2003) allow the probability of dropout at j to depend on the current (but possibly unobserved) response Y_j, but not on the future values $Y_{j+1}, \ldots, Y_J$ (see also Section 8.3.4). The constraint is more intuitive when given in terms of dropout time $U = 1 + \sum_j R_j$ ($= 1 + S$). It identifies all the unidentified conditional distributions in each pattern except for the one corresponding to current but unobserved value of the response, $p(y_j \mid y_1, \ldots, y_{j-1}, u = j)$ (here, 'current' is relative to dropout time). Beyond that, for each $j \geq 2$ and $k < j$,

$$p(y_j \mid y_1, \ldots, y_{j-1}, u = k) \quad = \quad p(y_j \mid y_1, \ldots, y_{j-1}, u \geq j) \qquad (8.19)$$

(recall that completers have $U = J + 1$). In terms of the follow-up time $S = \sum_j R_j$, the non-future dependence restriction leaves

$$p_{j-1}(y_j \mid y_1, \ldots, y_{j-1})$$

unidentified for $2 \leq j \leq J$, and imposes the constraint

$$p_k(y_j \mid y_1, \ldots, y_{j-1}) \quad = \quad p_{\geq j-1}(y_j \mid y_1 \ldots, y_{j-1}) \qquad (8.20)$$

for each $j > 2$ and $k < j - 1$. This constraint looks very similar to the MAR restriction in Theorem 8.1, except that this condition holds for $j > 2$ (not $j \geq 2$) and $k < j - 1$ (not $k < j$). Hence the MAR restriction is a special case. The non-future dependence constraints are illustrated in Table 8.5.

Table 8.5 *Schematic representation of non-future dependence missing value constraints for the pattern-mixture model with $J = 4$ and monotone dropout. Distributions below the dividing line are identified by observed data.*

	$j = 2$	$j = 3$	$j = 4$
$S = 1$	?	$p_{\geq 2}(y_3\|y_1, y_2)$	$p_{\geq 3}(y_4\|y_1, y_2, y_3)$
$S = 2$	$p_2(y_2\|y_1)$	?	$p_{\geq 3}(y_4\|y_1, y_2, y_3)$
$S = 3$	$p_3(y_2\|y_1)$	$p_3(y_3\|y_1, y_2)$	?
$S = 4$	$p_4(y_2\|y_1)$	$p_4(y_3\|y_1, y_2)$	$p_4(y_4\|y_1, y_2, y_3)$

A primary motivation for using this set of constraints is that in a sensitivity analysis, only one univariate distribution in each pattern is left unidentified; the MAR restriction provides a starting point for conducting sensitivity analysis within this class of restrictions. These restrictions are equivalent to the non-future dependence missing data mechanism introduced in the context of selection models in Section 8.3.4.

Identification via extrapolation

Unless the number of distinct dropout times is discrete and small (say 3 or 4), the number of constraints and/or sensitivity parameters potentially can become unmanageable, especially for saturated or nonparametric models, and structural constraints may be needed.

One approach is to assume that the mean response as a function of time follows a known function; this is the approach taken by several authors such as Wu and Bailey (1989), Hogan and Laird (1997a), and Fitzmaurice and Laird (2000). As an example, suppose the number of measurement occasions is large but common across subjects. Assume $(Y_1, \ldots, Y_J \mid U = u) \sim N(\boldsymbol{\mu}^{(u)}, \boldsymbol{\Sigma}^{(u)})$. One possible set of constraints is to assume the mean follows

$$\mu_j^{(u)} = f(t_j; \boldsymbol{\beta}^{(u)}),$$

where $f(t_j; \boldsymbol{\beta}^{(u)})$ has a known form such as $\beta_0^{(u)} + t_j \beta_1^{(u)}$, and the variance has some simplified form such as $\boldsymbol{\Sigma}^{(u)} = \boldsymbol{\Sigma}$. Because the variance matrix is assumed common across pattern, it will be identified. If the mean is linear over time within pattern, it will be fully identified for patterns with observations at two or more time points.

These models are of course restrictive, and there is no information in the data to verify the structural constraints. As with the discrete-time models, we can separate parameters $\boldsymbol{\omega}_E$ and $\boldsymbol{\omega}_O$ indexing the distributions $p(\boldsymbol{y}_{\text{mis}} \mid$

$y_{\text{obs}}, r, \omega_E$) and $p(y_{\text{obs}}, r \mid \omega_O)$. For example, instead of assuming that mean response within pattern follows a single line before and after dropout, we might assume that

$$f(t, \boldsymbol{\beta}^{(u)}) = \beta_0^{(u)} + \beta_1^{(u)} t + \Delta(t - u)_+,$$

where $a_+ = aI\{a > 0\}$ is the positive part of a. This larger model assumes that, conditionally on $U = u$, slope is $\beta_1^{(u)}$ for $t \leq u$ (prior to dropout), and $\beta_1^{(u)} + \Delta$ for $t > u$ (following dropout). Unlike the models described for the discrete case, setting $\Delta = 0$ does not imply MAR. But this model does admit a reparameterization $\boldsymbol{\xi}(\boldsymbol{\omega}) = (\boldsymbol{\xi}_S, \boldsymbol{\xi}_M)$, where $\boldsymbol{\xi}_S = \Delta$ is a sensitivity parameter. In Chapter 10, we illustrate this approach in detail using data from the pediatric AIDS trial.

8.4.3 Mixture models with discrete-time dropout

Continuous responses

To understand the PMM for continuous data in discrete time, and its connection to selection models, it is useful to revisit the model for the bivariate case (Little, 1994). We rely on specifications using mixtures of normal distributions, which tend to be mathematically tractable. For continuous responses, there is no reason to be confined to mixtures of normal distributions, although this area is largely open to further investigation.

In order to see the connection between mixture models and selection models, and the key differences between mixture models that assume MAR vs. those that allow MNAR, we first revisit Example 5.9. In the bivariate case, it is straightforward to impose the MAR constraint in a way that keeps the component distributions normal. However, for responses with dimension greater than two, pattern mixture models are more easily constructed by assuming normality within pattern for *observed data* distributions, but not necessarily for the full-data distributions.

Example 8.3. *Pattern mixture model for bivariate response with missing Y_2* (Example 5.9 revisited).
Consider a full-data model for the joint distribution of (Y_1, Y_2, R), such that $(Y_1, Y_2)^T$ is the bivariate full-data response, and R is a binary indicator of whether Y_2 is observed. Recall from Chapter 5 that the full-data PMM for bivariate normal data follows

$$\begin{aligned}(Y_1, Y_2)^T \mid R = r &\sim N(\boldsymbol{\mu}^{(r)}, \boldsymbol{\Sigma}^{(r)}) \\ R &\sim \text{Ber}(\phi),\end{aligned} \tag{8.21}$$

where, for pattern $r \in \{0, 1\}$, the parameters are $\{\mu_1^{(r)}, \mu_2^{(r)}, \sigma_{11}^{(r)}, \sigma_{22}^{(r)}, \sigma_{12}^{(r)}\}$. Recall also that for pattern r, we can reparameterize the model in terms of the marginal distribution of Y_1 and the conditional distributions of Y_2 given Y_1, such that

$$
\begin{aligned}
Y_1 \mid R = r &\sim N(\mu_1^{(r)}, \sigma_{11}^{(r)}) \\
Y_2 \mid Y_1, R = r &\sim N(\beta_0^{(r)} + \beta_1^{(r)} Y_1, \ \sigma_{2|1}^{(r)}),
\end{aligned}
$$

where

$$
\begin{aligned}
\phi^{(r)} &= (\mu_1^{(r)}, \sigma_{11}^{(r)}, \beta_0^{(r)}, \beta_1^{(r)}, \sigma_{2|1}^{(r)}) \\
&= g(\mu_1^{(r)}, \sigma_{11}^{(r)}, \mu_2^{(r)}, \sigma_{22}^{(r)}, \sigma_{12}^{(r)}).
\end{aligned}
$$

Clearly the parameters $\{\beta_0^{(0)}, \beta_1^{(0)}, \sigma_{2|1}^{(0)}\}$ cannot be identified from the observables, but they can be identified with parameter constraints. For simplicity, let us further suppose that $\Sigma^{(1)} = \Sigma^{(0)} = \Sigma$, so that variance components are equal across pattern. Because $\beta_1^{(r)} = \sigma_{21}^{(r)}/\sigma_{11}^{(r)}$ and $\sigma_{2|1}^{(r)} = \sigma_{22}^{(r)} - (\sigma_{21}^{(r)})^2/\sigma_{11}^{(r)}$, we have $\beta_1^{(0)} = \beta_1^{(1)} = \beta_1$ and $\sigma_{2|1}^{(0)} = \sigma_{2|1}^{(1)} = \sigma_{2|1}$. This still leaves $\beta_0^{(0)}$ (equivalently, $\mu_2^{(0)}$) unidentified.

From Theorem 8.1, MAR is satisfied when $p(y_2 \mid y_1, r = 1) = p(y_2 \mid y_1, r = 0)$. Equality of the conditional distributions implies equality of conditional means $E(Y_2 \mid Y_1 = y_1, R = r)$ for $r = 0, 1$ and for all y_1. Multivariate normality within pattern gives

$$
\begin{aligned}
E(Y_2 \mid Y_1 = y_1, R = r) &= \mu_2^{(r)} + \frac{\sigma_{21}}{\sigma_{11}}(y_1 - \mu_1^{(r)}) \\
&= \left\{\mu_2^{(r)} - \frac{\sigma_{21}}{\sigma_{11}}\mu_1^{(r)}\right\} + \frac{\sigma_{21}}{\sigma_{11}}y_1. \qquad (8.22)
\end{aligned}
$$

Hence equality of conditional means is satisfied when the intercept term in (8.22) is equal for $r = 0, 1$, i.e., when

$$
\mu_2^{(0)} - \beta_1 \mu_1^{(0)} = \mu_2^{(1)} - \beta_1 \mu_1^{(1)}.
$$

A more intuitive representation is

$$
\begin{aligned}
\mu_2^{(0)} &= \mu_2^{(1)} + \beta_1(\mu_1^{(0)} - \mu_1^{(1)}) \qquad (8.23) \\
&= E(Y_2 \mid Y_1 = \mu_1^{(0)}, R = 1),
\end{aligned}
$$

which is the predicted value, at $Y_1 = \mu_1^{(0)}$, of the missing Y_2 based on the regression of Y_2 on Y_1 among those with $(Y_1, Y_2)^{\mathrm{T}}$ observed. Under MAR, $\mu_2^{(0)}$ is entirely a function of identified parameters.

In Example 5.9, we saw that the selection model implied by this mixture of normals is

$$
\mathrm{logit}\{P(R = 1 \mid y_1, y_2)\} = \psi_0 + \psi_1 y_1 + \psi_2 y_2,
$$

where, relevant to this example,

$$\psi_2 = \frac{1}{\sigma_{2|1}^2}\left\{\mu_2^{(0)} - \mu_2^{(1)} - \beta_1(\mu_1^{(0)} - \mu_1^{(1)})\right\}.$$

By substituting the expression for $\mu_2^{(0)}$ from (8.23) into the expression for ψ_2 above, we get $\psi_2 = 0$. Hence MAR implies $\psi_2 = 0$. Not surprisingly, the converse also is true. Furthermore, it is straightforward to expand this model to allow MNAR while maintaining normality within pattern.

Referring to (8.23), let $\mu_{2:\text{MAR}}^{(0)} = \mu_2^{(1)} + \beta_1(\mu_1^{(0)} - \mu_1^{(1)})$ denote the identified value of $\mu_2^{(0)}$ under MAR. To embed this model in a larger one that allows MNAR, we can introduce a parameter Δ such that

$$\mu_2^{(0)} = \mu_{2:\text{MAR}}^{(0)} + \Delta.$$

Because this amounts to a shift in the marginal mean of Y_2 in pattern $R = 0$, bivariate normality of $(Y_1, Y_2)^{\text{T}}$ given $R = 0$ is preserved.   □

The mixture of normals model for $J = 2$ is instructive for understanding the connections between mixture models and their implied selection models; however, as we see in the examples that follow, mixture models constructed from multivariate normal component distributions with $J = 3$ cannot typically be constrained in a way that is consistent with an MAR assumption.

A better approach is to assume a normal distribution for the *observed* data within pattern, and then to impose MAR constraints that will identify the extrapolation distribution $p(\boldsymbol{y}_{\text{mis}} \mid \boldsymbol{y}_{\text{obs}}, \boldsymbol{r})$, which generally will not be normally distributed within pattern under MAR. The examples below illustrate two approaches to model construction. In Example 8.4, we assume observed responses within pattern are multivariate normal; in Example 8.5, we use a different parameterization to preserve conditional normality in the full-data distributions. The two parameterizations suggest different approaches to sensitivity analysis and MNAR formulations.

Example 8.4. *Pattern mixture model identified using interior family constraints.*
Here we use the normal distribution for observed data, and use the MAR assumption to identify the distribution of missing data. The case $J = 3$ is used to illustrate. Let $(Y_1, Y_2, Y_3)^{\text{T}}$ denote the full data. We assume monotone dropout, and we parameterize the model in terms of the number of observations, $S = \sum_j R_j$ (refer to Definition 5.3). The full-data distribution is factored as

$$p(y_1, y_2, y_3, s) = p_s(y_1, y_2, y_3)\, p(s),$$

where S follows a multinomial distribution with probabilities $\phi_s = P(S = s)$.

The observed data response distributions within each pattern are specified using normal distributions,

$$Y_1 \mid S = 1 \sim N(\mu^{(1)}, \sigma^{(1)}),$$
$$(Y_1, Y_2)^{\mathrm{T}} \mid S = 2 \sim N(\boldsymbol{\mu}^{(2)}, \boldsymbol{\Sigma}^{(2)}),$$
$$(Y_1, Y_2, Y_3)^{\mathrm{T}} \mid S = 3 \sim N(\boldsymbol{\mu}^{(3)}, \boldsymbol{\Sigma}^{(3)}).$$

Now consider identifying $p_s(\boldsymbol{y}_{\mathrm{mis}} \mid \boldsymbol{y}_{\mathrm{obs}})$ under MAR. Start with $S = 1$. We need to identify $p_1(y_2, y_3 \mid y_1)$, which can be factored as

$$p_1(y_2, y_3 \mid y_1) = p_1(y_2 \mid y_1) \, p_1(y_3 \mid y_2, y_1).$$

The MAR constraint (8.17) implies

$$p_1(y_2 \mid y_1) = p_{\geq 2}(y_2 \mid y_1),$$
$$p_1(y_3 \mid y_2, y_1) = p_3(y_3 \mid y_2, y_1).$$

The distribution $p_{\geq 2}(y_2 \mid y_1)$ is a *mixture* of the normal distributions from patterns $S = 2$ and $S = 3$,

$$p_{\geq 2}(y_2 \mid y_1) = \frac{\phi_2 \, p_2(y_2 \mid y_1) \, p_2(y_1) + \phi_3 \, p_3(y_2 \mid y_1) \, p_3(y_1)}{\phi_2 \, p_2(y_1) + \phi_3 \, p_3(y_1)}. \qquad (8.24)$$

Referring to Table 8.4, this corresponds to identification of $p_1(y_2 \mid y_1)$ as $\sum_{s=2}^{3} \Delta_{21s} \, p_s(y_2 \mid y_1)$, where

$$\Delta_{21s} = \frac{\phi_s p_s(y_1)}{\phi_2 p_2(y_1) + \phi_3 p_3(y_1)}. \qquad (8.25)$$

Hence the distribution $p_1(y_2 \mid y_1)$ is in general a *mixture* of normals, even though $p_2(y_2 \mid y_1)$ and $p_3(y_2 \mid y_1)$ are normal (by specification). If we impose the additional constraint that $p_2(y_2 \mid y_1) = p_3(y_2 \mid y_1)$, with both being normal, then $p_{\geq 2}(y_2 \mid y_1)$ clearly will be normal as well. $\qquad \square$

To see how the identification strategy for the model in Example 8.4 uses the interior family constraints, we focus again on those with $S = 1$, the distribution $p_1(y_2 \mid y_1)$ can be identified in general by the mixture

$$p_1(y_2 \mid y_1) = \Delta \, p_2(y_2 \mid y_1) + (1 - \Delta) \, p_3(y_2 \mid y_1). \qquad (8.26)$$

The parameter Δ is not identifiable from data and is a sensitivity parameter. Setting $\Delta = 1$ is a 'nearest-neighbor' identification scheme, equating the unidentified $p_1(y_2 \mid y_1)$ to the corresponding distribution from those with $S = 2$. Setting $\Delta = 0$ identifies $p_1(y_2 \mid y_1)$ by equating it to the corresponding distribution among those with complete follow-up; i.e., $p_3(y_2 \mid y_1)$. This is known as the complete case missing value (CCMV) restriction. Under MAR, as we see from (8.24) and (8.25),

$$\Delta = \frac{\phi_2 p_2(y_1)}{\phi_2 p_2(y_1) + \phi_3 p_3(y_1)}.$$

Apart from constraints like these, a sensitivity analysis can be based on varying Δ over $[0, 1]$, or a prior distribution can be assigned to Δ. Although most values of Δ technically represent MNAR models (the MAR constraint above being the exception), the full-data model space under this parameterization is indexed by a single parameter, and extrapolation distributions are confined to functions of the observed data distributions.

In the next example, we show how to parameterize the PMM to allow more global departures from MAR under monotone dropout. The *observed data* distribution is given as follows. At $j = 1$ we assign separate normal distributions to $Y_1 \mid S = s$ for $s = 1, \ldots, J$. Then at each time point $j \geq 2$, we assume the conditional distribution $p_{\geq j}(y_j \mid y_1, \ldots, y_{j-1})$ — i.e., the distribution of $Y_j \mid Y_1, \ldots, Y_{j-1}$ among those with $S \geq j$ — follows a normal distribution with mean and variance given by

$$E(Y_j \mid Y_1, \ldots, Y_{j-1}) = \alpha_0^{(\geq j)} + \sum_{l=1}^{j-1} \alpha_l^{(\geq j)} Y_l$$

$$\text{var}(Y_j \mid Y_1, \ldots, Y_{j-1}) = \tau_j^{(\geq j)}$$

and that

$$p_k(y_j \mid y_1, \ldots y_{j-1}) = p_{\geq j}(y_j \mid y_1, \ldots y_{j-1}), \ k \geq j.$$

Consequently, for any j, the joint distribution of $(Y_1, \ldots, Y_j)^{\mathrm{T}} \mid S = j$ is multivariate normal. For example,

$$
\begin{aligned}
p_3(y_1, y_2, y_3) &= p_3(y_3 \mid y_1, y_2) \, p_3(y_2 \mid y_1) \, p_3(y_1) \\
&= p_{\geq 3}(y_3 \mid y_1, y_2) \, p_{\geq 2}(y_2 \mid y_1) \, p_3(y_1),
\end{aligned}
$$

where the first line is just a factorization of the joint distribution given $S = 3$, and the second line replaces each factor with its specified model.

The *missing response data* distributions also are specified using normal distributions. For each $j \geq 2$, and for each $s < j$, we assume $p_s(y_j \mid y_1, \ldots, y_{j-1})$ is normal. It follows that the joint distribution $p_s(y_1, \ldots, y_j)$, for any s and j, is multivariate normal. Table 8.6 shows the identification strategy.

Imposing a normal distribution on the missing data is a stronger assumption than in Example 8.4; however, it does lend some consistency to the model specification by allowing each full-data distribution within pattern to be normal, and it expands the range of sensitivity analyses. Both attributes are illustrated in the following example.

Example 8.5. *Pattern mixture model based on normal distributions within pattern, with departure from MAR.*
Consider again the case $J = 3$ with no covariates. Following Table 8.6, identifiable observed data distributions are $p_1(y_1)$, $p_2(y_1)$, $p_3(y_1)$, $p_{\geq 2}(y_2 \mid y_1)$, and $p_{\geq 3}(y_3 \mid y_1, y_2)$ (which is equal to $p_3(y_3 \mid y_1, y_3)$). We specify these directly,

Table 8.6 *Schematic representation of mixture model having multivariate normal distribution within pattern, for $J = 4$. All densities in the table are normal with parameters depending on the subscript of $p(\cdot)$. Identified observed data distributions fall below the dividing line, and nonidentified missing distributions are above.*

	$j = 1$	$j = 2$	$j = 3$	$j = 4$
$S = 1$	$p_1(y_1)$	$p_1(y_2\|y_1)$	$p_1(y_3\|y_1, y_2)$	$p_1(y_4\|y_1, y_2, y_3)$
$S = 2$	$p_2(y_1)$	$p_{\geq 2}(y_2\|y_1)$	$p_2(y_3\|y_1, y_2)$	$p_2(y_4\|y_1, y_2, y_3)$
$S = 3$	$p_3(y_1)$	$p_{\geq 2}(y_2\|y_1)$	$p_{\geq 3}(y_3\|y_1, y_2)$	$p_3(y_4\|y_1, y_2, y_3)$
$S = 4$	$p_4(y_1)$	$p_{\geq 2}(y_2\|y_1)$	$p_{\geq 3}(y_3\|y_1, y_2)$	$p_4(y_4\|y_1, y_2, y_3)$

but make the further assumption that certain unidentified distributions are normal. The distribution of $(\boldsymbol{Y}_{\text{obs}} \mid S)$ is given by

$$
\begin{aligned}
Y_1 \mid S = 1 &\sim N(\mu^{(1)}, \sigma^{(1)}) \\
Y_1 \mid S = 2 &\sim N(\mu^{(2)}, \sigma^{(2)}) \\
Y_1 \mid S = 3 &\sim N(\mu^{(3)}, \sigma^{(3)}) \\
\left.\begin{array}{l} Y_2 \mid Y_1, S = 2 \\ Y_2 \mid Y_1, S = 1 \end{array}\right\} &\sim N(\alpha_0^{(\geq 2)} + \alpha_1^{(\geq 2)} Y_1,\ \tau_2^{(\geq 2)}) \\
Y_3 \mid Y_1, Y_2, S = 3 &\sim N(\beta_0^{(3)} + \beta_1^{(3)} Y_1 + \beta_2^{(3)} Y_2,\ \tau_3^{(3)}).
\end{aligned}
$$

To round out the full-data response distribution, we specify $(\boldsymbol{Y}_{\text{mis}} \mid S)$ as follows:

$$
\begin{aligned}
Y_2 \mid Y_1, S = 1 &\sim N(\alpha_0^{(1)} + \alpha_1^{(1)} Y_1,\ \tau_2^{(1)}) \\
Y_3 \mid Y_1, Y_2, S = s &\sim N(\beta_0^{(s)} + \beta_1^{(s)} Y_1 + \beta_2^{(s)} Y_2,\ \tau_3^{(s)}) \quad (s = 1, 2).
\end{aligned}
$$

The general MAR constraints require

$$
\begin{aligned}
p_1(y_2 \mid y_1) &= p_{\geq 2}(y_2 \mid y_1), & (8.27) \\
p_1(y_3 \mid y_1, y_2) &= p_3(y_3 \mid y_1, y_2), & (8.28) \\
p_2(y_3 \mid y_1, y_2) &= p_3(y_3 \mid y_1, y_2). & (8.29)
\end{aligned}
$$

Condition (8.27) can be satisfied by equating means and variances of $p_1(y_2 \mid y_1)$ and $p_{\geq 2}(y_2 \mid y_1)$,

$$
\begin{aligned}
\alpha_0^{(1)} + \alpha_1^{(1)} y_1 &= \alpha_0^{(\geq 2)} + \alpha_1^{(\geq 2)} y_1 \quad (\forall y_1 \in \mathbb{R}), \\
\tau_2^{(1)} &= \tau_2^{(\geq 2)}.
\end{aligned}
$$

The first equality above holds only when $\alpha_0^{(1)} = \alpha_0^{(\geq 2)}$ and $\alpha_1^{(1)} = \alpha_1^{(\geq 2)}$.

Similar constraints can be imposed on the parameters indexing $p_1(y_3 \mid y_1, y_2)$ and $p_2(y_3 \mid y_1, y_2)$ to fully satisfy MAR.

Writing the model in this way makes it fairly simple to embed the MAR specification in a large class of MNAR models indexed by parameters Δ_0, Δ_1, and Δ_2 that measure departures from MAR. For example, to characterize MNAR in terms of departures from (8.27), write

$$\left.\begin{aligned} \alpha_0^{(1)} &= \alpha_0^{(\geq 2)} + \Delta_0 \\ \alpha_1^{(1)} &= \alpha_1^{(\geq 2)} + \Delta_1 \\ \log \tau_2^{(1)} &= \log \tau_2^{(\geq 2)} + \Delta_2 \end{aligned}\right\} \qquad (8.30)$$

Assuming constraints (8.28) and (8.29) hold, dropout is MAR when $\Delta_0 = \Delta_1 = \Delta_2 = 0$. None of the Δ parameters appears in the observed data likelihood. In general, a separate Δ is needed for each model constraint, but in practice it is necessary to limit the dimensionality of these. Our analysis of the Growth Hormone study in Section 10.2 provides methods for doing so. □

In Chapter 9, we formalize this structure for fully Bayesian model specifications by writing (8.30) as a function

$$\boldsymbol{\xi}_{\mathrm{S}} = h(\boldsymbol{\xi}_{\mathrm{M}}, \boldsymbol{\Delta}),$$

where $\boldsymbol{\xi}_{\mathrm{S}}$ are the (nonidentified) sensitivity parameters in the full-data model, $\boldsymbol{\xi}_{\mathrm{M}}$ are (identified) parameters indexing the implied observed data model, and $\boldsymbol{\Delta}$ captures departures from MAR. In many cases, the h function represents the missing data mechanism, and makes explicit how assumptions or priors are being used to infer the full-data model (also see Rubin, 1977).

Examples 8.4 and 8.5 differ not only in how the observed data distribution is specified, but also how the missing data are extrapolated. In Example 8.4, the missing data distribution is identified using a mixture of normals constructed using a weighted average of observed data distributions; see (8.26). The range of possibilities for extrapolating missing data is confined to this structure; though it may be fairly limited in scope, it is simple for practical settings because of the small number of unidentified parameters (for $J = 3$, there is only one).

In Example 8.5, parametric distributions are assumed for the extrapolations. This imposes untestable distributional assumptions, but allows for considerable flexibility in extrapolating the missing data either by fixing values or assigning priors to parameters like $(\Delta_0, \Delta_1, \Delta_2)$ in (8.30).

A key consideration for model specification, including specification of priors or ranges for sensitivity parameters, is understanding the physical meaning of sensitivity parameters in context. This is discussed further in Chapter 9. In addition, in Section 10.2, we address specific related issues (model specification, dimensionality of sensitivity parameters, formulation and interpretation of priors, and calibration of sensitivity analyses) in a detailed analysis of data from the growth hormone study.

Binary responses with discrete-time dropout

Pattern mixture models can be used for binary data as well. Some key references here include Baker and Laird (1988), Ekholm and Skinner (1998), Birmingham and Fitzmaurice (2002), and Little and Rubin (2002). In this section we describe pattern mixture models for the bivariate case $J = 2$ and for higher-dimensional settings $(J > 2)$. For each model, we show how to impose MAR constraints using Theorem 8.1, and introduce parameters that index a larger class of MNAR models. Many of the key ideas that apply to continuous-data models also apply here.

Example 8.6. *Pattern mixture model for bivariate binary data.*
The bivariate case with missingness in Y_2 provides a useful platform for understanding model specification with binary responses. As with the continuous case, let $\boldsymbol{Y} = (Y_1, Y_2)^{\mathrm{T}}$ denote the full-data response, and let S denote the number of observed measurements (i.e., $S = 2$ if Y_2 is observed and $S = 1$ if it is missing). The entire full-data distribution $p(\boldsymbol{y}, s \mid \boldsymbol{\omega})$ can be enumerated as in Table 8.1, and viewed as a multinomial distribution with seven distinct parameters.

The pattern mixture model factors the joint distribution as

$$p(y_1, y_2 \mid s, \boldsymbol{\alpha})\, p(s \mid \boldsymbol{\phi}),$$

where, referring again to Table 8.1, $(\boldsymbol{\alpha}, \boldsymbol{\phi}) = \boldsymbol{g}(\boldsymbol{\omega})$ is just a reparameterization of $\boldsymbol{\omega}$. A simple PM specification is

$$
\begin{aligned}
Y_2 \mid Y_1, S = s &\sim \mathrm{Ber}(\mu_{2|1}^{(s)}) \\
Y_1 \mid S = s &\sim \mathrm{Ber}(\mu_1^{(s)})
\end{aligned}
\tag{8.31}
$$

for $s \in \{1, 2\}$, with $P(S = s) = \phi_s$. Regression models can be used to structure the parameters for the responses,

$$
\begin{aligned}
\mathrm{logit}(\mu_{2|1}^{(s)}) &= \alpha_0^{(s)} + \alpha_1^{(s)} y_1, \\
\mathrm{logit}(\mu_1^{(s)}) &= \beta^{(s)},
\end{aligned}
$$

for $s \in \{1, 2\}$. The seven unique parameters in this PMM comprise three regression parameters each for $k = 1, 2$, and the Bernoulli probability $P(S = 2) = \phi_2 = 1 - \phi_1$.

Without further assumptions, $\alpha_0^{(1)}$ and $\alpha_1^{(1)}$ cannot be identified; indeed, $\alpha_0^{(1)}$ and $\alpha_1^{(1)}$ are sensitivity parameters. In this simple model, MAR holds when $p_1(y_2 \mid y_1) = p_2(y_2 \mid y_1)$, and can be imposed by setting $(\alpha_0^{(1)}, \alpha_1^{(1)}) = (\alpha_0^{(2)}, \alpha_1^{(2)})$, yielding $\mu_{2|1}^{(1)} = \mu_{2|1}^{(2)}$.

For handling MNAR mechanisms, the regression formulation is favored over the the multinomial parameterization because it facilitates simple representations of departures from MAR in terms of differences between the $\boldsymbol{\alpha}$

parameters in the component distributions $p_k(y_2 \mid y_1)$; specifically, if we let

$$
\begin{aligned}
\alpha_0^{(1)} &= \alpha_0^{(2)} + \Delta_0, \\
\alpha_1^{(1)} &= \alpha_1^{(2)} + (\Delta_1 - \Delta_0),
\end{aligned}
\tag{8.32}
$$

then $(\Delta_0, \Delta_1) = (0, 0)$ represents MAR, and departures from MAR can be represented by non-zero values of Δ_0, Δ_1, or both.

For $y = 0, 1$, the parameter Δ_y is a log odds ratio comparing the odds that $Y_2 = 1$ between completers and dropouts, conditionally on $Y_1 = y$,

$$
\begin{aligned}
\Delta_y &= \log \left[\frac{E(Y_2 \mid S = 1, Y_1 = y)/\{1 - E(Y_2 \mid S = 1, Y_1 = y)\}}{E(Y_2 \mid S = 2, Y_1 = y)/\{1 - E(Y_2 \mid S = 2, Y_1 = y)\}} \right] \\
&= \log \left\{ \frac{\text{odds}(Y_2 \mid S = 1, Y_1 = y)}{\text{odds}(Y_2 \mid S = 2, Y_1 = y)} \right\}.
\end{aligned}
$$

$\square$

Another advantage to using the regression formulation is that its representation of MNAR extends readily to settings where $J > 2$. To illustrate, consider a binary response vector $\boldsymbol{Y} = (Y_1, \ldots, Y_J)^{\mathrm{T}}$. As above, denote follow-up time as as the number of observed measurements, $S = \sum_j R_j$. The full-data model can be specified in terms of the dropout-specific component distributions $p_s(y_1, \ldots, y_J)$ and the dropout model $P(S = s)$. Assume the component distributions, for $s = 1, \ldots, J$, follow

$$
\begin{aligned}
Y_1 \mid S = s &\sim \text{Ber}(\mu_1^{(s)}), \\
Y_j \mid Y_1, \ldots, Y_{j-1}, S = s &\sim \text{Ber}(\eta_j^{(s)}) \quad j = 2, \ldots, J.
\end{aligned}
$$

A general regression formulation for $\eta_j^{(s)}$ would clearly have too many parameters if it were to include main effects, two-way and higher-order interactions involving $Y_1, \ldots, Y_{j-1}$. For purposes of illustration, one can consider a simplified model where serial dependence has order 1. Using a logistic regression,

$$
\text{logit}(\eta_j^{(s)}) = \gamma_j^{(s)} + \theta_j^{(s)} y_{j-1}.
\tag{8.33}
$$

Even using this simplified structure, the full-data parameters $(\gamma_j^{(s)}, \theta_j^{(s)})$ cannot in general be identified for $j > s$; it turns out that they qualify as sensitivity parameters.

In the next example, we illustrate how to specify the PMM for $J = 3$ using first-order Markov models as the component distributions. In Section 10.3, we use a similar model to analyze data from the OASIS trial; there, we provide details on model specification under both MAR and MNAR, and guidance about specification of informative priors that capture assumptions about departures from MAR.

Example 8.7. *Pattern mixture model for binary response with $J = 3$.*
Here we consider a model for the joint distribution $p(y_1, y_2, y_3, s)$, with $S =$

$\sum_{j=1}^{3} R_j$ again denoting number of observed responses. For $s \in \{1, 2, 3\}$, we decompose the full-data distribution as a mixture

$$p(y_1, y_2, y_3) = \sum_{s=1}^{3} \phi_s \, p_s(y_1, y_2, y_3)$$

over the distribution of S, with

$$p_s(y_1, y_2, y_3) = p_s(y_1) \, p_s(y_2 \mid y_1) \, p_s(y_3 \mid y_2, y_1)$$

and $\phi_s = P(S = s)$. We further assume first-order serial dependence, so that $p_s(y_3 \mid y_2, y_1) = p_s(y_3 \mid y_2)$. Within pattern s,

$$\begin{aligned}
Y_1 \mid S = s &\sim \text{Ber}(\mu^{(s)}), \\
Y_2 \mid Y_1, S = s &\sim \text{Ber}(\eta_2^{(s)}), \\
Y_3 \mid Y_2, S = s &\sim \text{Ber}(\eta_3^{(s)}),
\end{aligned}$$

with

$$\begin{aligned}
\text{logit}(\mu^{(s)}) &= \beta^{(s)}, \\
\text{logit}(\eta_j^{(s)}) &= \gamma_j^{(s)} + \theta_j^{(s)} Y_{j-1} \quad (j = 2, 3). \quad (8.34)
\end{aligned}$$

The MAR constraint (8.17) is satisfied when these equalities hold:

$$\begin{aligned}
p_1(y_2 \mid y_1) &= p_{\geq 2}(y_2 \mid y_1) \\
p_1(y_3 \mid y_2) &= p_2(y_3 \mid y_2) = p_3(y_3 \mid y_2).
\end{aligned} \quad (8.35)$$

The first part of the MAR constraint can be imposed by assuming

$$Y_2 \mid Y_1, S \geq 2 \sim \text{Ber}(\eta_j^{(\geq 2)}),$$

with

$$\text{logit}(\eta_j^{(\geq 2)}) = \gamma_j^{(\geq 2)} + \theta_j^{(\geq 2)} Y_{j-1}, \quad (j = 2, 3) \quad (8.36)$$

and then equating regression coefficients across patterns by setting

$$\begin{aligned}
\gamma_2^{(1)} &= \gamma_2^{(\geq 2)} \\
\theta_2^{(1)} &= \theta_2^{(\geq 2)}.
\end{aligned} \quad (8.37)$$

In the absence of covariates, models (8.34) and (8.36) are compatible because with binary data, the mixture distribution $p_{\geq 2}(y_2 \mid y_1)$ obtained by averaging the component-specific logistic models for $p_2(y_2 \mid y_1)$ and $p_3(y_2 \mid y_1)$ is equivalent to the single logistic model for data aggregated over patterns 2 and 3.

The second part of the MAR constraint (8.35) that requires $p_1(y_3 \mid y_2) = p_2(y_3 \mid y_2) = p_3(y_3 \mid y_2)$ can be satisfied by setting

$$\begin{aligned}
\gamma_3^{(s)} &= \gamma_3^{(3)} \\
\theta_3^{(s)} &= \theta_3^{(3)}
\end{aligned}$$

for $s = 1, 2$.

As with the continuous data model in Example 8.5, the MAR model can easily be embedded in a larger class of models that allows MNAR. For example, referring to (8.37), we can write

$$
\begin{aligned}
\gamma_2^{(1)} &= \gamma_2^{(\geq 2)} + \Delta_0 \\
\theta_2^{(1)} &= \theta_2^{(\geq 2)} + (\Delta_1 - \Delta_0).
\end{aligned}
\tag{8.38}
$$

The parameters Δ_0 and Δ_1 correspond to log odds ratios that compare odds of $Y_2 = 1$ between those who have only one observed response ($S = 1$) and those who have two or more ($S \geq 2$). Specifically, notice that

$$
\begin{aligned}
\gamma_2^{(1)} &= \operatorname{logit}\{P(Y_2 = 1 \mid Y_1 = 0, S = 1)\}, \\
\gamma_2^{(\geq 2)} &= \operatorname{logit}\{P(Y_2 = 1 \mid Y_1 = 0, S \geq 2)\},
\end{aligned}
$$

so that

$$
\Delta_0 = \log\left\{\frac{\operatorname{odds}(Y_2 = 1 \mid Y_1 = 0, S = 1)}{\operatorname{odds}(Y_2 = 1 \mid Y_1 = 0, S \geq 2)}\right\}.
$$

Similarly, under the parameterization in (8.38), Δ_1 is the same log odds ratio, but conditioned on $Y_1 = 1$,

$$
\Delta_1 = \log\left\{\frac{\operatorname{odds}(Y_2 = 1 \mid Y_1 = 1, S = 1)}{\operatorname{odds}(Y_2 = 1 \mid Y_1 = 1, S \geq 2)}\right\}.
$$

More Δ parameters can be introduced along these lines, and the model configured such that MAR is implied by setting each Δ to zero. $\square$

In Chapter 9, we provide a more complete development of principles for finding sensitivity parameters, parameterizing untestable assumptions in terms of unidentified full-data model parameters. incorporating informative prior information, and conducting sensitivity analyses. Further consideration of this model and a detailed illustration using data from the OASIS Study is provided in Section 10.3.

8.4.4 Mixture models with continuous-time dropout

Our illustrations of mixture models to this point have focused on settings where dropout occurs in discrete time and for small numbers of measurement times. It is possible to apply these models in continuous-time settings as well. Here, we change notation for the event time from S to U, where $U > 0$ is a continuous positive-valued random variable representing time to dropout. The full-data model is still factored as

$$
p(\boldsymbol{y}, u \mid \boldsymbol{\omega}) = p(\boldsymbol{y} \mid u, \boldsymbol{\omega})\, p(u \mid \boldsymbol{\omega}),
$$

and the full-data response model is obtained by averaging over the dropout distribution

$$p(\boldsymbol{y} \mid \boldsymbol{\omega}) = \int p(\boldsymbol{y} \mid u, \boldsymbol{\omega}) \, p(u \mid \boldsymbol{\omega}) \, du.$$

For continuous data, a general formulation of this joint distribution uses a varying coefficient model, illustrated below, for the conditional distribution $p(\boldsymbol{y} \mid u, \boldsymbol{x}, \boldsymbol{\omega})$ (Hogan et al., 2004a). In the longitudinal data setting, the standard pattern mixture model for discrete-time dropout (Little, 1993, 1994) and the conditional linear model (Wu and Bailey, 1988) can be viewed as special cases of the varying coefficient model approach. Moreover, although the VCM approach is developed for continuous responses and is based on a mixture of normal distributions (used below to illustrate), in principle the approach can be used for other types of distributions (Su and Hogan, 2007).

In the mixture of normals case, we assume that the full-data response vector $\boldsymbol{Y} = (Y_1, \ldots, Y_J)^{\mathrm{T}}$ arises from a process $\{Y(t) : t \geq 0\}$ observed at times $t_1, \ldots, t_J$. The set of measurement times is assumed to vary by individual. Conditionally on dropout at time $U = u$, we assume $Y(t)$ has a mean function denoted by $\mu(t|u)$ and covariance given by $C(s, t|u)$; e.g., $C(s, t|u) = \sigma(s|u)\sigma(t|u)\rho(s, t|u)$ for $s \neq t$. We further assume that the process is conditionally normal (i.e., a Gaussian process), such that

$$Y(t) \mid U = u \quad \sim \quad N(\, \mu(t \mid u), C(t, t \mid u) \,).$$

To obtain $E\{Y(t)\}$, we integrate the conditional mean function over the distribution of dropout time,

$$E\{Y(t)\} \quad = \quad \int \mu(t \mid u) \, p(u) \, du.$$

A question arises as to the functional form of $\mu(t|u)$; in the VCM approach, it is assumed that $\mu(t|u)$ is a known function of t for any given u, but as a function of u it can be any smooth function. To illustrate, we use a simple formulation where $\mu(t|u)$ is linear in t, and the goal is to estimate the overall mean of $\{Y(t)\}$.

Example 8.8. *Mixture of varying coefficient models for estimating the intercept and slope under continuous-time dropout.*
Consider a full-data response $\boldsymbol{Y}_i$ that is generated by potentially observing the process $\{Y_i(t)\}$ at time points $t_{i1}, \ldots, t_{iJ_i}$, i.e.,

$$\boldsymbol{Y}_i = (Y_{i1}, \ldots, Y_{iJ_i})^{\mathrm{T}} = (Y_i(t_{i1}), \ldots, Y_i(t_{iJ_i}))^{\mathrm{T}}.$$

Dropout for individual i occurs at U_i. Let $\boldsymbol{x}_i(t) = \{x_{i1}(t), \ldots, x_{ip}(t)\}$ represent a p-dimensional covariate process, and let

$$\begin{aligned} \boldsymbol{X}_i &= (\, \boldsymbol{x}_i(t_{i1})^{\mathrm{T}}, \ldots, \boldsymbol{x}_i(t_{iJ_i})^{\mathrm{T}} \,)^{\mathrm{T}} \\ &= (\boldsymbol{x}_{i1}^{\mathrm{T}}, \ldots, \boldsymbol{x}_{iJ_i}^{\mathrm{T}})^{\mathrm{T}} \end{aligned}$$

denote a $J_i \times p$ covariate matrix.

A varying coefficient model for those who drop out at u is

$$Y_i \mid X_i, U_i = u \sim N(\ X_i\beta(u),\ \Sigma_i(u)\ ),$$

where $\beta(u)$ is a $p \times 1$ parameter vector comprised of functions of u; i.e., $\beta(u) = (\beta_1(u), \ldots, \beta_p(u))^{\mathrm{T}}$ and the (j, k)th element of $\Sigma_i(u)$ is $C(t_{ij}, t_{ik} \mid u)$. The functions can be specified using penalized splines (Ruppert et al., 2003; Crainiceanu et al., 2005).

For this example we assume $\Sigma_i(u) = \Sigma_i\{\phi(u)\} = \Sigma_i(\phi)$; that is, we assume the conditional covariance matrix is parameterized by the vector $\phi(u)$, but that $\phi(u) = \phi$ does not depend on dropout time. This assumption may be relaxed by modeling covariance parameters as smooth functions of u; see Chapter 6.

This model implies that conditional on dropout time,

$$E\{Y_i(t) \mid x_i(t), U = u\} = \mu_i(t|u) = x_i(t)\beta(u),$$

and therefore

$$\begin{aligned} E(Y_{ij} \mid X_i, U_i = u) &= x_i(t_{ij})\beta(u) \\ &= x_{ij}\beta(u). \end{aligned}$$

If $x_{ij} = (1, t_{ij})$, then $E(Y_{ij} \mid t_{ij}, U_i = u)$ follows a straight line as a function of t, but the intercept and slope parameters are functions of the dropout time u. For this model, the mean of $Y_i(t)$ itself is a straight line because

$$\begin{aligned} E(Y_{ij}) &= \int \{\beta_1(u) + \beta_2(u)t_{ij}\}\, p(u)\, du \qquad (8.39) \\ &= \beta_1 + \beta_2 t_{ij}, \end{aligned}$$

where $\beta_q = \int \beta_q(u)\, p(u)\, du$ for $q = 1, 2$. $\square$

An important feature of this model is that the distribution of dropout times, $p(u)$, can be left unspecified. For inference, mixtures of Dirichlet process models (cf. Section 3.6) can be used to ensure flexibility. Parametric distributions for $p(u)$ also can be used where appropriate, but in general leaving the distribution unspecified does not introduce significant additional complications. A detailed analysis of the Pediatric AIDS trial, including details on model and prior specification using a VCM, is given in Section 10.4.

For normally distributed data, the VCM approach reduces to the standard multivariate normal repeated measures regression — and consequently the missing data mechanism reduces to MAR — when $\beta(u) = \beta$ and $\Sigma(u) = \Sigma$.

When the regression coefficients are a nonconstant function of u, the working assumption for the VCM is that conditional on $U = u$, the regression coefficient remains the same both before and after u. This assumption is particularly important for time-varying covariates, the most common being time itself. In the simple model described in Example 8.8, it is assumed that among those dropping out at $U = u$, the slope $\beta_1(u)$ applies both prior to and after dropout. Hence the full-data mean of $Y(t)$ at a fixed time $t = t^*$ will be based on extrapolations for those who have dropped out prior to t^*.

It is possible to relax the assumption about constant slope (or covariate effect) both before and after dropout by adding one or more sensitivity parameters (cf. to the discussion of identification via extrapolation in Section 8.4.2); this is addressed in Su and Hogan (2007) and discussed further in our analysis in Section 10.4.

In principle, the VCM can be applied to settings where dropout time is continuous and responses are discrete. In that case, calculation of full-data functionals (such as the mean) requires integrating a nonlinear link function over the dropout distribution. This is in contrast to (8.39), where it is only necessary to integrate the functions $\beta(u)$ over $p(u)$.

8.4.5 Combinations of MAR and MNAR dropout

In many trials, subjects drop out for a variety of reasons, leading to situations where a single study may have both MNAR and MAR mechanisms. For example, in the schizophrenia clinical trial (Section 1.2), subjects dropped out for a variety of reasons, including adverse events, lack of treatment effect, and other reasons — including improvement in schizophrenia symptoms (see Table 1.1). Subjects who dropped out due to lack of treatment effect might be assumed to potentially be MNAR dropouts, while those who dropped out for other reasons might be assumed MAR dropouts (see Hogan and Laird, 1997a for further details).

Using mixture models, we can define patterns based on the follow-up time corresponding to MNAR dropout. Recall that for the schizophrenia data, the measurement times are $\mathcal{T} = \{1, 2, 3, 4, 6\}$; let S denote time followed up until an individual leaves the study for a reason that would be classified as MNAR. Here, $S \in \mathcal{T}$. If an individual leaves the study for a reason that is assumed MAR, then S is right-censored. For example, if an individual discontinues follow up after week 2 due to an adverse event that is unrelated to treatment efficacy, we observe $S > 2$ or equivalently $S \in \{3, 4, 6\}$.

Consider a pattern mixture model given by

$$\boldsymbol{Y} \mid S = s \sim N(\boldsymbol{x}\boldsymbol{\beta}^{(s)}, \boldsymbol{\Sigma})$$
$$S \sim \text{Mult}(\phi_1, \ldots, \phi_6),$$

where $\boldsymbol{x}$ is a design matrix reflecting treatment group and time trend (e.g., quadratic trend over time), and the multinomial parameters for S are constrained such that $\phi_5 \equiv 0$ (to reflect that measurements are not taken at week 5) and $\sum_s \phi_s = 1$. Let $\boldsymbol{\alpha} = \{\boldsymbol{\beta}^{(1)}, \ldots, \boldsymbol{\beta}^{(6)}, \boldsymbol{\Sigma}\}$, and $\boldsymbol{\phi} = (\phi_1, \ldots, \phi_6)^{\mathrm{T}}$.

Implementation of posterior sampling is best done using data augmentation, as opposed to working directly with the observed data likelihood. For individuals that drop out for reasons deemed MNAR, S is observed. For each individual, the data augmentation step draws $y^*_{i,\mathrm{mis}}$ from the distribution $p(y_{\mathrm{mis}} \mid y_{i,\mathrm{obs}}, s_i, \alpha)$, which is simply a conditional normal distribution in pattern $S = s_i$ for the pattern mixture model given above.

For individuals that discontinue follow up at time k for reasons deemed MAR, data augmentation proceeds in two steps. First, draw s^*_i from a multinomial distribution having probabilities $(\phi^*_{i1}, \ldots, \phi^*_{i6})$, where

$$
\begin{aligned}
\phi^*_{ij} &= p(S = j \mid y_{i,\mathrm{obs}}, S > k, \alpha, \phi) \\
&= \frac{\phi_j \, p(y_{i,\mathrm{obs}} \mid S = j, \alpha) \, I(j > k)}{\sum_{j=k+1}^{6} \phi_j \, p(y_{i,\mathrm{obs}} \mid S = j, \alpha)}.
\end{aligned}
$$

Note that $\phi^*_{ij} = 0$ for $j \leq k$. Next, we draw a new value of $y^*_{i,\mathrm{mis}}$ as above, this time conditioning on s^*_i and using the distribution $p(y_{\mathrm{mis}} \mid y_{i,\mathrm{obs}}, s^*_i, \alpha)$.

For a complete analysis of the schizophrenia data, see Hogan and Laird (1997a); in that paper, an EM algorithm is used, where the E step is very similar to the data augmentation step described here.

8.4.6 Mixture models or selection models?

With binary (or categorical) data, when the measurement occasions are discrete and dropout is the sole cause of missingness, the duality between mixture and selection models holds for any J, but of course the dimensionality increases exponentially: for J measurement occasions with J possible dropout times, the number of unique parameters is $J2^J - 1$, and any realistic analysis must rely on simplifying assumptions to reduce the dimension of the parameter space. This raises an obvious question of which factorization to use for the full-data model $p(y, r \mid \omega)$.

With a selection model, the simplifications must be made in terms of the full-data response distribution $p(y)$ and the selection mechanism $p(r \mid y)$. Possible strategies include limiting the association structure in the full-data response distribution (e.g., to include only two-way interactions), or limiting the selection mechanism such that dropout at t_j depends only on a small part of the observed history (say Y_j and Y_{j-1}). For purposes of sensitivity analysis, however, this can become problematic because unless $p(y)$ is specified nonparametrically, the full-data model parameter ω can be identified from the observed data, and the choice of an appropriate sensitivity parameter is often not possible. An alternative is a semiparametric formulation for the full-data response model Scharfstein et al. (2003), but this approach can present nontrivial technical complications.

By contrast, model simplifications in mixture models may be more feasible, despite the proliferation of nonidentified parameters. Sensible simplifying assumptions can be imposed on the observed data, while keeping the distribution of the missing data indexed by one or more nonidentified parameters. Our example using first-order dependence within pattern on mixtures of longitudinal binary data distributions is representative. There is a clear delineation between identified and nonidentified parameters and, importantly, simplifying assumptions that are used to constrain the observed data distribution can be empirically critiqued.

8.4.7 Covariate effects in mixture models

Although most of our examples do not involve drawing inference about more than one or two covariates, it is important to understand how covariate effects are computed and interpreted in mixture models. Because the full-data model is a mixture over component distributions corresponding to missing data pattern or dropout time, covariate effects must be interpreted in terms of the mixture distribution. For the PMM with identity link, covariate effects for the full data can sometimes have a simple representation as a weighted average over pattern-specific covariate effects, and the scale of the covariate effects is preserved; see Examples 8.9 and 8.10. This is generally not true with nonlinear link functions, as we illustrate in Example 8.11.

We focus on covariates $\boldsymbol{X}$ that are exogenous, either time-invariant (e.g., baseline characteristics, gender) or fixed functions of time (e.g., age), and whose *effects* are time invariant (see Roy and Lin (2002) for settings with stochastic time-varying covariates subject to missingness from dropout).

PMM with identity link function within pattern

Here we illustrate the computation of covariate effects using the identity link. In the first example, missingness does not depend on covariates, and in the second it does.

Example 8.9. *Mixture of regressions with identity link for bivariate response, where dropout does not depend on covariates.*
Consider the PMM for bivariate data where interest is in estimating the effect of covariates $\boldsymbol{X}$, represented in a $2 \times p$ matrix, on the bivariate outcome $\boldsymbol{Y} = (Y_1, Y_2)^{\mathrm{T}}$. Recall that the full-data model is factored as

$$p(y_1, y_2, r \mid \boldsymbol{x}, \boldsymbol{\omega}) = p(y_1, y_2 \mid r, \boldsymbol{x}, \boldsymbol{\omega})\, p(r \mid \boldsymbol{x}, \boldsymbol{\omega}).$$

In the case where missingness does not depend directly on covariates, we have $p(r \mid \boldsymbol{x}, \boldsymbol{\omega}) = p(r \mid \boldsymbol{\omega})$, which implies that the effect of $\boldsymbol{X}$ on $\boldsymbol{Y}$ can be fully characterized by its within-pattern effects. We further assume that the covariate effect is constant over time and can be captured in a time-independent parameter $\boldsymbol{\beta}$.

Regardless of the parametric distribution assigned to $p(y_1, y_2 \mid r, \boldsymbol{x}, \boldsymbol{\omega})$, an identity link within pattern implies the mean is linear in covariates, via

$$E(\boldsymbol{Y} \mid \boldsymbol{X} = \boldsymbol{x}, R = r) \quad = \quad \boldsymbol{x}\boldsymbol{\beta}^{(r)},$$

where $\boldsymbol{\beta}^{(r)}$ is a $p \times 1$ vector of regression parameters. With the two-component mixture, $R \sim \text{Ber}(\phi)$. Hence

$$
\begin{aligned}
E(\boldsymbol{Y} \mid \boldsymbol{x}) &= E_R\{E(\boldsymbol{Y} \mid \boldsymbol{x}, R)\} \\
&= \phi \boldsymbol{x} \boldsymbol{\beta}^{(1)} + (1 - \phi) \boldsymbol{x} \boldsymbol{\beta}^{(0)} \\
&= \boldsymbol{x}\{\phi \boldsymbol{\beta}^{(1)} + (1 - \phi) \boldsymbol{\beta}^{(0)}\},
\end{aligned}
$$

and the covariate effect is weighted average of pattern-specific coefficients, i.e.,

$$
\boldsymbol{\beta} = \phi \boldsymbol{\beta}^{(1)} + (1 - \phi) \boldsymbol{\beta}^{(0)}.
$$

These can be identified because the covariate is assumed to have a constant effect over time.   □

More generally, dropout may depend on covariates, in which case

$$
p(\boldsymbol{y} \mid \boldsymbol{x}) = \sum_{r \in \mathcal{R}} p(\boldsymbol{y} \mid \boldsymbol{r}, \boldsymbol{x}) \, p(\boldsymbol{r} \mid \boldsymbol{x}).
$$

Usually we are interested in $\boldsymbol{\mu}(\boldsymbol{x}) = E(\boldsymbol{Y} \mid \boldsymbol{X} = \boldsymbol{x})$, computed via

$$
\boldsymbol{\mu}(\boldsymbol{x}) = \sum_{r \in \mathcal{R}} \boldsymbol{\mu}^{(r)}(\boldsymbol{x}) \, p(\boldsymbol{r} \mid \boldsymbol{x}).
$$

The next example considers this more general setting using a discrete mixture over distinct dropout times.

Example 8.10. *Discrete-time mixture of regressions with identity link, where dropout depends on covariates.*
Denote the full-data response by $\boldsymbol{Y} = (Y_1, \ldots, Y_J)^{\mathrm{T}}$, with missing data indicators $\boldsymbol{R} = (R_1, \ldots, R_J)^{\mathrm{T}}$. For this example we consider only baseline covariates, collected in a $p-$dimensional row vector $\boldsymbol{X}$. Dropout is characterized using the follow-up time $S = \sum_j R_j$, with $S \in \{1, \ldots, J\}$, and the model is a mixture over $p(s \mid \boldsymbol{x})$.

The within-pattern regression model follows

$$
\mu_j^{(s)}(\boldsymbol{x}) = E(Y_j \mid S = s, \boldsymbol{X} = \boldsymbol{x}) = \boldsymbol{x} \boldsymbol{\beta}^{(s)},
$$

and follow-up time depends on covariates via

$$
S \mid \boldsymbol{X} = \boldsymbol{x} \quad \sim \quad \text{Mult}(\phi_1(\boldsymbol{x}), \ldots, \phi_J(\boldsymbol{x})),
$$

where $\phi_s(\boldsymbol{x}) = P(S = s \mid \boldsymbol{x})$. In practice the $\phi_s(\boldsymbol{x})$ could be represented using a saturated model if components of $\boldsymbol{X}$ are discrete and low dimensional (e.g., treatment group in a randomized trial); otherwise they would be specified using a model for multinomial distribution, such as relative risk regression. The mean response as a function of covariates is

$$
\mu_j(\boldsymbol{x}) = E(Y_j \mid \boldsymbol{X} = \boldsymbol{x}) = \sum_s \phi_s(\boldsymbol{x}) \, \boldsymbol{x} \boldsymbol{\beta}^{(s)}.
$$

The covariate *effect* is seen to depend on $\boldsymbol{x}$,

$$\frac{\partial \mu_j(\boldsymbol{x})}{\partial \boldsymbol{x}} = \sum_s \left\{ \boldsymbol{x} \frac{\partial \phi_s(\boldsymbol{x})}{\partial \boldsymbol{x}} + \phi_s(\boldsymbol{x}) \right\} \boldsymbol{\beta}^{(s)}. \tag{8.40}$$

For capturing the effect of fixed differences $\boldsymbol{x} - \boldsymbol{x}'$, we have

$$\mu_j(\boldsymbol{x}) - \mu_j(\boldsymbol{x}') = \sum_s \{ \boldsymbol{x} \phi_s(\boldsymbol{x}) - \boldsymbol{x}' \phi_s(\boldsymbol{x}') \} \boldsymbol{\beta}^{(s)}. \tag{8.41}$$

If missingness does not depend on covariates, then $p(s \mid \boldsymbol{x}) = p(s)$, which implies $\partial \phi_s(\boldsymbol{x}) / \partial \boldsymbol{x} = \boldsymbol{0}$ and $\phi(\boldsymbol{x}) - \phi(\boldsymbol{x}') = 0$ (for the discrete case). Hence (8.40) simplifies to $\sum_s \phi_s \boldsymbol{\beta}^{(s)}$ and (8.41) simplifies to $(\boldsymbol{x} - \boldsymbol{x}') \sum_s \phi_s \boldsymbol{\beta}^{(s)}$; each is a weighted average of pattern-specific regression coefficients. $\square$

PMM with nonlinear link functions within pattern

For PMM that are specified as mixtures of regression models, evaluating covariate effects is somewhat more complicated if nonlinear link functions are used (e.g., logistic or log). As with the previous examples, we specify regressions within patterns defined by follow-up time $S = \sum_j R_j$. Let $\mu_j^{(s)}(\boldsymbol{x}) = E(Y_j \mid \boldsymbol{X}_j = \boldsymbol{x}, S = s)$ denote the within-pattern mean, and assume

$$g\{\mu_j^{(s)}(\boldsymbol{x})\} = \boldsymbol{x} \boldsymbol{\beta}^{(s)},$$

where $g : \mathbb{R} \to \mathbb{R}$ is a smooth monotone link function.

In general the effect of $\boldsymbol{X}$ on the full-data mean of Y_j is

$$\begin{aligned} \frac{\partial \mu_j(\boldsymbol{x})}{\partial \boldsymbol{x}} &= \frac{\partial}{\partial \boldsymbol{x}} \sum_s \left\{ \mu_j^{(s)}(\boldsymbol{x}) \, \phi_s(\boldsymbol{x}) \right\} \\ &= \sum_s \left[\left\{ \frac{\partial}{\partial \boldsymbol{x}} g^{-1}(\boldsymbol{x} \boldsymbol{\beta}^{(s)}) \right\} \phi_s(\boldsymbol{x}) + g^{-1}(\boldsymbol{x} \boldsymbol{\beta}^{(s)}) \left\{ \frac{\partial}{\partial \boldsymbol{x}} \phi_s(\boldsymbol{x}) \right\} \right]. \end{aligned}$$

If dropout does not depend on covariates, then $\phi_s(\boldsymbol{x}) = \phi_s$ and

$$\frac{\partial \mu_j(\boldsymbol{x})}{\partial \boldsymbol{x}} = \sum_s \left\{ \frac{\partial}{\partial \boldsymbol{x}} g^{-1}(\boldsymbol{x} \boldsymbol{\beta}^{(s)}) \right\} \phi_s(\boldsymbol{x}). \tag{8.42}$$

Example 8.11. *Covariate effects in mixture of loglinear regression models.* Following the setup from above, if a log link is used within pattern, we have

$$g^{-1}(\boldsymbol{x} \boldsymbol{\beta}^{(s)}) = \exp(\boldsymbol{x} \boldsymbol{\beta}^{(s)})$$

and

$$\frac{\partial}{\partial \boldsymbol{x}} g^{-1}(\boldsymbol{x} \boldsymbol{\beta}^{(s)}) = \boldsymbol{\beta}^{(s)} \exp(\boldsymbol{x} \boldsymbol{\beta}^{(s)}).$$

Now assume $\phi_s(\boldsymbol{x}) = \phi_s$, i.e., dropout does not depend on covariates. For the mixture of loglinear models, (8.42) becomes

$$\frac{\partial \mu_j(\boldsymbol{x})}{\partial \boldsymbol{x}} \;=\; \sum_s \boldsymbol{\beta}^{(s)} \, \exp(\boldsymbol{x}\boldsymbol{\beta}^{(s)}) \, \phi_s.$$

Hence, even if dropout is independent of $\boldsymbol{x}$, the covariate effect still depends on $\boldsymbol{x}$ when the link function within pattern is nonlinear.

In fact, the effect of $\boldsymbol{x}$ on the full-data mean of Y_j is a weighted average of within-pattern regression coefficients $\boldsymbol{\beta}^{(s)}$, with weights that depend on $\boldsymbol{x}$. Inspection of (8.42) shows this will be true in general for nonlinear g. $\square$

In summary, important considerations in the computation of covariate effects from mixture models include (a) whether the mean is linear in covariates; (b) whether missingness depends on covariates, and (c) whether the covariate effects are time varying. Our focus here has been on (a) and (b), illustrating computations for settings where the covariate effects are time constant. When the link function is nonlinear, it can be difficult to capture the covariate effect succintly. To improve interpretability of covariate effects, Wilkins and Fitzmaurice (2006) and Roy and Daniels (2007) have introduced marginalized models, imposing constraints on the marginal mean similar to those used for marginalized transition models.

8.5 Shared parameter models

8.5.1 General structure

Shared parameter models were introduced in Section 5.9.3. In the most general case, a SPM takes the form

$$p(\boldsymbol{y}, \boldsymbol{r}, \boldsymbol{\omega}) = \int p(\boldsymbol{y}, \boldsymbol{r} \mid \boldsymbol{b}, \boldsymbol{\omega}) \, p(\boldsymbol{b} \mid \boldsymbol{\omega}) \, d\boldsymbol{b},$$

where $\boldsymbol{b}$ are subject-specific random effects. The chief characteristic of these models is that a single set of 'shared parameters' — usually random effects — applies to the joint distribution of $\boldsymbol{Y}$ and $\boldsymbol{R}$. In many cases it is assumed that $\boldsymbol{Y}$ and $\boldsymbol{R}$ are independent conditionally on $\boldsymbol{b}$, though this is not a requirement. Theoretically, all SPM can be represented either as a mixture model or selection model, but the functional form of the component distributions is typically not tractable.

In this example, we describe an SPM that can be used for a longitudinal response process with continuous time dropout. It uses a standard random effects model formulation for the full-data response distribution and a proportional hazards model for the dropout time.

Example 8.12. *Shared parameter model for normally distributed full-data response with continuous time dropout.*
Henderson et al. (2000) describe a general structure for the SPM whereby the

full-data response is characterized by a Gaussian process having between- and within-subject variation. Missingness is induced by dropout, which depends through a second model on individual-specific random effects characterizing between-subject variation of the responses.

The Henderson et al. specification assumes a continuous-time process $Y(t)$ that can be right-censored by dropout at time U; to make the notation agree with our conventions, let $Y_j = Y(t_j)$. Then their model is written as

$$
\begin{aligned}
Y_j \mid \boldsymbol{x}_j, Z_j &\sim N(\boldsymbol{x}_j\boldsymbol{\beta} + Z_j,\ \sigma^2) \\
h_U(t_j \mid Z_j) &= h_0(t_j)\exp(Z_j\gamma),
\end{aligned}
\tag{8.43}
$$

where Z_j is a realization at t_j of a *latent* Gaussian process $Z(t)$ that characterizes between-subject variation, and within-subject variation in the response model is characterized by a stationary Gaussian process having constant variance σ^2. The function h_U is the hazard of dropout, and h_0 is a baseline hazard. In this formulation, Z_j is viewed as the 'shared parameter'. The scalar parameter γ links the longitudinal process $Y(t)$ to the hazard of dropout. When $\gamma = 0$ we have MAR, and otherwise we have MNAR.

Contextually, this model can be motivated by the need to deal with responses measured with error, whereby $\boldsymbol{x}_j\boldsymbol{\beta} + Z_j$ is the error-free version of Y_j, with measurement error captured by a residual process.

Connections to the SPM formulations given in Section 5.9.3 can be seen by considering the simple case where $\boldsymbol{x}_j = (1, t_j)$ and $Z_j = \boldsymbol{x}_j\boldsymbol{b}$, where $\boldsymbol{b} = (b_0, b_1)^{\mathrm{T}} \sim N(\boldsymbol{0}, \boldsymbol{\Omega})$ are subject specific random effects for intercept and slope. Here the underlying error-free process follows a straight line over time. Then the model (8.43) is written as follows, using subject indices i to emphasize sources of variation:

$$
\begin{aligned}
Y_{ij} \mid \boldsymbol{x}_{ij}, \boldsymbol{b}_i &\sim N(\boldsymbol{x}_{ij}\boldsymbol{\beta} + \boldsymbol{x}_{ij}\boldsymbol{b}_i,\ \sigma^2) \\
h_U(t_{ij} \mid b_i) &= h_0(t_{ij})\exp\left\{(\boldsymbol{x}_{ij}\boldsymbol{b}_i)\gamma\right\}.
\end{aligned}
$$

The full-data likelihood for this model is proportional to

$$
p(\boldsymbol{y}, u \mid \boldsymbol{x}, \boldsymbol{\beta}, \boldsymbol{\Omega}, \sigma, \gamma) = \int p(\boldsymbol{y} \mid \boldsymbol{x}, \boldsymbol{b}, \boldsymbol{\beta}, \boldsymbol{\Omega}, \sigma)\, p(u \mid \boldsymbol{x}, \boldsymbol{b}, \gamma)\, p(\boldsymbol{b} \mid \boldsymbol{\Omega})\, d\boldsymbol{b},
$$

demonstrating that it is a shared parameter model with $\boldsymbol{Y} \perp\!\!\!\perp U \mid (\boldsymbol{b}, \boldsymbol{X})$. □

8.5.2 Pros and cons of shared parameter models

Shared parameter models are very effective for decomposing the variance for multivariate processes, and work in this area has been very effective at facilitating joint modeling of repeated measures and event times. Most commonly, shared parameter models are used either to (a) make adjustments for selection bias, when the main objective is drawing inference about the full-data distribution of repeated measures, or (b) model the hazard of an event time as a function of stochastic time-varying covariates.

With respect to handling dropout, these models can be effective for complex data structures (e.g., multivariate longitudinal responses, situations where observations are taken very frequently across time) or when the main outcome of interest can be conceptualized as a latent variable (for example severity of disease as measured by several indicators). In the latter case, the mechanism relating the full-data distribution of interest (the latent variable) is explicit.

A disadvantage of shared parameter models is that the functional dependence between full-data responses Y and dropout time U is not usually explicit; to obtain $p(u \mid y)$ or $p(y \mid u)$, the latent variables must be integrated out of the full-data model. Consequently, the missing data mechanism is not always transparent.

Another by-product of assuming a common latent structure for both the response and dropout distributions is that the hazard of dropout in a shared parameter model will generally depend on future observations, even after conditioning on past and current observations. To illustrate, consider the simple shared parameter model where the full-data response is $Y = (Y_1, Y_2, Y_3)^{\mathrm{T}}$, Y_1 is always observed, and hazard of dropout at times 2 and 3 depends on a common random effect. This model can be specified as

$$
\begin{aligned}
Y_j \mid b &\sim N(b, \sigma^2) &(j = 1, 2, 3) \\
b &\sim N(0, \tau^2) \\
R_j \mid R_{j-1} = 1, b &\sim \mathrm{Ber}(\ \Phi(\gamma b)\ ) &(j = 2, 3),
\end{aligned}
$$

where $\Phi(\cdot)$ is the cdf of a standard normal distribution and γ is a scalar parameter. The hazard of dropout at time 2 as a function of y is

$$
\begin{aligned}
P(R_2 = 1 \mid y) &= \frac{\int P(R_2 = 1 \mid b)\, p(y \mid b)\, p(b)\, db}{p(y)} \\
&= \frac{\int P(R_2 = 1 \mid b)\, p(b \mid y)\, p(y)\, db}{p(y)} \\
&= \int \Phi(\gamma b)\, p(b \mid y)\, db \\
&= E_{b \mid Y}\{\Phi(\gamma b) \mid y\}.
\end{aligned}
\tag{8.44}
$$

Because $\sum_j y_j$ is a sufficient statistic for b in $p(b \mid y)$, the expectation (8.44) depends on $\sum_j y_j$ and therefore on y_3. The exception is when $\gamma = 0$ (MCAR), in which case $\Phi(\gamma b) = \Phi(0) = \frac{1}{2}$ is a constant.

The latent variable structure also can make it difficult to separate parameters indexing $p(y_{\mathrm{mis}} \mid y_{\mathrm{obs}}, u)$ from those indexing $p(y_{\mathrm{obs}}, u)$, and it is therefore hard to embed an MAR specification in a larger class of models for assessing sensitivity to departures from MAR. Finally, shared parameter models frequently rely for identification on distributional assumptions about the latent variable b, which governs both observed and missing data. These assumptions are frequently motivated by convenience rather than by context.

To summarize, shared parameter models are very useful for characterizing joint distributions of repeated measures and event times, and can be particularly useful as a method of data reduction when the dimension of Y is high. Nonetheless, their application to the problem of making full-data inference from incomplete longitudinal data should be made with caution and with an eye toward justifying the required assumptions. Sensitivity analysis is an open area of research for these models.

8.6 Model selection and model fit in nonignorable models

Unlike with the ignorable models described in Chapter 6, model comparison and assessment of fit for nonignorable models must consider the missing data mechanism $p(r \mid y, \psi(\omega))$ itself. Model checking and model selection are therefore based on the full-data model $p(y, r \mid \omega)$, and not just the full-data response model $p(y \mid \theta)$.

Likelihood-based criteria, like the DIC, will now be based on the fit of the full-data model to the observed data, (y_{obs}, r). Poor fit indicates the full-data modeling assumptions are *not* consistent with the observed data.

None of the metrics and checks for comparing and assessing fit of models provide information about the feasibility of the implicit or explicit assumptions about the missing data. They only provide information about how modeling assumptions, priors, and specific missing data assumptions as an ensemble fit the *observed* data (y_{obs}, r). Thus, the following criteria and checks can only be used to assess specific missing data assumptions within the context of a specific fully parametric model for the full data. We can therefore use these criteria to compare the fit of different parametric models within the same class (e.g., selection models) or between classes (e.g., a parametric selection model vs. a pattern mixture model).

8.6.1 Deviance information criterion (DIC)

For nonignorable dropout, the full-data likelihood is proportional to $p(y, r \mid \omega)$. In Chapter 6, we discussed two forms of the DIC: one based on the *observed data response* likelihood, DIC_O, and one based on the posterior predictive expectation of the *full-data response* likelihood, DIC_F. We develop corresponding criteria here based on the observed data likelihood and the full-data likelihood, respectively. DIC_O takes the form

$$\text{DIC}_O = -4E_{\omega}\{\ell(\omega \mid y_{\text{obs}}, r)\} + 2\ell(\overline{\omega} \mid y_{\text{obs}}, r),$$

where $\ell = \log L$ and $\overline{\omega} = E(\omega \mid \boldsymbol{y}_{\mathrm{obs}}, \boldsymbol{r})$. The observed data likelihood $L(\omega \mid \boldsymbol{y}_{\mathrm{obs}}, \boldsymbol{r})$ satisfies

$$L(\omega \mid \boldsymbol{y}_{\mathrm{obs}}, \boldsymbol{r}) \propto \int p(\boldsymbol{y}, \boldsymbol{r} \mid \omega) d\boldsymbol{y}_{\mathrm{mis}}, \qquad (8.45)$$

and is typically not available in closed form for selection models (SM) and shared parameter models (SPM).

When the observed data likelihood is available in closed form, the DIC can often be computed entirely in WinBUGS. We provide examples of this on the book webpage.

DIC_F is based on the full-data likelihood. We first construct the DIC based on the full-data model (assuming complete response data),

$$
\begin{aligned}
\mathrm{DIC}_{\mathrm{full}}(\boldsymbol{y}_{\mathrm{obs}}, \boldsymbol{y}_{\mathrm{mis}}, \boldsymbol{r}) \;=\;& -4E_{\omega}\left\{\ell(\omega \mid \boldsymbol{y}_{\mathrm{obs}}, \boldsymbol{y}_{\mathrm{mis}}, \boldsymbol{r})\right\} \\
& + 2\ell\{\overline{\omega}(\boldsymbol{y}_{\mathrm{mis}}) \mid \boldsymbol{y}_{\mathrm{obs}}, \boldsymbol{y}_{\mathrm{mis}}, \boldsymbol{r}\},
\end{aligned}
$$

where $\overline{\omega}(\boldsymbol{y}_{\mathrm{mis}}) = E(\omega \mid \boldsymbol{y}_{\mathrm{obs}}, \boldsymbol{y}_{\mathrm{mis}}, \boldsymbol{r})$. To obtain DIC_F, we take the expectation of $\mathrm{DIC}_{\mathrm{full}}(\boldsymbol{y}_{\mathrm{obs}}, \boldsymbol{y}_{\mathrm{mis}}, \boldsymbol{r})$ with respect to the posterior predictive distribution $p(\boldsymbol{y}_{\mathrm{mis}} \mid \boldsymbol{y}_{\mathrm{obs}}, \boldsymbol{r})$,

$$
\begin{aligned}
\mathrm{DIC}_F \;=\;& E_{\boldsymbol{y}_{\mathrm{mis}}}\left\{\mathrm{DIC}_{\mathrm{full}}(\boldsymbol{y}_{\mathrm{obs}}, \boldsymbol{y}_{\mathrm{mis}}, \boldsymbol{r})\right\} \\
=\;& -4E_{\boldsymbol{y}_{\mathrm{mis}}}\left[E_{\omega}\{\ell(\omega \mid \boldsymbol{y}_{\mathrm{obs}}, \boldsymbol{y}_{\mathrm{mis}}, \boldsymbol{r})\}\right] \\
& + 2E_{\boldsymbol{y}_{\mathrm{mis}}}\left\{\ell(\overline{\omega}(\boldsymbol{y}_{\mathrm{mis}}) \mid \boldsymbol{y}_{\mathrm{obs}}, \boldsymbol{y}_{\mathrm{mis}}, \boldsymbol{r})\right\}. \qquad (8.46)
\end{aligned}
$$

Computing DIC_F requires the evaluation of $\overline{\omega}(\boldsymbol{y}_{\mathrm{mis}}) = E(\omega \mid \boldsymbol{y}_{\mathrm{obs}}, \boldsymbol{y}_{\mathrm{mis}}, \boldsymbol{r})$ (in the second term of (8.46)) at each value of $\boldsymbol{y}_{\mathrm{mis}}$ sampled during the data augmentation step of the sampling algorithm. We provide suggestions for computing $\overline{\omega}(\boldsymbol{y}_{\mathrm{mis}})$ in the setting of different nonignorable models, similar to the approaches proposed in Chapter 6 in the next section.

Once the expectations, $\overline{\omega}(\boldsymbol{y}_{\mathrm{mis}})$ are computed, DIC_F can be computed with little difficulty. The first term in (8.46) is the expectation of the full-data log likelihood with respect to $p(\boldsymbol{y}_{\mathrm{mis}}, \omega \mid \boldsymbol{y}_{\mathrm{obs}}, \boldsymbol{r})$. Except for SPMs, for which the full-data likelihood can often not be written in closed form, the first term can be estimated as

$$\frac{1}{M} \sum_{m=1}^{M} \ell(\omega^{(m)} \mid \boldsymbol{y}_{\mathrm{obs}}, \boldsymbol{y}_{\mathrm{mis}}^{(m)}, \boldsymbol{r}),$$

where $(\boldsymbol{y}_{\mathrm{mis}}^{(m)}, \omega^{(m)})$ are the draws from the data-augmented Gibbs sampling algorithm that was used to sample from the posterior distribution of ω; we can compute this term in WinBUGS. The second term, given $\overline{\omega}(\boldsymbol{y}_{\mathrm{mis}})$, can be computed using the same draws as the first term via

$$\frac{1}{M} \sum_{m=1}^{M} \ell(\overline{\omega}(\boldsymbol{y}_{\mathrm{mis}}^{(m)}) \mid \boldsymbol{y}_{\mathrm{obs}}, \boldsymbol{y}_{\mathrm{mis}}^{(m)}, \boldsymbol{r}).$$

Both DIC_F and DIC_O can be used to compare the fit of the full-data models to the observed data, $(\boldsymbol{y}_{\mathrm{obs}}, \boldsymbol{r})$. Next, we present more details on implementation and computation of both forms of the DIC for selection models, pattern mixture models, and shared parameter models.

Computation for selection models

Computation of DIC_O in selection models is complicated by the fact that the observed data log likelihood will typically *not* be available in closed form (see, e.g., Diggle and Kenward model in Example 8.1). However, it can, in general, be computed using a reweighting approach. Recall the form of the observed data likelihood given in (8.45) and assume distinct (and *a priori* independent) parameters for the missing data mechanism and the full-data response model. It can be shown that the observed data likelihood for a selection model factorization is

$$
\begin{aligned}
L(\boldsymbol{\omega} \mid \boldsymbol{y}_{\mathrm{obs}}, \boldsymbol{r}) \quad &\propto \quad \int p(\boldsymbol{r} \mid \boldsymbol{y}; \boldsymbol{\psi}) \, p(\boldsymbol{y} \mid \boldsymbol{\theta}) \, d\boldsymbol{y}_{\mathrm{mis}} \\
&= \quad p(\boldsymbol{y}_{\mathrm{obs}} \mid \boldsymbol{\theta}) \, E_{\boldsymbol{y}_{\mathrm{mis}} \mid \boldsymbol{y}_{\mathrm{obs}}, \boldsymbol{\theta}} \{ p(\boldsymbol{r} \mid \boldsymbol{y}_{\mathrm{mis}}, \boldsymbol{y}_{\mathrm{obs}}, \boldsymbol{\psi}) \}. \quad (8.47)
\end{aligned}
$$

The expectation in (8.47) can be computed using Monte Carlo integration by averaging draws from $p(\boldsymbol{y}_{\mathrm{mis}} \mid \boldsymbol{y}_{\mathrm{obs}}, \boldsymbol{\theta})$, the predictive distribution, conditional on $\boldsymbol{\theta}$, based on *only* the full-data response model. However, a separate Monte Carlo integration would need to be done for every value of $\boldsymbol{\theta}$ from the Gibbs sampler output of $p(\boldsymbol{\theta} \mid \boldsymbol{y}_{\mathrm{obs}}, \boldsymbol{r})$. As a more computationally practical alternative, we recommend the following approach. First, obtain a Monte Carlo sample of $\boldsymbol{y}_{\mathrm{mis}}$ for a likely value of $\boldsymbol{\theta}$, say $\boldsymbol{\theta}^{\star} = E(\boldsymbol{\theta} \mid \boldsymbol{y}_{\mathrm{obs}}, \boldsymbol{r})$; denote this sample as $\{\boldsymbol{y}_{\mathrm{mis}}^{(l)} : l = 1, \ldots, L\}$. Then, reweight (cf. Chapter 6) this sample to estimate the expectation of interest as a function of $\boldsymbol{\theta}$,

$$
E_{\boldsymbol{y}_{\mathrm{mis}} \mid \boldsymbol{y}_{\mathrm{obs}}, \boldsymbol{\theta}} \{ p(\boldsymbol{r} \mid \boldsymbol{y}_{\mathrm{mis}}, \boldsymbol{y}_{\mathrm{obs}}, \boldsymbol{\psi}) \} = \frac{\sum_{l=1}^{L} w_l p(\boldsymbol{r} \mid \boldsymbol{y}_{\mathrm{mis}}^{(l)}, \boldsymbol{y}_{\mathrm{obs}}, \boldsymbol{\psi})}{\sum_{l=1}^{L} w_l},
$$

with weights $w_l = w_l(\boldsymbol{\theta})$ given by

$$
w_l = \frac{p(\boldsymbol{y}_{\mathrm{mis}}^{(l)} \mid \boldsymbol{y}_{\mathrm{obs}}, \boldsymbol{\theta})}{p(\boldsymbol{y}_{\mathrm{mis}}^{(l)} \mid \boldsymbol{y}_{\mathrm{obs}}, \boldsymbol{\theta}^{\star})}.
$$

For DIC_F, we do not need to compute the observed data likelihood. The computational complexity for DIC_F, as discussed earlier, is the computation of $\overline{\boldsymbol{\omega}}(\boldsymbol{y}_{\mathrm{mis}}) = E(\boldsymbol{\omega} \mid \boldsymbol{y}_{\mathrm{obs}}, \boldsymbol{y}_{\mathrm{mis}}, \boldsymbol{r})$ for every value of $\boldsymbol{y}_{\mathrm{mis}}$ in the MCMC sample. We can do this using a similar reweighting strategy to that proposed in Chapter 6. First, run one additional Gibbs sampler with $\boldsymbol{y}_{\mathrm{mis}}$ fixed at a likely value, say $\boldsymbol{y}_{\mathrm{mis}}^{\star} = E(\boldsymbol{y}_{\mathrm{mis}} \mid \boldsymbol{y}_{\mathrm{obs}}, \boldsymbol{r})$; denote this sample as $\{\boldsymbol{\omega}^{(l)} : l = 1, \ldots, L\}$. Then, compute $\overline{\boldsymbol{\omega}}(\boldsymbol{y}_{\mathrm{mis}}^{(m)}) = E(\boldsymbol{\omega} \mid \boldsymbol{y}_{\mathrm{obs}}, \boldsymbol{y}_{\mathrm{mis}}^{(m)}, \boldsymbol{r})$ for each sampled value $\boldsymbol{y}_{\mathrm{mis}}^{(m)}$

from the original MCMC sample using

$$\overline{\boldsymbol{\omega}}(\boldsymbol{y}_{\mathrm{mis}}^{(m)}) = \frac{\sum_{l=1}^{L} w_l \boldsymbol{\omega}^{(l)}}{\sum_{l=1}^{L} w_l} \tag{8.48}$$

with weights w_l given by

$$w_l = \frac{p(\boldsymbol{\theta}^{(l)}, \boldsymbol{\psi}^{(l)}, \boldsymbol{y}_{\mathrm{mis}}^{(m)}, \boldsymbol{y}_{\mathrm{obs}}, \boldsymbol{r})}{p(\boldsymbol{\theta}^{(l)}, \boldsymbol{\psi}^{(l)}, \boldsymbol{y}_{\mathrm{mis}}^{\star}, \boldsymbol{y}_{\mathrm{obs}}, \boldsymbol{r})}. \tag{8.49}$$

Expressions of the form $p(\boldsymbol{\theta}^{(l)}, \boldsymbol{\psi}^{(l)}, \boldsymbol{y}_{\mathrm{mis}}, \boldsymbol{y}_{\mathrm{obs}}, \boldsymbol{r})$ in w_l can be replaced by $p(\boldsymbol{y} \mid \boldsymbol{\theta}^{(l)}) \, p(\boldsymbol{r} \mid \boldsymbol{y}, \boldsymbol{\psi}^{(l)})$ because $p(\boldsymbol{\theta}^{(l)}, \boldsymbol{\psi}^{(l)})$ cancel out. Recall here that m indexes the original MCMC sample.

Computation for pattern mixture models

In a pattern mixture formulation, the full-data likelihood is factored as $p(\boldsymbol{y} \mid \boldsymbol{r}, \boldsymbol{\alpha}) p(\boldsymbol{r} \mid \boldsymbol{\phi})$. So the observed data likelihood, which needs to be computed for DIC_O, is proportional to

$$p(\boldsymbol{r} \mid \boldsymbol{\phi}) \int p(\boldsymbol{y} \mid \boldsymbol{r}, \boldsymbol{\alpha}) d\boldsymbol{y}_{\mathrm{mis}}, \tag{8.50}$$

which can be computed in closed form here as long as the pattern-specific distributions $p(\boldsymbol{y} \mid \boldsymbol{r}, \boldsymbol{\alpha})$ offer a closed form for $\int p(\boldsymbol{y} \mid \boldsymbol{r}, \boldsymbol{\alpha}) d\boldsymbol{y}_{\mathrm{mis}}$. The mixture models proposed in Examples 8.4, 8.5, and 8.8 all offer a closed form for (8.50). In such cases, DIC_O can be computed in WinBUGS.

For DIC_F, we need to compute

$$\overline{\boldsymbol{\omega}}(\boldsymbol{y}_{\mathrm{mis}}) = E(\boldsymbol{\alpha}, \boldsymbol{\phi} \mid \boldsymbol{y}_{\mathrm{mis}}, \boldsymbol{y}_{\mathrm{obs}}, \boldsymbol{r})$$

for all values of $\boldsymbol{y}_{\mathrm{mis}}$ in the data-augmented Gibbs sampler output for the posterior, $\{\boldsymbol{\omega}^{(m)}, \boldsymbol{y}_{\mathrm{mis}}^{(m)} : m = 1, \ldots, M\}$, where $\boldsymbol{\omega}^{(m)} = (\boldsymbol{\alpha}^{(m)}, \boldsymbol{\phi}^{(m)})$. One approach, similar to that suggested for selection models, is the following: First, run one additional Gibbs sampler with $\boldsymbol{y}_{\mathrm{mis}}$ fixed at (a likely value), say $\boldsymbol{y}_{\mathrm{mis}}^{\star} = E(\boldsymbol{y}_{\mathrm{mis}} \mid \boldsymbol{y}_{\mathrm{obs}}, \boldsymbol{r})$; denote this sample as $\{\boldsymbol{\omega}^{(l)} : l = 1, \ldots, L\}$. Note that this sample is drawn from the full-data posterior where we have filled in $\boldsymbol{y}_{\mathrm{mis}}$ with $\boldsymbol{y}_{\mathrm{mis}}^{\star}$. Then, to compute $\overline{\boldsymbol{\omega}}(\boldsymbol{y}_{\mathrm{mis}}^{(m)})$ for the values $\{\boldsymbol{y}_{\mathrm{mis}}^{(m)} : m = 1, \ldots, M\}$ from the original data-augmented Gibbs sampler run, we can use (8.48) with weights

$$w_l = \frac{p(\boldsymbol{\alpha}^{(l)}, \boldsymbol{\phi}^{(l)}, \boldsymbol{y}_{\mathrm{mis}}^{(m)}, \boldsymbol{y}_{\mathrm{obs}}, \boldsymbol{r})}{p(\boldsymbol{\alpha}^{(l)}, \boldsymbol{\phi}^{(l)}, \boldsymbol{y}_{\mathrm{mis}}^{\star}, \boldsymbol{y}_{\mathrm{obs}}, \boldsymbol{r})}. \tag{8.51}$$

As with (8.49), we can make the substitution

$$p(\boldsymbol{\alpha}^{(l)}, \boldsymbol{\phi}^{(l)}, \boldsymbol{y}_{\mathrm{mis}}, \boldsymbol{y}_{\mathrm{obs}}, \boldsymbol{r}) = p(\boldsymbol{y} \mid \boldsymbol{\alpha}^{(l)}, \boldsymbol{r}) \, p(\boldsymbol{r} \mid \boldsymbol{\phi}^{(l)}).$$

Computation for shared parameter models

Shared parameter models pose an additional problem for computation of both forms of the DIC. The joint distribution $p(\boldsymbol{y}, \boldsymbol{r} \mid \boldsymbol{\omega})$,

$$p(\boldsymbol{y}, \boldsymbol{r} \mid \boldsymbol{\omega}) = \int p(\boldsymbol{y}, \boldsymbol{r} \mid \boldsymbol{b}, \boldsymbol{\omega}) \, p(\boldsymbol{b} \mid \boldsymbol{\omega}) \, d\boldsymbol{b},$$

typically cannot be computed in closed form (i.e., cannot analytically integrate out the shared parameters). This joint distribution is required to compute both forms of the DIC. To evaluate $p(\boldsymbol{y}, \boldsymbol{r} \mid \boldsymbol{\omega})$, a Monte Carlo integration could be done at each iteration to compute the integrated (over $\boldsymbol{b}$) likelihood. However, this would be extremely inefficient. A more computationally feasible approach might be to use Laplace approximations (Tierney and Kadane, 1986) to approximate the likelihood integrated over $\boldsymbol{b}$.

Strategies for computing DIC_O and DIC_F are similar to those for selection models and pattern mixture models but with the additional complication that the full-data model itself is typically not available in closed form. Neither of these criteria can be computed in WinBUGS.

However, to compare the fit among different shared parameter models, we can avoid the integration over the shared parameters $\boldsymbol{b}$ by using the likelihood conditional on the shared parameters, which is proportional to $p(\boldsymbol{y}, \boldsymbol{r} \mid \boldsymbol{b}, \boldsymbol{\omega})$ (cf. discussion of DIC with random effects models in Chapter 3). The DIC_O based on this likelihood can be computed in WinBUGS.

Summary

As discussed in Chapter 6, the relative merits and operating characteristics of these two forms of the DIC with incomplete data need careful study, including the computational efficiency of the reweighting-based approaches. For some models, the DIC can be computed completely in WinBUGS. For others, parts of the DIC need to computed using simple user-written functions in a package like R. Comparison of nonignorable models (which require specification of the missing data mechanism) to ignorable models (which do not) is an area that needs further study.

8.6.2 *Posterior predictive checks*

In Chapters 3 and 6 we recommended posterior predictive checks to assess the fit of the specific aspects of a parametric model to $\boldsymbol{y}_{\text{obs}}$. In the setting of nonignorable missingness, we typically want to assess the fit of the full-data model to features of the observed data $(\boldsymbol{y}_{\text{obs}}, \boldsymbol{r})$. We therefore sample from the posterior predictive distribution

$$p(\boldsymbol{y}_{\text{rep}}, \boldsymbol{r}_{\text{rep}} \mid \boldsymbol{y}_{\text{obs}}, \boldsymbol{r}).$$

Given that we now model the missing data mechanism, we can compute data summaries based on replicates of observed data, not just replicates of complete data. For individual i, we define replicates of the observed data as

$$\boldsymbol{y}_{i,\text{obs,rep}} \quad = \quad \{y_{ij,\text{rep}} : r_{ij,\text{rep}} = 1\},$$

i.e., the components $y_{ij,\text{rep}}$ of the replicated complete datasets for which the corresponding missing data indicator $r_{ij,\text{rep}}$ is equal to 1. These replicated 'observed' datasets do not have a one-to-one correspondence with the observed data. We further clarify with a simple example.

Consider complete data (Y_1, Y_2, R), where $R = 0$ corresponds to Y_2 being missing. When we sample from the posterior predictive distribution, $\boldsymbol{r}_{\text{rep}} \neq \boldsymbol{r}$. As a result, we will have components $r_{ij,\text{rep}}$ of $\boldsymbol{r}_{\text{rep}}$ that take a value of one when the corresponding component r_{ij} of $\boldsymbol{r}$ is zero. So, even though we have an 'observed' $y_{ij,\text{rep}}$, there is not necessarily a corresponding $y_{ij,\text{obs}}$ with which to compare it.

There are two solutions to this problem. The first would be to do the predictive checks based on replicates of the complete data, as in Chapter 6. The second solution would be to construct statistics that attach indicators $I\{r_{ij} = r_{ij,\text{rep}} = 1\}$ to the pairs y_{ij} and $y_{ij,\text{rep}}$; that is, at each iteration, compute the checks using only those components of $\boldsymbol{y}_{\text{obs}}$ and $\boldsymbol{y}_{\text{rep}}$ where $r_{ij,\text{rep}} = 1$.

The power of these two approaches to detect model departures needs further study. Both approaches will result in less power than if we actually had complete data. The first approach loses power by 'filling in' the missing data based on the model, thus biasing the checks in favor of the model. The second approach loses power by not using all the observed data. Under ignorable dropout, we advocated the first approach based on completed datasets. One of the reasons for this was that under ignorability, the missing data mechanism $p(\boldsymbol{r} \mid \boldsymbol{y}, \psi)$ is not specified. However, in nonignorable models, we specify the missing data mechanism so this reason is no longer valid and the second approach can be implemented.

Checks of the full-data response model will be similar to those discussed in Chapter 6, where we assessed the fit of aspects of the full-data response model under ignorable dropout. Although our primary interest is typically the fit of the full-data model to the observed responses, $\boldsymbol{y}_{\text{obs}}$, we should also check the fit of the missing data mechanism. To assess the fit, we can compute statistics that measure the agreement between the observed missing data indicators $\boldsymbol{r}$ and the replicates $\boldsymbol{r}_{\text{rep}}$. Because our focus is on monotone missingness (dropout), we might consider a statistic such as

$$h\{T(\boldsymbol{r}_i), T(\boldsymbol{r}_{i,\text{rep}})\} = \sum_j r_{ij} - \sum_j r_{ij,\text{rep}}.$$

If the missing data mechanism fits well, we would expect this distribution to have a median around zero.

8.7 Further reading

Marginalized mixture models for marginal covariate effects

Wilkins and Fitzmaurice (2006) developed what they termed a 'hybrid model' for nonignorable longitudinal binary data. Specifically, they model the joint distribution of (y, r) using a canonical log-linear model (Fitzmaurice and Laird, 1993) that allows for direct specification of the marginal means for the longitudinal binary data and for the dropout indicators. However, within this setup, it can be hard to specify sensible parsimonious longitudinal dependence. Roy and Daniels (2007) extended earlier work by Roy (2003) on grouping dropout times into patterns using a latent class approach in the framework of mixture models. Using ideas from Heagerty (1999), they construct the model to allow direct specification of the marginal means (marginalized over the random effects and the dropout distribution).

Semiparametric selection models

For continuous longitudinal responses, nonparametric specifications for the full-data response model are typically not practical. To offer some flexibility in the context of random effects models, the random effects distributions can be specified nonparametrically using mixtures of Dirichlet process models (Kleinman and Ibrahim, 1998b) as described in Chapter 3.

Flexible modeling of the missing data mechanism

Work on more flexible estimation of the 'form' of the missing data mechanism has been undertaken by several authors. Lee and Berger (2001) have proposed Bayesian nonparametric estimation of the missing data mechanism in the simple setting of a univariate y and no covariates using Dirichlet process priors. Scharfstein and Irizarry (2003) entered baseline covariates x into the missing data mechanism nonparametrically (using generalized additive models). However, full extension of these ideas to the longitudinal setting with covariates is nontrivial.

Model Checking

An alternative approach to posterior predictive checks for model checking is to examine residuals for the observed data standardized with means and variances conditional on being observed. See Dobson and Henderson (2003) for a discussion.

CHAPTER 9

Informative Priors
and Sensitivity Analysis

9.1 Overview

Exploring sensitivity to unverifiable missing data assumptions, characterizing the uncertainty about these assumptions, and incorporating subjective beliefs about the distribution of missing responses, are key elements in any analysis of incomplete data. The Bayesian paradigm provides a natural venue for accomplishing these aims through the construction and integration of prior information and beliefs into the analysis.

9.1.1 General approach

Recall that the extrapolation factorization of the full-data model is

$$p(\boldsymbol{y}_{\mathrm{mis}}, \boldsymbol{y}_{\mathrm{obs}}, \boldsymbol{r} \mid \boldsymbol{\omega}) \;=\; p(\boldsymbol{y}_{\mathrm{mis}} \mid \boldsymbol{y}_{\mathrm{obs}}, \boldsymbol{r}, \boldsymbol{\omega}_{\mathrm{E}}) \, p(\boldsymbol{y}_{\mathrm{obs}}, \boldsymbol{r} \mid \boldsymbol{\omega}_{\mathrm{O}}), \quad (9.1)$$

where $\boldsymbol{\omega}_{\mathrm{E}}$ indexes the conditional distribution of missing responses, given observed data (the extrapolation distribution), and $\boldsymbol{\omega}_{\mathrm{O}}$ indexes the distribution of observables. Our use of the term *sensitivity analysis* refers to *assessment of sensitivity of model-based inferences to assumptions that cannot be verified or checked with data*. In the missing data setting, inference about the full-data distribution requires unverifiable assumptions about the extrapolation distribution $p(\boldsymbol{y}_{\mathrm{mis}} \mid \boldsymbol{y}_{\mathrm{obs}}, \boldsymbol{r}, \boldsymbol{\omega}_{\mathrm{E}})$.

Without assumptions such as a parametric model for the full-data response, or constraints such as MAR for the missing data mechanism, the observed data provide no information about the extrapolation distribution.

The general strategy espoused here is to work with the subset of sensitivity parameters $\boldsymbol{\xi}_{\mathrm{S}}$ that satisfy the conditions laid out in Definition 8.1. The sensitivity parameters are then used to encode prior beliefs about the missing data mechanisms, either by fixing their values at some constant, examining inferences across a range of constants, or by assigning an appropriate prior distribution.

The key idea is to separate what is identified from what is not, and to use prior distributions — possibly including point mass priors — to inform unidentified parts of the full-data model. Hence the priors are viewed as an integral part of the full-data model. In that sense we are responding to the observations and recommendations of Diggle (1999) that

> ... *scientists are inveterate optimists. They ask questions of their data that their data cannot answer. Statisticians can either refuse to answer such questions, or they can explain what is needed over and above the data to yield an answer and be properly cautious in reporting their results.*

An important objective therefore is to make the priors understandable and transparent to consumers of the model-based inferences, and to make them amenable to incorporation of external information (e.g., from historical data or elicited from experts). Indeed, Scharfstein et al. (1999) observe that

> ... *the biggest challenge in conducting sensitivity analysis is the choice of one or more sensibly parameterized functions whose interpretation can be communicated to subject matter experts with sufficient clarity ... .*

This chapter outlines the strategy for model parameterization and prior specification for both mixture and selection models. Generally speaking, parameterizations for sensitivity analysis tend to be more straightforward with mixture models; however, semiparametric selection models can be used for this purpose as well. We illustrate using several examples of each type of model.

9.1.2 Global vs. local sensitivity analysis

Although we are discussing ideas that are more general than sensitivity analysis *per se*, it is important to distinguish between *local* and *global* sensitivity analysis. Our concern is with the latter, but it is helpful to distinguish the two with a brief comparison.

One type of local sensitivity analysis calls for parameters indexing departures from MAR to be varied in a neighborhood of MAR. In some circumstances, as discussed in Section 8.3, the parameters being varied are actually identified by the observed data. This is especially true for models where the full-data response $p(\boldsymbol{y} \mid \boldsymbol{\omega})$ is assumed to follow a parametric model, and the selection mechanism $p(\boldsymbol{r} \mid \boldsymbol{y}, \boldsymbol{\omega})$ has an assumed functional form.

To illustrate concretely, consider again the bivariate setting where Y_2 may be missing, and suppose

$$(Y_1, Y_2)^{\mathrm{T}} \sim N(\boldsymbol{\mu}, \boldsymbol{\Sigma})$$
$$R \mid Y_1, Y_2 \sim \mathrm{Ber}(\pi),$$

with

$$\mathrm{logit}(\pi) = \psi_0 + \psi_1 y_1 + \psi_2 y_2. \tag{9.2}$$

MAR holds when $\psi_2 = 0$; however, as discussed in Chapter 8, the parameter ψ_2 is identified, even though Y_2 is missing for some individuals.

For models that can be identified under MNAR by virtue of parametric assumptions, imposing the MAR constraint (e.g., setting $\psi_2 = 0$ above) may yield substantially different fit to observed data when compared to the fit under MNAR. Equivalently, the observed data likelihood differs between the MAR and MNAR specifications. Empirical illustrations can be found in Diggle and Kenward (1994) and Baker et al. (2003). This phenomenon motivates a *local* sensitivity approach for assessing sensitivity of inferences in a neighborhood of MAR (Troxel et al., 2004; Zhang and Heitjan, 2006). Referring to (9.2), the approach is to examine changes in inference about a full-data parameter such as μ as a function of ψ_2 in a neighborhood of $\psi_2 = 0$. Nandram and Choi (2002a,b) take a similar approach, using priors on identified selection parameters such as ψ_2.

Another type of local sensitivity examines the influence of individual data points on model-based inference (Verbeke et al., 2001), using methods similar to influence analysis in regression.

By contrast, *global sensitivity analysis* enables the analyst to examine inferences about the full data over a class of full-data models that (a) are indexed by one or more nonidentifiable parameters and (b) have identical or very similar observed-data likelihoods.

Two approaches can be identified. The first is to begin with a model for the observed data, and expand it to admit one or more missing data mechanisms that are consistent with the observed data distribution. A very general approach, based on model expansion in terms of the missing data mechanism, is developed in Robins (1997) and comprehensively described in van der Laan and Robins (2003). Application to categorical data models can be found in Vansteelandt et al. (2006), and application to longitudinal responses appears in Scharfstein et al. (1999). For more general discussion of model expansion and model uncertainty from a Bayesian viewpoint, see Draper (1995) and Gustafson (2006).

A second but closely related approach to global sensitivity analysis begins with specification of a full-data distribution, followed by examination of inferences across a range of values for one or more unidentified parameters. In the Bayesian setup, priors would be placed on the unidentified parameters. This approach can be traced at least to Rubin (1977), who assumes a full-data distribution that is a mixture of normal distributions over respondents and nonrespondents, and uses covariate information to partially inform the distribution of nonrespondents. Assumptions about the differences between respondents and nonrespondents are expressed as prior distributions.

Other approaches based on specification of a full-data model with non-identified parameters embedded have been developed mainly for categorical data settings, as in Forster and Smith (1998). The work by Nandram and Choi (2002a,b) is similar in spirit, but the 'sensitivity parameters' are identifiable. Copas and Eguchi (2005) give a general treatment that extends to problems such as unmeasured confounding, but does not restrict attention to using nonidentified parameters for sensitivity analysis.

Our development in this chapter relies mainly the second approach described above. Inference is based on a specific model for the full-data distribution, where one or more of its parameters are sensitivity parameters in the sense of Definition 8.1. The sensitivity parameters are informed by (possibly point mass) prior distributions that characterize assumptions about the missing data mechanism or the missing data itself. Our focus is on how to specify and parameterize models so that are amenable either to (a) global sensitivity analysis or (b) incorporation of informative prior information that, essentially, fully characterizes the posterior distribution of the sensitivity parameters.

For mixture models, the material described here builds on Rubin (1977) whose work examined informative priors for full-data models of survey data. For selection models, we build on the ideas in Scharfstein et al. (1999) and Scharfstein et al. (2003), who develop methods for semiparametric inference about full-data parameters using incomplete (longitudinal) data. For categorical responses, there is little qualitative difference between taking a selection model or mixture model approach other than the interpretation of the sensitivity parameters that they admit. For continuous responses, selection models present a number of technical challenges for applied settings.

9.2 Some principles

In what follows, we articulate several principles for parameterizing models for sensitivity analysis and incorporation of informative prior distributions. These principles will be illustrated for both mixture and selection models in the examples below.

1. Parameterize the model in terms of a sensitivity parameter ξ_S that satisfies Definition 8.1. Consequently,

   (a) assumptions about the missing data mechanism are fully encoded by (possibly point mass) prior distributions;

   (b) one can move smoothly through the space of full-data models with MNAR missingness by varying the value of the sensitivity parameters;

   (c) changing the value of the sensitivity parameter does not affect the observed data likelihood, and hence does not affect fit to the observed data

2. Use an appropriately constructed prior distribution on the sensitivity parameters to reflect certainty or uncertainty about the missing data assumptions and other nonidentifiable aspects of the model.

3. Anchor or 'center' full-data models at MAR, such that the MAR assumption coincides with a fixed point on the range of the sensitivity parameters ξ_S (e.g., $\xi_S = 0$). The effect of the missing data assumptions can then be viewed in terms of departures from MAR.

The prior for $\boldsymbol{\xi}_S$ may be informed by a credible source of external information, such as expert opinion, historical data, or some combination of the two. The Bayesian setup is ideal for this because we can, at least in principle, quantify uncertainty about assumptions through these priors.

A common practice for assessing sensitivity to missing data assumptions is to compare inferences about a parameter of interest under several different full-data models that are not compatible in their fit to the observed data. For example, one might compare inferences under a selection model, a pattern mixture model, and a shared parameter model. It is also common to compare nested models that have different assumptions about the missing data mechanism, but also have different observed data likelihoods (for example by fitting model (9.2) and comparing it to the nested model with $\psi_2 = 0$). By contrast, principles 1.(b) and 1.(c) emphasize that model parameterizations should allow for exploration of sensitivity to missing data assumptions along a continuum of model specifications that have similar or identical fits to the observed data.

9.3 Parameterizing the full-data model

Recall that any full-data model has an associated extrapolation factorization (9.1). The component $p(\boldsymbol{y}_{\mathrm{mis}} \mid \boldsymbol{y}_{\mathrm{obs}}, \boldsymbol{r}, \boldsymbol{\omega}_{\mathrm{E}})$ is the extrapolation distribution, or the distribution of the missing data given the observed data; without parametric assumptions, it cannot be identified by observed data. The component $p(\boldsymbol{y}_{\mathrm{obs}}, \boldsymbol{r} \mid \boldsymbol{\omega}_{\mathrm{O}})$ is the observed data distribution and is proportional in $\boldsymbol{\omega}$ to the observed data likelihood.

We seek a parameterization $\boldsymbol{\xi}(\boldsymbol{\omega}) = (\boldsymbol{\xi}_S, \boldsymbol{\xi}_{\mathrm{M}})$ such that $\boldsymbol{\xi}_{\mathrm{M}}$ is identified by observed data and $\boldsymbol{\xi}_S$ is a sensitivity parameter. Sensitivity analysis and elicitation of informative priors is then be based on functions of $\boldsymbol{\xi}_S$ alone (although this may not be possible in all full-data models).

To make inferences under MNAR, we construct a prior for $\boldsymbol{\xi}_S$ and draw inference about the full-data distribution under this prior (Rubin, 1977). In practice we will frequently introduce a redundant parameter (or vector of parameters) $\boldsymbol{\Delta}$ that captures explicitly the departures from MAR. The prior on $\boldsymbol{\xi}_S$ can then be expressed in terms of $\boldsymbol{\Delta}$ and $\boldsymbol{\xi}_{\mathrm{M}}$. The following example provides a concrete illustration.

Example 9.1. *Mixture model parameterization in terms of* $(\boldsymbol{\xi}_S, \boldsymbol{\xi}_M)$.
Consider a mixture model for univariate full-data response Y such that

$$
\begin{aligned}
Y \mid R = 1 &\sim N(\mu^{(1)}, \sigma^2) \\
Y \mid R = 0 &\sim N(\mu^{(0)}, \sigma^2) \\
R &\sim \mathrm{Ber}(\phi),
\end{aligned}
$$

where, as usual, $R = 1$ indicates that Y is observed and $R = 0$ indicates missingness. Suppose the target of inference is $\theta = E(Y)$; in terms of model parameters,

$$\theta = \phi\mu^{(1)} + (1 - \phi)\mu^{(0)}.$$

This model assumes observed and missing responses differ in their mean but share a common variance. The constraint on the variance is a *modeling assumption* that cannot be verified by observed data. However, we can still parameterize the model in terms of a sensitivity parameters ξ_S that satisfies Definition 8.1.

The vector of full-data parameters is

$$\boldsymbol{\omega} = (\mu^{(0)}, \mu^{(1)}, \sigma, \phi).$$

Define $\boldsymbol{\xi}(\boldsymbol{\omega}) = (\xi_S, \boldsymbol{\xi}_M)$ such that

$$\xi_S = \mu^{(0)}$$
$$\boldsymbol{\xi}_M = (\mu^{(1)}, \sigma, \phi).$$

The sensitivity parameter ξ_S does not appear in the observed data likelihood; hence, the model has equivalent fit to the observed data across all values of ξ_S. It is easy to show that MAR holds when $\mu^{(0)} = \mu^{(1)}$; or, in terms of the sensitivity parameter, when $\xi_S = \mu^{(1)}$. To 'center' the model at MAR, we introduce a function h such that

$$\begin{aligned} \xi_S &= h(\boldsymbol{\xi}_M, \Delta) \\ &= \mu^{(1)} + \Delta. \end{aligned} \tag{9.3}$$

MAR holds when $\Delta = 0$. Departures from MAR are represented by varying Δ away from 0. Because Δ does not appear in the observed-data likelihood, varying the missing data mechanism by moving smoothly through values of Δ in (9.3) has no effect on fit to the observed data. The effect of missing data assumptions on inferences about $\theta = E(Y)$ can be seen by expressing θ as a function of Δ; i.e.,

$$\begin{aligned} \theta(\Delta) &= \phi\mu^{(1)} + (1 - \phi)\mu^{(0)} \\ &= \phi\mu^{(1)} + (1 - \phi)(\mu^{(1)} + \Delta) \\ &= \mu^{(1)} + (1 - \phi)\Delta. \end{aligned}$$

$\square$

The 'sensitivity' to departures from MAR — or indeed from any missing data mechanism characterized by $\Delta = \Delta^*$ — can be measured in terms of the derivative $\partial\theta(\Delta)/\partial\Delta$ evaluated near $\Delta = \Delta^*$. In Example 9.1, $\theta(\Delta)$ is linear and its derivative is the proportion of missing data, $1 - \phi$; however, the same principle applies in more complex models (see also Troxel et al., 2004 and Zhang and Heitjan, 2006).

In the frequentist setting, a typical sensitivity analysis (see, e.g., Scharfstein et al., 1999) fits the full-data model at fixed values over a range of Δ, and examines how the inference of interest varies with Δ. For example, if the target parameter is θ, the sensitivity analysis might comprise point estimates and confidence intervals for $\theta(\Delta)$ over a range for Δ. Viewed in the Bayesian context, this corresponds to summarizing conditional posteriors, where conditioning is done on fixed values of Δ. It can also be viewed as examining sensitivity to the prior, where all the priors under consideration are point masses.

More generally, one can use priors for Δ that are centered at a specific value and convey uncertainty about the missing data assumption by assigning nonzero prior variance. This approach is used by Nandram and Choi (2002b) in selection models with binary data, and by Rubin (1977) in mixture models for univariate response. In the next two sections, we provide details on how to specify priors in both selection and mixture models, using the principles articulated above.

9.4 Specifying priors

Let $h(\boldsymbol{\xi}_{\mathrm{M}}, \boldsymbol{\Delta})$ be a one-to-one function such that

$$\boldsymbol{\xi}_{\mathrm{S}} = h(\boldsymbol{\xi}_{\mathrm{M}}, \boldsymbol{\Delta}),$$

where $\boldsymbol{\Delta}$ captures the information about the missing data mechanism. To anchor the model at MAR, there must exist some $\boldsymbol{\Delta}_0$ such that $h(\boldsymbol{\xi}_{\mathrm{S}}, \boldsymbol{\Delta}_0)$ implies MAR. Typically we can parameterize the model such that $\boldsymbol{\Delta}_0 = \mathbf{0}$; examples can be found in Chapter 8 (Examples 8.4, 8.5, and 8.6); see also Example 9.1.

It is frequently useful to specify the joint prior $p(\boldsymbol{\xi}_{\mathrm{S}}, \boldsymbol{\xi}_{\mathrm{M}})$ as

$$p(\boldsymbol{\xi}_{\mathrm{S}}, \boldsymbol{\xi}_{\mathrm{M}}) = p(\boldsymbol{\xi}_{\mathrm{S}} \mid \boldsymbol{\xi}_{\mathrm{M}}) \, p(\boldsymbol{\xi}_{\mathrm{M}}), \tag{9.4}$$

where $p(\boldsymbol{\xi}_{\mathrm{S}} \mid \boldsymbol{\xi}_{\mathrm{M}})$ captures assumptions about the missing data mechanism. In terms of the sensitivity parameter $\boldsymbol{\Delta}$, we can re-express $p(\boldsymbol{\xi}_{\mathrm{S}} \mid \boldsymbol{\xi}_{\mathrm{M}})$ as

$$p(\boldsymbol{\xi}_{\mathrm{S}} \mid \boldsymbol{\xi}_{\mathrm{M}}) = p(\boldsymbol{\xi}_{\mathrm{S}} \mid \boldsymbol{\xi}_{\mathrm{M}}, \boldsymbol{\Delta}) \, p(\boldsymbol{\Delta} \mid \boldsymbol{\xi}_{\mathrm{M}}), \tag{9.5}$$

where

$$p(\boldsymbol{\xi}_{\mathrm{S}} \mid \boldsymbol{\xi}_{\mathrm{M}}, \boldsymbol{\Delta}) = I\{\boldsymbol{\xi}_{\mathrm{S}} = h(\boldsymbol{\xi}_{\mathrm{M}}, \boldsymbol{\Delta})\}$$

is a point mass prior; that is, $\boldsymbol{\xi}_{\mathrm{S}}$ is a deterministic function of $(\boldsymbol{\xi}_{\mathrm{M}}, \boldsymbol{\Delta})$ as above.

The first part of the prior in (9.5) simply reparameterizes $\boldsymbol{\xi}_{\mathrm{S}}$ in terms of departures from MAR; the second part, $p(\boldsymbol{\Delta} \mid \boldsymbol{\xi}_{\mathrm{M}})$, encodes the missing data mechanism *and* the uncertainty about it. A point mass prior

$$p(\boldsymbol{\Delta} \mid \boldsymbol{\xi}_{\mathrm{M}}) = I\{\boldsymbol{\Delta} = \boldsymbol{\Delta}^*\}$$

for a fixed value $\boldsymbol{\Delta}^*$ encodes a particular missing data mechanism such as

MAR with absolute certainty. For example, if $\boldsymbol{\Delta} = \mathbf{0}$ implies MAR, then specific MNAR mechanisms can be assumed by setting $p(\boldsymbol{\Delta} \mid \boldsymbol{\xi}_{\mathrm{M}}) = I\{\boldsymbol{\Delta} = \boldsymbol{\Delta}^*\}$ for some $\boldsymbol{\Delta}^* \neq \mathbf{0}$.

Alternatively, priors that convey uncertainty about the missing data mechanism can be used. A non-degenerate prior such as

$$\boldsymbol{\Delta} \mid \boldsymbol{\xi}_{\mathrm{M}} \sim N(\boldsymbol{d}, \boldsymbol{D}),$$

where $\boldsymbol{d}$ and $\boldsymbol{D}$ are hyperparameters giving the prior mean and variance for $\boldsymbol{\Delta}$, can be used to center the missing data mechanism at $\boldsymbol{\Delta} = \boldsymbol{d}$, with prior uncertainty quantified through a variance matrix $\boldsymbol{D}$.

More generally, when $\boldsymbol{\Delta} = \boldsymbol{\Delta}_0$ implies MAR, we can use the prior mean and variance as follows (using the case of scalar Δ):

MAR with no uncertainty	$E(\Delta \mid \boldsymbol{\xi}_{\mathrm{M}}) = \Delta_0$	$\mathrm{var}(\Delta \mid \boldsymbol{\xi}_{\mathrm{M}}) = 0$
MAR with uncertainty	$E(\Delta \mid \boldsymbol{\xi}_{\mathrm{M}}) = \Delta_0$	$\mathrm{var}(\Delta \mid \boldsymbol{\xi}_{\mathrm{M}}) > 0$
MNAR with no uncertainty	$E(\Delta \mid \boldsymbol{\xi}_{\mathrm{M}}) = \Delta^* \neq \Delta_0$	$\mathrm{var}(\Delta \mid \boldsymbol{\xi}_{\mathrm{M}}) = 0$
MNAR with uncertainty	$E(\Delta \mid \boldsymbol{\xi}_{\mathrm{M}}) = \Delta^* \neq \Delta_0$	$\mathrm{var}(\Delta \mid \boldsymbol{\xi}_{\mathrm{M}}) > 0$

To illustrate, we revisit Example 9.1.

Example 9.2. *Prior specifications for univariate mixture of normals* (continuation of Example 9.1).
Recall that the full-data model follows the mixture distribution

$$\begin{aligned}
Y \mid R = 1 &\sim N(\mu^{(1)}, \sigma^2) \\
Y \mid R = 0 &\sim N(\mu^{(0)}, \sigma^2) \\
R &\sim \mathrm{Ber}(\phi),
\end{aligned}$$

and that the full-data parameters are partitioned as

$$\begin{aligned}
\boldsymbol{\xi}_{\mathrm{M}} &= (\mu^{(1)}, \sigma, \phi) \\
\xi_{\mathrm{S}} &= \mu^{(0)}.
\end{aligned}$$

Let

$$\xi_{\mathrm{S}} = h(\boldsymbol{\xi}_{\mathrm{M}}, \Delta) = \mu^{(1)} + \Delta \tag{9.6}$$

so that $\Delta = 0$ coincides with MAR.

Following (9.4) and (9.5), we can factor the prior as

$$\begin{aligned}
p(\boldsymbol{\xi}_{\mathrm{M}}, \boldsymbol{\xi}_{\mathrm{S}}, \Delta) &= p(\boldsymbol{\xi}_{\mathrm{S}} \mid \boldsymbol{\xi}_{\mathrm{M}}, \Delta) \, p(\Delta \mid \boldsymbol{\xi}_{\mathrm{M}}) \, p(\boldsymbol{\xi}_{\mathrm{M}}) \\
&= p(\mu^{(0)} \mid \mu^{(1)}, \sigma, \phi, \Delta) \, p(\Delta \mid \mu^{(1)}, \sigma, \phi) \, p(\mu^{(1)}, \sigma, \phi).
\end{aligned}$$

To center the model at MAR with uncertainty about MAR governed by σ, we can use the priors

$$\begin{aligned}
\mu^{(0)} \mid \mu^{(1)}, \sigma, \phi, \Delta &\sim I\{\mu^{(0)} = \mu^{(1)} + \Delta\} \\
\Delta \mid \mu^{(1)}, \sigma, \phi &\sim N(0, \sigma^2)
\end{aligned} \tag{9.7}$$

and let $p(\mu^{(1)}, \sigma, \phi) = p(\mu^{(1)})p(\sigma)p(\phi)$, with vague priors for each.

To assume MAR with absolute certainty, we replace the $N(0, \sigma^2)$ prior in (9.7) with the point mass $I\{\Delta = 0\}$. We see from (9.6) that departures from MAR are measured in terms of $\mu^{(1)} - \mu^{(0)}$; i.e., differences in mean response between observed and missing Y. MNAR mechanisms can be represented in (9.7) by using a distribution or point mass centered away from zero. For example, setting

$$\Delta \mid \mu^{(1)}, \sigma, \phi \quad \sim \quad I\{\Delta = \sigma\}$$

states that $\mu^{(0)}$ differs from $\mu^{(1)}$ by one standard deviation. Fixed constants also can be used, as can prior distributions with nonzero variance. $\square$

In the next two sections, we discuss how to apply this framework in mixture models and in selection models.

9.5 Pattern mixture models

9.5.1 General parameterization

The factorization that defines pattern mixture models makes them a natural class of models for partitioning $\boldsymbol{\omega}$ into its identified and nonidentified components. To see this, recall that the PMM factorization is

$$p(\boldsymbol{y}, \boldsymbol{r} \mid \boldsymbol{\omega}) = p(\boldsymbol{y} \mid \boldsymbol{r}, \boldsymbol{\alpha}(\boldsymbol{\omega})) \, p(\boldsymbol{r} \mid \boldsymbol{\phi}(\boldsymbol{\omega})). \tag{9.8}$$

From here, we assume $\boldsymbol{\omega} = (\boldsymbol{\alpha}, \boldsymbol{\phi})$ and drop dependence of $\boldsymbol{\alpha}$ and $\boldsymbol{\phi}$ on $\boldsymbol{\omega}$. Then,

$$\begin{aligned} p(\boldsymbol{y}, \boldsymbol{r} \mid \boldsymbol{\alpha}, \boldsymbol{\phi}) &= p(\boldsymbol{y} \mid \boldsymbol{r}, \boldsymbol{\alpha}) \, p(\boldsymbol{r} \mid \boldsymbol{\phi}) \\ &= p(\boldsymbol{y}_{\mathrm{mis}} \mid \boldsymbol{y}_{\mathrm{obs}}, \boldsymbol{r}, \boldsymbol{\alpha}) \, p(\boldsymbol{y}_{\mathrm{obs}} \mid \boldsymbol{r}, \boldsymbol{\alpha}) \, p(\boldsymbol{r} \mid \boldsymbol{\phi}). \end{aligned} \tag{9.9}$$

Notice that the extrapolation model depends only on parameters $\boldsymbol{\alpha}$ indexing the mixture components $p(\boldsymbol{y} \mid \boldsymbol{r}, \boldsymbol{\alpha})$. This property makes the reparameterization of $\boldsymbol{\omega} = (\boldsymbol{\alpha}, \boldsymbol{\phi})$ as $\boldsymbol{\xi}(\boldsymbol{\omega}) = (\boldsymbol{\xi}_{\mathrm{S}}, \boldsymbol{\xi}_{\mathrm{M}})$ relatively straightforward in many practical settings. To illustrate in a simple case, we use the mixture of bivariate normals from Examples 5.9 and 5.11.

Example 9.3. *Sensitivity parameterizations for mixture of bivariate normals.* From Example 5.11, recall that the full data are $\boldsymbol{Y} = (Y_1, Y_2)^{\mathrm{T}}$ and $R = I\{Y_2 \text{ observed}\}$. Assume

$$\begin{aligned} \boldsymbol{Y} \mid R = r &\sim N(\boldsymbol{\mu}^{(r)}, \boldsymbol{\Sigma}^{(r)}), \\ R &\sim \mathrm{Ber}(\phi). \end{aligned}$$

Within pattern r, we can reparameterize the distribution of $\boldsymbol{Y}$ as

$$\begin{aligned} Y_2 \mid Y_1, R = r &\sim N(\beta_0^{(r)} + \beta_1^{(r)} Y_1, \sigma_{2|1}^{(r)}), \\ Y_1 \mid R = r &\sim N(\mu_1^{(r)}, \sigma_{11}^{(r)}). \end{aligned}$$

Hence, in terms of (9.9),

$$\boldsymbol{\alpha} = \left\{ \beta_0^{(r)}, \beta_1^{(r)}, \sigma_{2|1}^{(r)}, \mu_1^{(r)}, \sigma_{11}^{(r)} : r = 0, 1 \right\}.$$

Now define $\boldsymbol{\xi}(\boldsymbol{\omega}) = \boldsymbol{\xi}(\boldsymbol{\alpha}, \phi) = (\boldsymbol{\xi}_S, \boldsymbol{\xi}_M)$ such that

$$\boldsymbol{\xi}_S = (\beta_0^{(0)}, \beta_1^{(0)}, \sigma_{2|1}^{(0)}),$$

$$\boldsymbol{\xi}_M = (\beta_0^{(1)}, \beta_1^{(1)}, \sigma_{2|1}^{(1)}, \mu_1^{(0)}, \mu_1^{(1)}, \sigma_{11}^{(1)}, \sigma_{11}^{(0)}, \phi).$$

Notice that $\boldsymbol{\xi}_S$ is exclusively a function of $\boldsymbol{\alpha}$.

The model can be centered at MAR as follows. Let $\boldsymbol{\Delta} = (\Delta_0, \Delta_1, \Delta_2)^{\mathrm{T}}$ with $\Delta_2 > 0$, and define the function $\boldsymbol{h}(\boldsymbol{\xi}_M, \boldsymbol{\Delta})$ as

$$
\begin{aligned}
h_1(\boldsymbol{\xi}_M, \boldsymbol{\Delta}) &= \beta_0^{(1)} + \Delta_0 \\
h_2(\boldsymbol{\xi}_M, \boldsymbol{\Delta}) &= \beta_1^{(1)} + \Delta_1 \\
h_3(\boldsymbol{\xi}_M, \boldsymbol{\Delta}) &= \Delta_2 \sigma_{2|1}^{(1)}.
\end{aligned}
$$

By setting $\boldsymbol{\xi}_S = \boldsymbol{h}(\boldsymbol{\xi}_M, \boldsymbol{\Delta})$, we have

$$
\begin{aligned}
\beta_0^{(0)} &= \beta_0^{(1)} + \Delta_0 \\
\beta_1^{(0)} &= \beta_1^{(1)} + \Delta_1 \\
\sigma_{2|1}^{(0)} &= \Delta_2 \sigma_{2|1}^{(1)}.
\end{aligned}
$$

Setting $(\Delta_0, \Delta_1, \Delta_2) = (0, 0, 1)$ yields MAR. Priors on $\boldsymbol{\Delta}$ can be used to encode alternate missing data assumptions. $\qquad\square$

9.5.2 Using model constraints to reduce dimensionality of sensitivity parameters

In the previous example, a relatively simple model yields 3 sensitivity parameters. As the dimension increases, the number of sensitivity parameters will grow quickly. In the next example, we show how to use a simple modeling constraint to reduce the dimension of $\boldsymbol{\Delta}$ to 1. Our analysis of the Growth Hormone data in Chapter 10 takes a similar approach for a trivariate normal distribution.

Example 9.4. *Parameterization and prior specifications for mixture of bivariate normals having common variance structure.*
Continuing with Example 9.3, if we constrain $\boldsymbol{\Sigma}^{(r)} = \boldsymbol{\Sigma}$, part of the full-data is identified by parameter constraints. It is easy to show that

$$
\begin{aligned}
\beta_1^{(1)} &= \beta_1^{(0)} &= \beta_1 \\
\sigma_{2|1}^{(1)} &= \sigma_{2|1}^{(0)} &= \sigma_{2|1} \\
\sigma_{11}^{(1)} &= \sigma_{11}^{(0)} &= \sigma_{11}.
\end{aligned}
$$

Hence,

$$\xi_{\mathrm{S}} = \beta_0^{(0)}$$
$$\boldsymbol{\xi}_{\mathrm{M}} = (\beta_0^{(1)}, \beta_1, \sigma_{2|1}, \sigma_{11}, \phi).$$

To center the model at MAR, let $\xi_{\mathrm{S}} = \beta_0^{(0)} = h(\boldsymbol{\xi}_{\mathrm{M}}, \Delta)$, where

$$h(\boldsymbol{\xi}_{\mathrm{M}}, \Delta) = \beta_0^{(1)} + \Delta.$$

The prior specification can follow (9.4) and (9.5); i.e.,

$$p(\xi_{\mathrm{S}}, \boldsymbol{\xi}_{\mathrm{M}}, \Delta) = p(\xi_{\mathrm{S}} \mid \boldsymbol{\xi}_{\mathrm{M}}, \Delta)\, p(\Delta \mid \boldsymbol{\xi}_{\mathrm{M}})\, p(\boldsymbol{\xi}_{\mathrm{M}}).$$

For the first factor, we have

$$\beta_0^{(0)} \mid \Delta, \boldsymbol{\xi}_{\mathrm{M}} \sim I\{\beta_0^{(0)} = \beta_0^{(1)} + \Delta\}.$$

It may be useful (but not necessary) to calibrate the prior for Δ given $\boldsymbol{\xi}_{\mathrm{M}}$ using identified model parameters; for example,

$$\Delta \mid \boldsymbol{\xi}_{\mathrm{M}} \sim N(d, \sigma_{2|1}).$$

If $d = 0$, the prior is centered at MAR. The conditional variance $\sigma_{2|1} = \mathrm{var}(Y_2 \mid Y_1)$ is chosen because the sensitivity parameter $\xi_{\mathrm{S}} = \beta_0^{(0)}$ characterizes the conditional mean $E(Y_2 \mid Y_1, R = 0)$. Finally, priors for components of $\boldsymbol{\xi}_{\mathrm{M}}$ can be specified in the usual ways (e.g., by assuming independence between components and assigning vague or flat priors).   □

The general strategy illustrated by Example 9.4 — to specify a constrained full-data model in conjunction with a parameterization that admits sensitivity parameters — is used for analysis of the Growth Hormone study and the OASIS study in Chapter 10.

9.6 Selection models

The extrapolation factorization (9.1) is not generally easy to derive in closed form for parametric selection models. In most *parametric* selection models (Section 8.3), all full-data parameters are identified, including the parameters ω_{E} of the extrapolation distribution. The parameters ω_{E} are complex functions of the missing data mechanism parameters $\boldsymbol{\psi}$ and the full-data response model parameters $\boldsymbol{\theta}$, and it is not generally possible to find a reparameterization $\boldsymbol{\xi}(\omega) = \boldsymbol{\xi}(\boldsymbol{\theta}, \boldsymbol{\psi}) = (\boldsymbol{\xi}_{\mathrm{S}}, \boldsymbol{\xi}_{\mathrm{M}})$ that satisfies the criteria of Definition 8.1.

An alternative is to use *semiparametric* selection models. By semiparametric we mean that the full-data distribution $p(\boldsymbol{y} \mid \boldsymbol{\theta})$ is left unspecified, but the missing data mechanism $p(\boldsymbol{r} \mid \boldsymbol{y}, \boldsymbol{\psi})$ may have a parametric component. If the distribution $p(\boldsymbol{y} \mid \boldsymbol{\theta})$ has discrete support (as when $\boldsymbol{Y}$ is categorical), then semiparametric models will usually admit sensitivity parameters. When

Y is continuous, the situation is somewhat more complicated, but candidate sensitivity parameters can often still be found.

In either case, model parameterization has in important bearing on prior specification and posterior inference. We use three examples to illustrate. The first two illustrate the importance of model parameterization with binary response. The third shows how to parameterize a selection model for a continuous univariate response.

Example 9.5. *Prior specification in nonparametric and semiparametric selection models for longitudinal binary data.*
In this example we revisit the nonparametric selection model for bivariate binary data with missingness in the second observation; this model was introduced in Section 8.3.7. Let $Y = (Y_1, Y_2)^{\mathrm{T}}$ denote the full-data response, with $R = 1$ if Y_2 is observed and $R = 0$ if missing. The entire full-data distribution $p(y, r \mid \omega)$ can be enumerated as in Table 9.1. The parameters $\omega_{00}^{(1)}, \ldots, \omega_{11}^{(1)}$, corresponding to $R = 1$, all are identifiable from the observed data. The sums

$$\begin{aligned} \omega_{1+}^{(0)} &= \omega_{00}^{(0)} + \omega_{01}^{(0)} \\ \omega_{0+}^{(0)} &= \omega_{10}^{(0)} + \omega_{11}^{(0)}, \end{aligned}$$

corresponding to $R = 0$, also are identifiable. Let

$$\omega_{\mathrm{I}} = (\omega_{00}^{(1)}, \omega_{01}^{(1)}, \omega_{10}^{(1)}, \omega_{11}^{(1)}, \omega_{0+}^{(0)}, \omega_{1+}^{(0)}),$$

denote the set of *identified* parameters in this multinomial distribution. Before moving to the discussion of the selection model, note that this model admits a set of sensitivity parameters. Referring to Definition 8.1, if we set $\boldsymbol{\xi}_{\mathrm{M}} = \omega_{\mathrm{I}}$, then either $\boldsymbol{\xi}_{\mathrm{S}} = (\omega_{00}^{(0)}, \omega_{10}^{(0)})$ or $\boldsymbol{\xi}_{\mathrm{S}} = (\omega_{01}^{(0)}, \omega_{11}^{(0)})$ meets the criteria for a

Table 9.1 *Multinomial parameterization of full-data distribution for bivariate binary data with possibly missing Y_2.*

R	Y_1	Y_2	$p(y_1, y_2, r \mid \omega)$
0	0	0	$\omega_{00}^{(0)}$
0	0	1	$\omega_{01}^{(0)}$
0	1	0	$\omega_{10}^{(0)}$
0	1	1	$\omega_{11}^{(0)}$
1	0	0	$\omega_{00}^{(1)}$
1	0	1	$\omega_{01}^{(1)}$
1	1	0	$\omega_{10}^{(1)}$
1	1	1	$\omega_{11}^{(1)}$

sensitivity parameter. Essentially, the full-data distribution has two degrees of freedom unaccounted for by observed data.

The corresponding selection model can be written as

$$
\begin{aligned}
Y_1 &\sim \mathrm{Ber}(\theta_1) \\
Y_2 \mid Y_1 &\sim \mathrm{Ber}(\theta_{2\mid 1}) \\
R \mid Y_1, Y_2 &\sim \mathrm{Ber}(\pi),
\end{aligned}
$$

with

$$
\begin{aligned}
\mathrm{logit}(\theta_1) &= \alpha \\
\mathrm{logit}(\theta_{2\mid 1}) &= \beta_0 + \beta_1 y_1 \\
\mathrm{logit}(\pi) &= \psi_0 + \psi_1 y_1 + \psi_2 y_2 + \psi_3 y_1 y_2.
\end{aligned}
$$

Priors are usually specified as

$$
p(\boldsymbol{\psi}, \boldsymbol{\theta}) = p(\boldsymbol{\psi})p(\boldsymbol{\theta}), \tag{9.10}
$$

where here, $\boldsymbol{\theta} = (\alpha, \boldsymbol{\beta})$ (Scharfstein et al., 2003). It is often further assumed that

$$
p(\boldsymbol{\psi}) = p(\psi_0, \psi_1)p(\psi_2, \psi_3), \tag{9.11}
$$

i.e., the sensitivity parameters in the MDM are *a priori* independent of the other MDM parameters. These independent priors do not (in general) imply that $p(\psi_2, \psi_3, \boldsymbol{\omega}_\mathrm{I})$ — the corresponding joint prior on the sensitivity parameters (ψ_2, ψ_3) and the identified parameters $\boldsymbol{\omega}_\mathrm{I}$ — can be factored as $p(\psi_2, \psi_3)p(\boldsymbol{\omega}_\mathrm{I})$. The prior dependence between (ψ_2, ψ_3) and $\boldsymbol{\omega}_\mathrm{I}$ has implications for posterior inference. In particular, the prior and posterior for the sensitivity parameters, (ψ_2, ψ_3) are not (in general) equivalent unlike in mixture models. Before exploring the *a priori* dependence and the implications for posterior inference, we note that this *a priori* dependence is desirable. Scharfstein et al. (2003) (p. 501) argue that it makes little sense to construct (or elicit) independent priors on $\boldsymbol{\omega}_\mathrm{I}$ and (ψ_2, ψ_3). But, by specifying independent priors on $\boldsymbol{\theta}$ and $\boldsymbol{\psi}$, the dependence between $\boldsymbol{\omega}_\mathrm{I}$ and (ψ_2, ψ_3) is induced.

To see how $\boldsymbol{\omega}_\mathrm{I}$ and (ψ_2, ψ_3) are *a priori* dependent, we explore the reparameterization of the model from $(\boldsymbol{\omega}_\mathrm{I}, \psi_2, \psi_3)$ to $(\boldsymbol{\theta}, \psi_0, \psi_1, \psi_2, \psi_3)$. The function to move between these parameterizations is indexed by (ψ_2, ψ_3),

$$
\boldsymbol{\omega}_\mathrm{I} = \boldsymbol{h}_{\psi_2, \psi_3}(\boldsymbol{\theta}, \psi_0, \psi_1).
$$

We derive this function for the simple cross-sectional setting of a binary response. Consider the selection model

$$
\begin{aligned}
Y &\sim \mathrm{Ber}(\theta) \\
\mathrm{logit}\{P(R = 1 \mid y)\} &= \psi_0 + \psi_1 y,
\end{aligned}
$$

where $R = 1$ means Y is observed. In this model, ψ_1 is the sensitivity parameter. We denote the identified parameters as $\boldsymbol{\omega}_I = (\theta_1, \alpha)$, where

$$\theta_1 = P(Y = 1 \mid R = 1)$$
$$\alpha = P(R = 1).$$

With a little algebra, it can be shown that

$$\begin{pmatrix} \theta \\ \psi_0 \end{pmatrix} = \boldsymbol{h}_{\psi_1}(\theta_1, \alpha) = \begin{pmatrix} \alpha\theta_1 + \dfrac{1 - \alpha}{1 + \theta_1 \exp(-\psi_1)/(1 - \theta_1)} \\ \operatorname{logit}(\alpha) + \log\{(1 - \theta_1) + \exp(-\psi_1)\theta_1\} \end{pmatrix}.$$

The *a priori* dependence can be seen explicitly by deriving the Jacobian using these results. For a related discussion, see Gustafson (2006).

As a result, *a priori* independence between the full data response model parameters and the missing data mechanism parameters given in (9.10) and between the sensitivity parameters and the other parameters in the missing data mechanism given in (9.11) does not imply *a priori* independence between $\boldsymbol{\omega}_I$ and (ψ_2, ψ_3) (the sensitivity parameters). This lack of *a priori* independence between $\boldsymbol{\omega}_I$ and (ψ_2, ψ_3) implies that the posterior for the sensitivity parameters is not equal to the prior; i.e.,

$$p(\psi_2, \psi_3 \mid \boldsymbol{y}_{\text{obs}}, \boldsymbol{r}) \quad \neq \quad p(\psi_2, \psi_3).$$

However, it is often still the case that

$$p(\psi_2, \psi_3 \mid \boldsymbol{y}, \boldsymbol{r}) \quad \approx \quad p(\psi_2, \psi_3). \tag{9.12}$$

To see why the equality between prior and posterior does not hold here, we write

$$p(\psi_2, \psi_3 \mid \boldsymbol{y}_{\text{obs}}, \boldsymbol{r}) \quad \propto \quad \int p(\boldsymbol{y}_{\text{obs}}, \boldsymbol{r} \mid \psi_2, \psi_3, \boldsymbol{\omega}_I) p(\boldsymbol{\omega}_I \mid \psi_2, \psi_3) p(\psi_2, \psi_3) d\boldsymbol{\omega}_I$$
$$= \quad \int p(\boldsymbol{y}_{\text{obs}}, \boldsymbol{r} \mid \boldsymbol{\omega}_I) p(\boldsymbol{\omega}_I \mid \psi_2, \psi_3) p(\psi_2, \psi_3) d\boldsymbol{\omega}_I$$
$$= \quad p(\boldsymbol{y}_{\text{obs}}, \boldsymbol{r} \mid \psi_2, \psi_3) p(\psi_2, \psi_3).$$

The first equality holds given that $\boldsymbol{\omega}_I$ parameterizes the observed data distribution. When we integrate out $\boldsymbol{\omega}_I$, we are left with an observed data distribution that depends on (ψ_2, ψ_3). However, in a nonparametric selection model, it will typically be the case that

$$\frac{p(\boldsymbol{y}_{\text{obs}}, \boldsymbol{r} \mid \psi_2, \psi_3)}{p(\boldsymbol{y}_{\text{obs}}, \boldsymbol{r})} \quad \approx \quad 1.$$

That is, $p(\boldsymbol{y}_{\text{obs}}, \boldsymbol{r} \mid \psi_2, \psi_3)$ is almost constant with respect to (ψ_2, ψ_3). How well (9.12) holds for different semiparametric specifications is an area of ongoing work.   □

We illustrate the approximate equality given in (9.12) using the OASIS data (described in Section 1.6) in the next example.

Example 9.6. *Prior dependence in nonparametric selection models using data from the OASIS study.*

For simplicity, we use the first and second observations in the OASIS smoking cessation trial. We fit the nonparametric selection model from Example 9.5,

$$
\begin{aligned}
Y_1 &\sim \text{Ber}(\theta_1) \\
Y_2 \mid Y_1 &\sim \text{Ber}(\theta_{2|1}) \\
R \mid Y_1, Y_2 &\sim \text{Ber}(\pi),
\end{aligned}
$$

with

$$
\begin{aligned}
\text{logit}(\theta_1) &= \alpha \\
\text{logit}(\theta_{2|1}) &= \beta_0 + \beta_1 y_1 \\
\text{logit}(\pi) &= \psi_0 + \psi_1 y_1 + \psi_2 y_2 + \psi_3 y_1 y_2,
\end{aligned}
$$

separately for each treatment. Diffuse normal priors on $(\boldsymbol{\theta}, \psi_0, \psi_1)$ and independent $N(0,1)$ priors on ψ_2 and ψ_3 are used. The normal priors on ψ_2 and ψ_3 correspond to an MAR missing data mechanism with uncertainty. Neither ψ_2 nor ψ_3 appear in the observed data likelihood, but they are not *a priori* independent of the identified parameters (see Example 9.5). However, for nonparametric selection models, we expect the approximate equivalence to hold, as in (9.12).

Table 9.2 shows posterior summaries of ψ_2 and ψ_3 from using the $N(0,1)$ priors described above. Posterior means and variances are very close to 0 and 1, respectively, supporting the approximate equivalence given in (9.12). The posterior mean and credible interval of the odds ratio for treatment effect are 1.3 (.3, 3.3). If instead of the normal priors, we use MAR point mass priors (no uncertainty), the posterior mean and credible interval of the odds ratio are 1.3 (.5, 2.9). As expected, the posterior credible intervals are wider under the $N(0,1)$ priors.   □

The final example of semiparametric selection models illustrates their construction for a continuous univariate response.

Table 9.2 *Posterior means and standard deviations for the sensitivity parameters in the nonparametric selection model.*

Treatment	Parameter	Posterior Mean	Posterior SD
ET	ψ_2	.01	1.00
	ψ_3	.06	1.00
ST	ψ_2	.03	.99
	ψ_3	−.01	.99

Example 9.7. *Sensitivity analysis with semiparametric selection models for cross-sectional continuous response.*

Denote the response as Y and the missing indicator as R. Suppose we specify the following semiparametric selection model using a mixture of Dirichlet processes model (cf. Section 3.6):

$$
\begin{aligned}
\text{logit}\{P(R_i = 1 \mid y_i, \boldsymbol{\psi})\} &= \psi_0 + \psi_1 y_i \\
Y_i \mid \mu_i, \sigma^2 &\sim N(\mu_i, \sigma^2) \\
\mu_i &\sim G \\
G &\sim DP(G_0, \alpha),
\end{aligned}
\tag{9.13}
$$

where $\boldsymbol{\theta} = (G, \sigma^2)$ and (G_0, α) are fixed hyperparameters. In Scharfstein et al. (2003), a similar model was specified with a prior of the form (9.10). Clearly, if no distributional assumptions are made about the full-data response $\boldsymbol{y}$, the data will provide no information on the missing data mechanism parameter ψ_1, and the MDP on the distribution of the full-data response allows ψ_1 to be essentially unidentified; by contrast, it *is* identified when using a parametric model for the full-data response. Here, ψ_1 is a sensitivity parameter. □

Semiparametric selection models can provide a viable alternative to mixture models for sensitivity analyses. The underlying factorization into the full-data response model and the missing data mechanism facilitates elicitation of priors for parameters indexing the missing data mechanism (e.g., ψ_1 in (9.13)). See Scharfstein et al. (2003) and Scharfstein et al. (2006) for examples.

The main stumbling block to implementation of these models in general is the feasibility of their specification for longitudinal data with many time points, especially if the responses are continuous (cf. Chapter 8). The computational challenges also are nontrivial. For complex longitudinal settings, the full-data response model can be specified semiparametrically (Daniels and Scharfstein, 2007). Construction of sensitivity analysis and informative priors will still be valid as long as (9.12) holds.

9.7 Further reading

Model uncertainty and incomplete data

Copas and Eguchi (2005) also emphasize the importance of characterizing uncertainty in incomplete data problems, but more from the perspective of mis-specifying the full-data model. To account for model uncertainty, they suggest rules for adjusting standard errors and confidence intervals for parameters of interest. Work by Forster and Smith (1998) is closely related, and deals with categorical data. Recent work by Gustafson (2006) describes model expansion and model contraction for handling full-data models that are only partially identified by observed data. Illustrations related to measurement error and unmeasured confounding are provided, both of which can be viewed as missing data problems.

CHAPTER 10

Case Studies: Nonignorable Missingness

10.1 Overview

This chapter provides three detailed case studies that illustrate analyses under the missing not at random assumption. For each of our three examples, we give details on specifying both the full-data model and the appropriate prior distributions. The analyses shown here were implemented using WinBUGS software; the code is available from the book Web site.

In Section 10.2, pattern mixture models are used to analyze data from two arms of the Growth Hormone Study described in Section 1.3. We fit the model using both a multivariate normal model (assuming ignorability) and a pattern mixture model (assuming nonignorable MAR). Departures from MAR in the mixture model are incorporated by introducing sensitivity parameters and constructing appropriate prior distributions for them.

In Section 10.3, both pattern mixture models and selection models are used to analyze data from the OASIS Study, described in Section 1.6. We examine the role of model specification by comparing inferences using both selection and pattern mixture models. Analyses under MNAR are illustrated using a pattern mixture model with informative priors elicited from experts, and using a parametric selection model that identifies the parameters governing departures from MAR. The latter analysis highlights some limitations of using parametric selection models with informative priors and sensitivity analysis.

The third case study, in Section 10.4, uses a mixture of varying coefficient models to analyze longitudinal CD4 counts from the Pediatric AIDS trial, described in Section 1.7. The model used here permits continuous dropout times, and assumes the CD4 intercepts and slopes depend on dropout time via smooth but unspecified functions. The mixture of VCM analysis is compared to the standard random effects model under MAR.

233

10.2 Growth Hormone study: Pattern mixture models and sensitivity analysis

10.2.1 Overview

The Growth Hormone Study is described in detail in Section 1.3. In short, the trial examines effect of various combinations of growth hormone and exercise on 1-year changes in quadriceps strength in a cohort of 160 individuals. Measurements are taken at baseline, 6 months, and 12 months. Of the 160 randomized, only 111 (69%) had complete follow-up. The observed data appear in Table 1.2.

To illustrate various types of models, we confine attention to the arms using exercise plus placebo (EP; $n = 40$) and exercise plus growth hormone (EG; $n = 38$). On arm EP, 7 individuals (18%) had only one measurement, 2 (5%) had two measurements, and 31 (78%) had three. Missingness is more prevalent on the EG arm, with 12 (32%) having only one measurement, 4 (11%) having two, and 22 (58%) having three. Missingness is caused by dropout and follows a monotone pattern (Table 1.2).

The objective is to compare mean quadriceps strength at 12 months between the two groups. We use a pattern mixture approach. The pattern mixture model is first specified under MAR; we then elaborate the specification to allow MNAR using two additional nonidentified parameters. We illustrate how to conduct and interpret a sensitivity analysis using posteriors over a range of point mass priors for the sensitivity parameters. In addition, we show how to construct and use informative priors to obtain a single summary about the treatment effect.

The relevant variables for the full-data model are:

$$
\begin{aligned}
(Y_1, Y_2, Y_3)^{\mathrm{T}} &= \text{quad strength measures at months 0, 6, 12,} \\
(R_1, R_2, R_3)^{\mathrm{T}} &= \text{response indicators (1 = observed, 0 = missing).} \\
Z &= \text{treatment group indicator (1 = EG, 0 = EP),}
\end{aligned}
$$

We define patterns based on observed follow-up time $S = \sum_j R_j$. The objective is to compare mean quad strength between the two arms at month 12; the treatment effect is denoted by

$$
\theta = E(Y_3 \mid Z = 1) - E(Y_3 \mid Z = 0).
$$

10.2.2 Multivariate normal model under ignorability

For reference, we compare the mixture model results to an ignorable model where we assume the full-data distribution is multivariate normal within treatment arm. This analysis was done in Section 7.2.

10.2.3 Pattern mixture model specification

In this section we provide the full-data model specification, show how to constrain the model under MAR, and show how to parameterize the model in terms of sensitivity parameters that capture departures from MAR.

Our analysis is based on the pattern mixture model specified in Example 8.5. To avoid too many subscripts, we suppress treatment subscripts z until they are needed; hence the model below applies separately to each treatment arm.

The full-data distribution is a mixture over follow-up times, with components (patterns) defined by $S \in \{1, 2, 3\}$. Nonidentified components are marked by $\star$; parameters for these do not appear in the observed data likelihood. The complete model specification is as follows:

$$Y_1 \mid S = 1 \sim N(\mu^{(1)}, \sigma^{(1)})$$
$$Y_1 \mid S = 2 \sim N(\mu^{(2)}, \sigma^{(2)})$$
$$Y_1 \mid S = 3 \sim N(\mu^{(3)}, \sigma^{(3)})$$

$$\star \quad Y_2 \mid Y_1, S = 1 \sim N(\alpha_0^{(1)} + \alpha_1^{(1)} Y_1, \tau_2^{(1)})$$
$$\left. \begin{array}{l} Y_2 \mid Y_1, S = 2 \\ Y_2 \mid Y_1, S = 3 \end{array} \right\} \sim N(\alpha_0^{(\geq 2)} + \alpha_1^{(\geq 2)} Y_1, \tau_2^{(\geq 2)})$$

$$\star \quad Y_3 \mid Y_1, Y_2, S = 1 \sim N(\beta_0^{(1)} + \beta_1^{(1)} Y_1 + \beta_2^{(1)} Y_2, \tau_3^{(1)})$$
$$\star \quad Y_3 \mid Y_1, Y_2, S = 2 \sim N(\beta_0^{(2)} + \beta_1^{(2)} Y_1 + \beta_2^{(2)} Y_2, \tau_3^{(2)})$$
$$Y_3 \mid Y_1, Y_2, S = 3 \sim N(\beta_0^{(3)} + \beta_1^{(3)} Y_1 + \beta_2^{(3)} Y_2, \tau_3^{(3)})$$

$$S \sim \text{Mult}(\phi). \tag{10.1}$$

The multinomial parameter is $\phi = (\phi_1, \phi_2, \phi_3)$ with $\phi_s = P(S = s)$ for $s \in \{1, 2, 3\}$ and $\sum_s \phi_s = 1$.

Although the model specification looks cumbersome, it follows the structure $p(\boldsymbol{y}_{\text{mis}} \mid \boldsymbol{y}_{\text{obs}}, s)\, p(\boldsymbol{y}_{\text{obs}} \mid s)\, p(s)$, where the components marked with $\star$ comprise $p(\boldsymbol{y}_{\text{mis}} \mid \boldsymbol{y}_{\text{obs}}, s)$, and those not marked comprise $p(\boldsymbol{y}_{\text{obs}} \mid s)$ and $p(s)$.

10.2.4 MAR constraints for pattern mixture model

Under MAR, the distribution of missing data is identified using the constraints listed in Theorem 8.1. Let $p_s(y)$ denote $p(y \mid S = s)$. For the missing data in pattern $S = 1$, the MAR constraints imply

$$\begin{array}{rcl} p_1(y_2 \mid y_1) & = & p_{(\geq 2)}(y_2 \mid y_1), \\ p_1(y_3 \mid y_1, y_2) & = & p_3(y_3 \mid y_1, y_2); \end{array}$$

and for pattern $S = 2$, we have

$$p_2(y_3 \mid y_1, y_2) \;\; = \;\; p_3(y_3 \mid y_1, y_2).$$

By assuming the unidentified components of our model are normally distributed, MAR is satisfied by equating means and variances. For example, to identify $p_1(y_2 \mid y_1)$, we set

$$\alpha_0^{(1)} + \alpha_1^{(1)} y_1 \;\; = \;\; \alpha_0^{(\geq 2)} + \alpha_1^{(\geq 2)} y_1,$$
$$\tau_2^{(1)} \;\; = \;\; \tau_2^{(\geq 2)}.$$

In order for the first constraint to hold for all possible values of y_1, we require

$$\alpha_0^{(1)} \;\; = \;\; \alpha_0^{(\geq 2)},$$
$$\alpha_1^{(1)} \;\; = \;\; \alpha_1^{(\geq 2)}. \qquad (10.2)$$

Similarly, to identify $p_1(y_3 \mid y_1, y_2)$ and $p_2(y_3 \mid y_1, y_2)$, we equate regression parameters and variance components as follows, for $s = 1, 2$:

$$\beta_0^{(s)} \;\; = \;\; \beta_0^{(3)},$$
$$\beta_1^{(s)} \;\; = \;\; \beta_1^{(3)},$$
$$\beta_2^{(s)} \;\; = \;\; \beta_2^{(3)},$$
$$\tau_3^{(s)} \;\; = \;\; \tau_3^{(3)}. \qquad (10.3)$$

All told, 22 constraints are needed to impose the MAR assumption for this model; for each treatment arm, there are eight constraints on regression parameters and three on variance components.

10.2.5 Parameterizing departures from MAR

We can reparameterize the model in terms of sensitivity parameters to embed the MAR constraint in a larger class of MNAR models for the full-data distribution. Note first that the full-data model can easily be reparameterized as $\boldsymbol{\xi}(\boldsymbol{\omega}) = (\boldsymbol{\xi}_{\mathrm{S}}, \boldsymbol{\xi}_{\mathrm{M}})$, where elements of $\boldsymbol{\xi}_{\mathrm{S}}$ are parameters from distributions in (10.1) marked with $\star$, and elements of $\boldsymbol{\xi}_{\mathrm{M}}$ are the remaining parameters. Specifically,

$$\boldsymbol{\xi}_{\mathrm{S}} \;\; = \;\; \begin{cases} \alpha_0^{(1)}, \alpha_1^{(1)}, \tau_2^{(1)} \\ \beta_0^{(s)}, \beta_1^{(s)}, \beta_2^{(s)}, \tau_3^{(s)} & s = 1, 2 \end{cases}$$

$$\boldsymbol{\xi}_{\mathrm{M}} \;\; = \;\; \begin{cases} \mu^{(s)}, \sigma^{(s)} & s = 1, 2, 3 \\ \alpha_0^{(\geq 2)}, \alpha_1^{(\geq 2)}, \tau_2^{(\geq 2)} \\ \beta_0^{(3)}, \beta_1^{(3)}, \beta_2^{(3)}, \tau_3^{(3)} \\ \phi_1, \phi_2, \phi_3. \end{cases}$$

Moreover, $\boldsymbol{\xi}_S$ meets the criteria for a *sensitivity parameter* given in Definition 8.1.

Following the strategy laid out in Section 9.3, departures from MAR can be captured by reparameterizing $\boldsymbol{\xi}_S$ in terms of $\boldsymbol{\xi}_M$ and a set of parameters $\boldsymbol{\Delta}$ via $\boldsymbol{\xi}_S = \boldsymbol{h}(\boldsymbol{\xi}_M, \boldsymbol{\Delta})$. Components of $\boldsymbol{h}$ are most easily described in several parts. First, expanding on regression parameter constraints in (10.2), we have

$$
\begin{aligned}
\alpha_0^{(1)} &= h_1(\boldsymbol{\xi}_M, \boldsymbol{\Delta}) = \alpha_0^{(\geq 2)} + \Delta_{\alpha_0}^{(1:2)}, \\
\alpha_1^{(1)} &= h_2(\boldsymbol{\xi}_M, \boldsymbol{\Delta}) = \alpha_1^{(\geq 2)} + \Delta_{\alpha_1}^{(1:2)},
\end{aligned}
\tag{10.4}
$$

where the superscript (1:2) on Δ denotes a relation between parameters from patterns 1 and (≥ 2) (using a slight abuse of notation).

The next six components $h_3, \ldots, h_8$ of $\boldsymbol{h}$ reparameterize the β coefficients in $\boldsymbol{\xi}_S$ as

$$
\begin{aligned}
\beta_0^{(s)} &= \beta_0^{(3)} + \Delta_{\beta_0}^{(s:3)} \\
\beta_1^{(s)} &= \beta_1^{(3)} + \Delta_{\beta_1}^{(s:3)} \\
\beta_2^{(s)} &= \beta_2^{(3)} + \Delta_{\beta_2}^{(s:3)}
\end{aligned}
\tag{10.5}
$$

for $s = 1, 2$. The Δ parameters here are interpreted similarly to (10.4). For example, $\Delta_{\beta_0}^{(1:3)}$ is the difference between $\beta_0^{(1)}$, which is not identified, and $\beta_0^{(3)}$, which is.

Finally, h_9, h_{10}, h_{11} deal with variance components:

$$
\begin{aligned}
\tau_2^{(1)} &= \Delta_{\tau_2}^{(1:2)} \tau_2^{(\geq 2)} \\
\tau_3^{(1)} &= \Delta_{\tau_3}^{(1:3)} \tau_3^{(3)} \\
\tau_3^{(2)} &= \Delta_{\tau_3}^{(2:3)} \tau_3^{(3)}.
\end{aligned}
\tag{10.6}
$$

The full collection of Δ parameters is denoted

$$
\boldsymbol{\Delta} = (\boldsymbol{\Delta}_\alpha, \boldsymbol{\Delta}_\beta, \boldsymbol{\Delta}_\tau),
$$

where

$$
\begin{aligned}
\boldsymbol{\Delta}_\alpha &= (\Delta_{\alpha_0}^{(1:2)}, \Delta_{\alpha_1}^{(1:2)}), \\
\boldsymbol{\Delta}_\beta &= (\Delta_{\beta_0}^{(1:3)}, \Delta_{\beta_1}^{(1:3)}, \Delta_{\beta_2}^{(1:3)}, \Delta_{\beta_0}^{(2:3)}, \Delta_{\beta_1}^{(2:3)}, \Delta_{\beta_2}^{(2:3)}), \\
\boldsymbol{\Delta}_\tau &= (\Delta_{\tau_2}^{(1:2)}, \Delta_{\tau_3}^{(1:3)}, \Delta_{\tau_3}^{(2:3)}).
\end{aligned}
$$

The MNAR model includes MAR as a special case: setting $\boldsymbol{\Delta}_\alpha = 0$, $\boldsymbol{\Delta}_\beta = 0$, and $\boldsymbol{\Delta}_\tau = 1$ yields MAR.

10.2.6 Constructing priors

Following (9.4) and (9.5), we factor the prior on $(\boldsymbol{\xi}_S, \boldsymbol{\xi}_M, \boldsymbol{\Delta})$ as

$$p(\boldsymbol{\xi}_S, \boldsymbol{\xi}_M, \boldsymbol{\Delta}) = p(\boldsymbol{\xi}_S \mid \boldsymbol{\xi}_M, \boldsymbol{\Delta}) \, p(\boldsymbol{\Delta} \mid \boldsymbol{\xi}_M) \, p(\boldsymbol{\xi}_M),$$

where

$$p(\boldsymbol{\xi}_S \mid \boldsymbol{\xi}_M, \boldsymbol{\Delta}) = I\{\boldsymbol{\xi}_S = \boldsymbol{h}(\boldsymbol{\xi}_M, \boldsymbol{\Delta})\}$$

is a point mass, and $p(\boldsymbol{\Delta} \mid \boldsymbol{\xi}_M)$ reflects prior beliefs about the departures from MAR. For example, the prior

$$p(\boldsymbol{\Delta}_\alpha, \boldsymbol{\Delta}_\beta, \boldsymbol{\Delta}_\tau \mid \boldsymbol{\xi}_M) = I\{\boldsymbol{\Delta}_\alpha = \mathbf{0}, \boldsymbol{\Delta}_\beta = \mathbf{0}, \boldsymbol{\Delta}_\tau = \mathbf{1}\} \qquad (10.7)$$

assumes MAR with absolute certainty.

To move beyond the MAR assumption, we can make alternate choices for $p(\boldsymbol{\Delta} \mid \boldsymbol{\xi}_M)$. A sensitivity analysis can be conducted by examining posterior inferences about treatment effect (or any parameter of interest) over a bounded set of values (say $\mathscr{D}$) for the $\boldsymbol{\Delta}$ parameters. Concretely, this involves summarizing posterior treatment effects based on the set

$$\{p(\boldsymbol{\Delta} \mid \boldsymbol{\xi}_M) = I\{\boldsymbol{\Delta} = \boldsymbol{\Delta}^*\} \ : \ \boldsymbol{\Delta}^* \in \mathscr{D}\}$$

of point mass priors. A practical question arises as to the choice of $\mathscr{D}$ because its dimension can be high (in this case, there are 22 sensitivity parameters). In the analysis that follows, we first draw inferences using the MAR prior, and then illustrate the implementation of a sensitivity analysis where $\mathscr{D}$ is calibrated using the posterior of $\boldsymbol{\xi}_M$ under MAR.

10.2.7 Analysis using point mass MAR prior

In this analysis we use the prior (10.7) for $\boldsymbol{\Delta}$ and place diffuse priors on $\boldsymbol{\xi}_M$ as follows (for simplicity we do not include superscripts and subscripts — e.g., all μ parameters have the same prior; refer to (10.1) for specifics):

$$
\begin{aligned}
\mu &\sim N(0, 10^6), \\
\alpha, \beta &\sim N(0, 10^4), \\
\sigma^2 &\sim N(0, 20^2) \, I\{0 < \sigma^2 < 10^4\}, \\
\tau^2 &\sim N(0, 20^2) \, I\{0 < \tau^2 < 10^4\}, \\
\boldsymbol{\phi} &\sim \text{Dirichlet}(1, 1, 1).
\end{aligned}
$$

Posterior inference was implemented using two parallel Markov chains with 25,000 iterations each. We discarded results from the first 5000 per chain (burn-in) and based posterior inference on the remaining 40,000 draws.

Inference for the mean for each treatment group at each time point is summarized in Table 10.1. For comparison we also include inferences from a standard multivariate normal specification, described and fit in Section 7.2. As expected, the posterior means are very similar; posterior variability is slightly higher in the mixture model, which can be attributed to the larger number of parameters.

Table 10.1 *Growth hormone trial: posterior mean (s.d.) for quadriceps strength at each time point, stratified by treatment group. MVN = multivariate normal distribution assumed for joint distribution of responses; MAR = missing at random constraints; MNAR-1 = Δ assumed common across treatment groups; MNAR-2 = Δ assumed different by treatment group. Both MNAR analyses use uniform priors for Δ bounded away from zero. See Section 10.2.8 for details.*

| Treatment | Month | MVN | Pattern Mixture Models | | |
			MAR	MNAR-1	MNAR-2
EP	0	65 (4.2)	66 (4.6)	66 (4.6)	66 (4.6)
	6	81 (4.4)	82 (4.7)	80 (4.9)	80 (4.9)
	12	73 (3.7)	73 (4.0)	70 (4.3)	69 (4.6)
EG	0	69 (4.2)	69 (4.4)	69 (4.4)	69 (4.4)
	6	81 (6.0)	81 (6.4)	78 (6.8)	78 (6.8)
	12	78 (6.3)	78 (6.7)	73 (7.4)	72 (8.1)
Difference at 12 mos.		5.7 (7.3)	5.4 (7.8)	3.1 (8.2)	2.6 (9.3)

10.2.8 Analyses using MNAR priors

The pattern mixture model can be expanded to allow for MNAR through appropriate specification of the prior $p(\Delta \mid \xi_M)$. This prior will convey information about the degree to which the distribution of missing responses differs from that of observed responses. In general, priors for components of Δ_α and Δ_β will shift the posterior means away from their MAR values, with $\Delta > 0$ indicating that the mean response is higher for missing observations than for observed ones, relative to MAR.

Under this parameterization of the PMM, the conditional variance parameters τ from nonidentified distributions do not appear in the posterior distribution of the marginal mean parameters; hence priors for components of Δ_τ will not affect posterior inference about marginal means. This reduces the number of sensitivity parameters from 22 to 16.

Specification and calibration of MNAR priors

There are two obstacles that must be overcome in specifying MNAR priors: the high dimension of the parameter space for $\boldsymbol{\Delta}$, and the need to make a specific choice of prior distribution. Even if a series of point mass priors will be used to conduct a sensitivity analysis, we must specify the range of values to include.

We focus on departures from MAR in terms of the intercept sensitivity parameters $(\Delta_{\alpha_0}^{(1:2)}, \Delta_{\beta_0}^{(1:3)}, \Delta_{\beta_0}^{(2:3)})$, reducing the number of sensitivity parameters to six (three for each treatment group). The justification for this choice is based mainly on interpretability; each represents a difference in the conditional mean of Y_j given previous Y's between patterns. For example,

$$\Delta_{\alpha_0}^{(1:2)} \;=\; E(Y_2 \mid Y_1, S = 1) - E(Y_2 \mid Y_1, S \geq 2)$$

represents the difference in $E(Y_2 \mid Y_1)$ between those with missing and observed Y_2. Similarly,

$$\Delta_{\beta_0}^{(1:3)} \;=\; E(Y_3 \mid Y_2, Y_1, S = 1) - E(Y_3 \mid Y_2, Y_1, S = 3),$$
$$\Delta_{\beta_0}^{(2:3)} \;=\; E(Y_3 \mid Y_2, Y_1, S = 2) - E(Y_3 \mid Y_2, Y_1, S = 3).$$

As a final constraint, we assume $\Delta_{\beta_0}^{(1:3)} = \Delta_{\beta_0}^{(2:3)}$, and denote the common parameter by $\Delta_{\beta_0}^{(\bullet:3)}$. The implication of our final constraint is that, relative to MAR, the difference in $E(Y_3 \mid Y_1, Y_2)$ between those with observed and missing Y_3 is common across missing data patterns $S = 1$ and $S = 2$. Hence the reduced set of Δ parameters for sensitivity analysis and incorporation of informative priors is denoted by

$$\boldsymbol{\Delta} = (\Delta_{\alpha_0}^{(1:2)}, \Delta_{\beta_0}^{(\bullet:3)}). \tag{10.8}$$

There are four sensitivity parameters in all: one set for each treatment group.

Recall that the treatment effect of interest is

$$\theta \;=\; E(Y_3 \mid Z = 1) - E(Y_3 \mid Z = 0),$$

with $\theta > 0$ if quad strength on EG is greater than on EP. With respect to summarizing posterior inference about θ under MNAR, each of our analyses below has two components: *First*, we examine sensitivity to departures from MAR in terms of posterior mean and posterior probabilities for treatment effect over a set of values $\boldsymbol{\Delta} \in \mathscr{D}$. A sensitivity analysis summarizing the range of posterior inferences about mean treatment effect over a range of missing data mechanisms is carried out by summarizing the posterior mean $E(\theta \mid \boldsymbol{y}_{\text{obs}}, \boldsymbol{z}, \boldsymbol{\Delta})$ and posterior probability $P(\theta > 0 \mid \boldsymbol{y}_{\text{obs}}, \boldsymbol{z}, \boldsymbol{\Delta})$ over all $\boldsymbol{\Delta} \in \mathscr{D}$. In Analysis 1 we give ranges for these quantities over $\mathscr{D}$, and in Analysis 2 we use contour plots to identify subsets of $\mathscr{D}$ where inference about treatment effect might differ substantially from MAR. *Second*, we use informative priors for $\boldsymbol{\Delta}$ to generate an overall inference about θ. In our case, we assume $\boldsymbol{\Delta}$ is uniformly distributed over a domain $\mathscr{D}$.

In general, components of $\boldsymbol{\Delta}$ can range over the entire real line. It is therefore necessary to calibrate or otherwise bound its domain $\mathscr{D}$. The *direction* of departure from MAR can typically be informed by context. In our analyses below, we assume dropouts have lower mean quad strength than non-dropouts. In terms of the model, our assumption is that for $j = 2, 3$, the conditional mean of quad strength Y_j given the past is lower for those with missing Y_j compared to those with observed Y_j.

The *scale* of departure from MAR can be determined in any number of ways. Our approach is to use the variability in the observed data. Because the components of $\boldsymbol{\Delta}$ in (10.8) correspond to intercepts in the conditional distributions $p_{\geq 2}(y_2 \mid y_1)$ and $p_3(y_3 \mid y_1, y_2)$, a natural metric for scaling $\mathscr{D}$ is the set of residual variances for identified conditional distributions in the pattern mixture model (10.1). For example, by setting $\mathscr{D} = \mathscr{D}(\boldsymbol{\tau})$, where

$$\mathscr{D}(\boldsymbol{\tau}) \;=\; \left[-\sqrt{\tau_2^{(\geq 2)}}, \, 0\right] \times \left[-\sqrt{\tau_3^{(3)}}, \, 0\right],$$

departures from MAR are assumed to lie within one standard deviation for the conditional distributions $p_{\geq 2}(y_2 \mid y_1)$ and $p_3(y_3 \mid y_1, y_2)$.

Priors for $\boldsymbol{\Delta}$ can be calibrated in the same fashion. Following with this example, we can assume $p(\boldsymbol{\Delta} \mid \boldsymbol{\xi}_{\mathrm{M}}) = p(\boldsymbol{\Delta} \mid \boldsymbol{\tau})$, and set

$$\boldsymbol{\Delta} \mid \tau_2^{(\geq 2)}, \tau_3^{(3)} \;\sim\; \mathrm{Unif}\{\mathscr{D}(\boldsymbol{\tau})\}.$$

In our analyses, we approximate this approach by plugging in posterior means for relevant components of $\boldsymbol{\tau}$ to determine $\mathscr{D}$. Details are provided below.

Analysis 1: Assume common $\boldsymbol{\Delta}$ between treatments

Sensitivity analyses and informative priors used in the first analysis will be based on the two-dimensional parameter $\boldsymbol{\Delta} = (\Delta_{\alpha_0}^{(1:2)}, \Delta_{\beta_0}^{(\bullet:3)}) \in \mathscr{D}_1$. We set the maximum range of departures from MAR roughly equal to the highest residual standard deviation between the treatment-specific regressions for $p(y_2 \mid y_1)$ and $p(y_3 \mid y_1, y_2)$ based on the observed data. Using the results in Table 10.2, where the residual standard deviation ranges from 14 to 23 for the regression $(Y_2 \mid Y_1)$ and from 8.9 to 15 for $(Y_3 \mid Y_1, Y_2)$, we set $\mathscr{D}_1 = [-20, 0] \times [-15, 0]$.

To help with interpretation, it is useful to understand the effect of departures from MAR on the marginal means $\mu_2 = E(Y_2)$ and $\mu_3 = E(Y_3)$. For clarity we again suppress treatment indices. Consider first μ_2; in the pattern mixture model,

$$\begin{aligned} \mu_2 &= E(Y_2) \\ &= \sum_{s=1}^{3} \phi_s E(Y_2 \mid S = s). \end{aligned} \tag{10.9}$$

Table 10.2 *Growth hormone trial: posterior mean (posterior SD) of regression parameters for pattern mixture model (10.1) fit under MAR constraints.*

			Treatment Group	
Model	Covariate	Parameter	EP($z = 0$)	EG ($z = 1$)
$Y_2 \mid Y_1$	Int.	$\alpha_{0z}^{(\geq 2)}$	23 (6.9)	14 (15)
	Y_1	$\alpha_{1z}^{(\geq 2)}$	.89 (.10)	.97 (.19)
		$\tau_{2z}^{(\geq 2)}$	14	23
$Y_3 \mid Y_1, Y_2$	Int.	$\beta_{0z}^{(3)}$	11 (5.5)	-5.3 (12)
	Y_1	$\beta_{1z}^{(3)}$	.21 (.13)	.45 (.19)
	Y_2	$\beta_{2z}^{(3)}$	.59 (.12)	.65 (.14)
		$\tau_{3z}^{(3)}$	8.9	15

In this summation, $E(Y_2 \mid S = 1)$ is not identified and is therefore a function of sensitivity parameters; specifically,

$$
\begin{aligned}
E(Y_2 \mid S = 1) &= E_{Y_1 \mid S = 1}\{E(Y_2 \mid Y_1, S = 1)\} \\
&= E(\alpha_0^{(1)} + \alpha_1^{(1)} Y_1 \mid S = 1) \\
&= \alpha_0^{(1)} + \alpha_1^{(1)} \mu^{(1)} \\
&= (\alpha_0^{(\geq 2)} + \Delta_{\alpha_0}^{(1:2)}) + (\alpha_1^{(\geq 2)} + \Delta_{\alpha_1}^{(1:2)}) \mu^{(1)}.
\end{aligned}
$$

Recall that we have constrained $\Delta_{\alpha_1}^{(1:2)} = 0$, so that

$$
E(Y_2 \mid S = 1) = \alpha_0^{(\geq 2)} + \Delta_{\alpha_0}^{(1:2)} + \alpha_1^{(\geq 2)} \mu^{(1)}.
$$

Referring back to (10.9), write $E(Y_2) = \mu_2 = \mu_2(\Delta_{\alpha_0}^{(1:2)})$ to emphasize its dependence on the sensitivity parameter. The effect of departures from MAR on μ_2 is captured by the difference

$$
\mu_2(\Delta_{\alpha_0}^{(1:2)}) - \mu_2(0) = \phi_1 \Delta_{\alpha_0}^{(1:2)}. \tag{10.10}
$$

Hence the contribution of the sensitivity parameter to the shift in μ_2 is proportional to the fraction dropping out at $S = 1$.

As with $E(Y_2)$, the marginal mean $E(Y_3)$ is the weighted average

$$
\begin{aligned}
\mu_3 &= E(Y_3) \\
&= \sum_{s=1}^{3} \phi_s E(Y_3 \mid S = s).
\end{aligned}
$$

Because neither $E(Y_3 \mid S = 1)$ nor $E(Y_3 \mid S = 2)$ is identified, μ_3 depends on both sensitivity parameters; hence we write $\mu_3 = \mu_3(\Delta_{\alpha_0}^{(1:2)}, \Delta_{\beta_0}^{(\bullet:3)})$.

Calculations similar to those above can be used to show that

$$\mu_3(\Delta_{\alpha_0}^{(1:2)}, \Delta_{\beta_0}^{(\bullet:3)}) - \mu_3(0,0) = (\phi_1 + \phi_2)\Delta_{\beta_0}^{(\bullet:3)} + \phi_1\beta_2^{(3)}\Delta_{\alpha_0}^{(1:2)}. \quad (10.11)$$

Expressions (10.10) and (10.11) can be used in conjunction with output from Table 10.2 to understand the maximum effect of departures from MAR, based on the set $\mathcal{D}_1$. At the boundary value $(\Delta_{\alpha_0}^{(1:2)}, \Delta_{\beta_0}^{(\bullet:3)}) = (-20, -15)$, we see that for $z = 0$ (EP),

$$
\begin{aligned}
\mu_2(-20) - \mu_2(0) &= \tfrac{7}{40}(-20) \\
&= -3.5, \\
\mu_3(-20, -15) - \mu_3(0, 0) &= (\tfrac{7}{40} + \tfrac{2}{40})(-15) + \tfrac{7}{40}(.59)(-20) \\
&= -6.6, \quad (10.12)
\end{aligned}
$$

and for $z = 1$ (EG),

$$
\begin{aligned}
\mu_2(-20) - \mu_2(0) &= \tfrac{12}{38}(-20) \\
&= -6.3, \\
\mu_3(-20, -15) - \mu_3(0, 0) &= (\tfrac{12}{38} + \tfrac{4}{38})(-15) + \tfrac{12}{38}(.65)(-20) \\
&= -10.4. \quad (10.13)
\end{aligned}
$$

Hence departures from MAR may lead to mean quad strength that is up to 6.6 units lower for EP and 10.4 units lower for EG, relative to MAR.

When the sensitivity parameters are equivalent by treatment group, as they are here, the dropout proportion influences which treatment group will have greater sensitivity to departures from MAR. Because dropout rate is higher on the EG arm, the gradient for μ_{31} away from MAR is steeper than for μ_{30}; the posterior mean of μ_{31} ranges from 78 to 68; and for μ_{30}, the range is 73 to 68. Posterior treatment effect over $\mathcal{D}_1$ ranges from 5 to 1, with associated posteriors for $P(\theta > 0)$ taking values from .74 to .56. None of the posteriors over the set $\mathcal{D}_1$ leads to the conclusion that mean quad strength on EG is greater than on EP.

If prior belief about the values for Δ is confined to the fact that it falls within $\mathcal{D}$, and if each $\Delta \in \mathcal{D}_1$ is assumed to have equal *a priori* probability, then we can assign a uniform prior over $\mathcal{D}_1$ and summarize treatment effects using a single posterior distribution. Results from this approach are summarized in Table 10.1 under MNAR-1. As expected, posterior marginal means have shifted downward. Uncertainty about Δ is reflected in the increased posterior SD for parameters that rely on Δ (means at months 6, 12), but importantly, not for fully identified parameters (means at month 0). Posterior mean treatment difference is 3.1 (PSD = 8.2), which is closer to zero but does not change the qualitative conclusion under MAR that quad strength on the EG arm is similar to EP. Given the sensitivity analysis above, this is not surprising.

Analysis 2: Assume Δ is treatment specific

In the previous analysis, we assumed a common Δ across the treatment arms. However, it may be more reasonable to assume that the degree of departure from MAR differs on the two arms. Recall that our previous analysis was based on the sensitivity parameters

$$\Delta^{(1:2)}_{\alpha_0} = E(Y_2 \mid Y_1, S = 1) - E(Y_2 \mid S \geq 2)$$

$$\Delta^{(\bullet:3)}_{\beta_0} = E(Y_3 \mid Y_1, Y_2, S = 2) - E(Y_3 \mid Y_2, Y_1, S = 3)$$
$$= E(Y_3 \mid Y_1, Y_2, S = 1) - E(Y_3 \mid Y_2, Y_1, S = 3).$$

In this analysis, we allow these parameters to vary by treatment, but to keep the dimension of Δ equal to 2, we combine them by assuming

$$\Delta^{(1:2)}_{\alpha_0(z)} = \Delta^{(\bullet:3)}_{\beta_0(z)} = \Delta_z,$$

for $z = 0, 1$.

It is again helpful to quantify the effect of Δ_z on the shift in marginal means; arguing as before, it is straightforward to show that the maximum effects of departures from MAR on the marginal means $\mu_{2z} = E(Y_2 \mid Z = z)$ and $\mu_{3z} = E(Y_3 \mid Z = z)$ are

$$\mu_{2z}(\Delta_z) - \mu_{2z}(0) = \Delta_z \phi_{1z}$$
$$\mu_{3z}(\Delta_z) - \mu_{3z}(0) = \Delta_z \{\phi_{2z} + \phi_{1z}(1 + \beta_2^{(3)})\}.$$

We set $\mathscr{D}_2 = [0, 20] \times [0, 20]$, based on the largest residual standard deviation over both treatments in Table 10.2.

Using calculations similar to (10.12) and (10.13), the following maximum shifts in μ_{2z} and μ_{3z} relative to MAR are

$$\left. \begin{array}{rcl} \mu_{20}(-20) - \mu_{20}(0) &=& -3.5 \\ \mu_{30}(-20) - \mu_{30}(0) &=& -6.3 \end{array} \right\} \quad z = 0 \ (EP)$$

$$\left. \begin{array}{rcl} \mu_{21}(-20) - \mu_{21}(0) &=& -6.6 \\ \mu_{31}(-20) - \mu_{31}(0) &=& -12.5 \end{array} \right\} \quad z = 1 \ (EG).$$

Figure 10.1 shows contours of the posterior mean of θ and of $P(\theta > 0)$, conditional on fixed values of $(\Delta_0, \Delta_1) \in \mathscr{D}_2$. The contour plot of $P(\theta > 0)$ indicates the specific departures from MAR that would change the qualitative conclusions about treatment effect. In particular, the part of the posterior distribution having $P(\theta > 0) > .9$ is in the region where Δ_1 is near zero and Δ_0 is between -20 and -15; i.e., where dropouts on EG are roughly MAR, but dropouts on EP have quad strength much lower than expected relative to MAR.

As with Analysis 1, we can use an informative prior on $\boldsymbol{\Delta}$. For illustration we use a uniform prior on $\mathscr{D}_2 = [-20, 0] \times [-20, 0]$; the posterior is summarized in Table 10.1. The results largely agree with Analysis 1, but posterior variances are higher because of the wider range for $\boldsymbol{\Delta}$.

Posterior predictive model checking

Although the distributions $p(\boldsymbol{y}_{\text{mis}} \mid \boldsymbol{y}_{\text{obs}}, s)$ and $p(\boldsymbol{y}_{\text{obs}}, s)$ are fully separated, we have made a number of parametric and distributional assumptions about $p(\boldsymbol{y}_{\text{obs}}, s)$; primarily, we have assumed multivariate normality and linearity of associations. Unlike assumptions about $p(\boldsymbol{y}_{\text{mis}} \mid \boldsymbol{y}_{\text{obs}}, s)$, those applied to $p(\boldsymbol{y}_{\text{obs}}, s)$ can be critiqued. We use Pearson's χ^2 statistic (3.19) to assess fit of the identified distributions distributions in (10.1) to the observed data using the posterior predictive approach described in Chapter 8.

Specifically, to assess the fit of $p(y_1)$, we compute

$$T_1(\boldsymbol{y}_1; \boldsymbol{\omega}) = \frac{1}{n} \sum_{i=1}^{n} Q_{i1}(\boldsymbol{\omega}),$$

where

$$Q_{i1}(\boldsymbol{\omega}) = \frac{\{y_{i1} - E(Y_{i1} \mid \boldsymbol{\omega})\}^2}{\text{var}(Y_{i1} \mid \boldsymbol{\omega})}$$

with

$$
\begin{aligned}
E(Y_{i1} \mid \boldsymbol{\omega}) &= \sum_{s=1}^{3} \phi_s \, \mu^{(s)}, \\
\text{var}(Y_{i1} \mid \boldsymbol{\omega}) &= E\{\text{var}(Y_1 \mid S, \boldsymbol{\omega})\} + \text{var}\{E(Y_1 \mid S, \boldsymbol{\omega})\} \\
&= \sum_{s=1}^{3} \phi_s \left[\sigma^{(s)} + \{\mu^{(s)} - E(Y_{i1} \mid \boldsymbol{\omega})\}^2 \right].
\end{aligned}
$$

To assess the fit of $p(y_2 \mid y_1, S \geq 2)$ and $p(y_3 \mid y_1, y_2, S = 3)$, we use $T_2(\boldsymbol{y}; \boldsymbol{\omega})$ and $T_3(\boldsymbol{y}; \boldsymbol{\omega})$, respectively, where, for $j = 2, 3$,

$$T_j(\boldsymbol{y}; \boldsymbol{\omega}) = \frac{1}{n_{\geq j}} \sum_{i=1}^{n_{\geq j}} Q_{ij}(\boldsymbol{\omega}).$$

As above,

$$Q_{ij}(\boldsymbol{\omega}) = \frac{\{y_{ij} - E(Y_{ij} \mid y_{i1}, \dots, y_{i,j-1}, S \geq j, \boldsymbol{\omega})\}^2}{V_j(\boldsymbol{\omega})},$$

where

$$E(Y_{ij} \mid y_{i1}, \ldots, y_{i,j-1}, S \geq j, \boldsymbol{\omega}) = \sum_{s=j}^{3} \phi_s \, E(Y_{ij} \mid y_{i1}, \ldots, y_{i,j-1}, S = s)$$

$$= \sum_{s=j}^{3} \phi_s \left(\beta_0^{(j)} + \sum_{l=1}^{j-1} \beta_l^{(j)} Y_{il} \right)$$

and

$$V_j(\boldsymbol{\omega}) = \mathrm{var}(Y_j \mid y_1, \ldots, y_{j-1}, S \geq j, \boldsymbol{\omega}) = \tau_j^{(\geq j)}.$$

We normalize the data summary $T_j(\boldsymbol{y}; \boldsymbol{\omega})$ using $n_{\geq j}$, the total number of subjects in patterns greater than or equal to j. The reason for doing this is that the total number in these patterns in the replicated dataset will vary with each sample from the posterior predictive distribution and will differ from the observed number of subjects in those patterns in the original dataset.

Posterior predictive probabilities based on $T_j(\cdot; \boldsymbol{\omega})$ were computed as described in Chapter 8. The probabilities for these three checks were .71, .56, and .56, indicating no evidence of wide departures from the observed data model.

10.2.9 Summary of pattern mixture analysis

Our strategy here has been to construct the full-data distribution as a PMM. The general model allows missingness to be MNAR, with MAR as a special case. The degree of departure from MAR is quantified by a set of parameters $\boldsymbol{\Delta}$ measuring departures from MAR. The $\boldsymbol{\Delta}$ parameters are completely nonidentified, so the fit of the model to observed data is identical across the assumed missing data mechanisms. We assess the fit of the observed data model using posterior predictive checks.

Several key issues must be addressed when implementing a PMM analysis. First, the general model has a large number of sensitivity parameters. In practical settings, the dimension must be reduced; we have given suggestions for doing this.

Second, analyses based on examining conditional posteriors at fixed values of $\boldsymbol{\Delta}$ over a predetermined range $\mathscr{D}$ can be useful, but specification of $\mathscr{D}$ is subjective. Our approach has been to calibrate $\mathscr{D}$ using relevant posterior distributions under the MAR constraint. Other approaches are certainly possible and will depend on context. It is important to ensure that choices for $\mathscr{D}$ do not extrapolate the missing data outside of a 'reasonable' range (e.g., negative values of quadriceps strength).

Third, it must be kept in mind that the contour plots generated for Analyses 1 and 2 represent conditional posteriors (or posteriors over an infinite number of point mass priors), and that ideally, a final inference should be

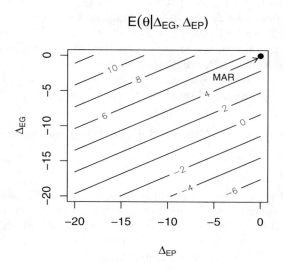

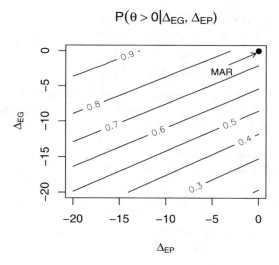

Figure 10.1 *Growth hormone trial: contours of posterior mean treatment effect θ (difference in quad strength) and of posterior probability $P(\theta > 0)$ as function of* **Δ** *from Analysis 2.*

based on a well-motivated prior distribution for the nonidentified sensitivity parameters. The priors for $\boldsymbol{\Delta} \mid \boldsymbol{\xi}_{\mathrm{M}}$ used in Analyses 1 and 2 are informed solely by the choice of $\mathscr{D}$, and all possible values in $\mathscr{D}$ are weighted equally in our example here. In practice, however, it is likely that more information is known about departures from MAR, whether from previous studies or from expert opinion. In the case study presentation given in Section 10.3, we illustrate the use of informative priors based on elicited expert opinion.

10.3 OASIS Study: Selection models, mixture models, and elicited priors

10.3.1 Overview

The OASIS trial is described in detail in Section 1.6. OASIS was a two-arm randomized trial designed to reduce smoking rates in alcoholics. Smoking status was assessed at 1, 3, 6, and 12 months following randomization. For each individual, the full data comprise the 4×1 response vector of smoking outcomes $\boldsymbol{Y} = (Y_1, Y_2, Y_3, Y_4)^{\mathrm{T}}$, obtained at months 1, 3, 6, and 12 following baseline; the corresponding vector of response indicators $\boldsymbol{R} = (R_1, R_2, R_3, R_4)^{\mathrm{T}}$; and treatment group Z (1 if randomized to enhanced intervention, 0 if standard intervention).

The objective is to compare smoking rates at month 12 between those randomized to $Z = 1$ vs. $Z = 0$. Inference is made in terms of the odds ratio

$$\varphi = \frac{\mathrm{odds}(Y_4 = 1 \mid Z = 1)}{\mathrm{odds}(Y_4 = 1 \mid Z = 0)}.$$

This trial has intermittent missingness; follow-up time $S = S(\boldsymbol{R})$ is defined as the last time point at which data are observed (all individuals are observed at the first time point):

$$S = \begin{cases} 1 & \text{if } \boldsymbol{R} = (1, 0, 0, 0) \\ 2 & \text{if } \boldsymbol{R} = (1, 1, 0, 0) \\ 3 & \text{if } \boldsymbol{R} \in \{(1, 1, 1, 0), (1, 0, 1, 0)\} \\ 4 & \text{if } \boldsymbol{R} \in \{(1, 1, 1, 1), (1, 0, 1, 1), (1, 1, 0, 1), (1, 0, 0, 1)\}. \end{cases}$$

To handle intermittent missing values, we assume missingness is MAR conditionally on S; that is, we assume

$$p(\boldsymbol{y}_{\mathrm{mis}} \mid \boldsymbol{y}_{\mathrm{obs}}, s, z, \boldsymbol{r}) = f(\boldsymbol{y}_{\mathrm{mis}} \mid \boldsymbol{y}_{\mathrm{obs}}, s, z),$$

or that $\boldsymbol{Y}_{\mathrm{mis}}$ is independent of $\boldsymbol{R}$ given $(\boldsymbol{Y}_{\mathrm{obs}}, S, Z)$.

Figure 10.2 displays the smoking rate by pattern from observed data. For both treatment arms, those who complete the study ($S = 4$) show lower smoking rates than dropouts. Furthermore, for $S \in \{2, 3\}$, smoking rate appears to increase preceding dropout in ET arm. Later, for the pattern mixture analysis, we combine the relatively few subjects with $S = 2$ or $S = 3$ into a single pattern that, using a slight abuse of notation, is labeled as $S = (2, 3)$.

We conduct several analyses using (parametric) selection models and using mixture models. For the latter, we describe the construction of an informative prior that will be used for inference under MNAR.

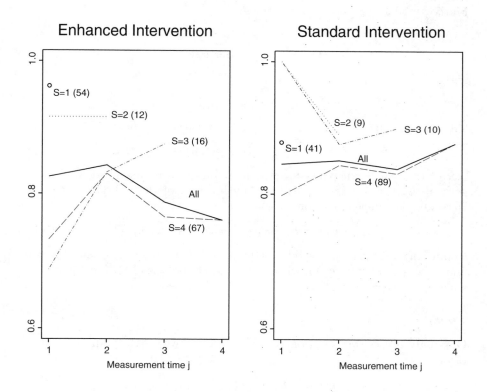

Figure 10.2 *OASIS study: proportion smoking by dropout pattern S at each month, stratified by treatment arm (number of subjects per pattern in parentheses).*

10.3.2 Selection model specification

A selection model factorization of the full-data model follows $p(\boldsymbol{y} \mid \boldsymbol{\omega}) \, p(\boldsymbol{r} \mid \boldsymbol{y}, \boldsymbol{\omega})$. We use a *parametric* selection model to infer treatment effect under both MAR and MNAR. The selection model being used here is parametric because, we impose constraints on the full-data distribution, and on the missing data mechanism (similar to the constraints used in Example 8.2). We focus attention here on the effect of priors on the parameters governing departures from MAR, and the degree to which they may be informed by observed data.

Full-data model

For the first factor, we assume the full-data response follows a first-order Markov model, separately by treatment,

$$
\begin{aligned}
Y_1 &\sim \mathrm{Ber}(\theta_1) \\
Y_j \mid Y_{j-1} &\sim \mathrm{Ber}(\theta_{j|j-1}),
\end{aligned}
$$

with

$$
\mathrm{logit}(\theta_{j|j-1}) = \beta_{0j} + \beta_1 y_{j-1}.
$$

The missing data mechanism is specified in terms of the hazard of dropout,

$$
h_S(t_j \mid \boldsymbol{y}) \;=\; P(S = j \mid S \geq j, \boldsymbol{y}), \quad j = 1, 2, 3, 4
$$

(recall that all participants have at least one measurement, so $S \geq 1$). In general, $h_S(t_j \mid \boldsymbol{Y})$ can depend on any function of the full-data vector $\boldsymbol{Y}$. Here we allow dependence on Y_{j-1} and Y_j via

$$
\mathrm{logit}\{h_S(t_j \mid \boldsymbol{y})\} = \psi_0 + \psi_1 y_{j-1} + \psi_2 y_j + \psi_3 y_j y_{j-1}. \tag{10.14}
$$

Conditional on Y_{j-1}, the log odds of dropout at t_j is $\psi_0 + \psi_2 y_j$ when $Y_{j-1} = 0$, and is $(\psi_0 + \psi_1) + (\psi_2 + \psi_3) y_j$ when $Y_{j-1} = 1$. Dropout is MAR (and ignorable) when $\psi_2 = \psi_3 = 0$.

Priors

Our analysis uses various priors on (ψ_2, ψ_3). First, we use the point mass MAR prior

$$
p(\psi_2, \psi_3) \;=\; I\{\psi_2 = \psi_3 = 0\}.
$$

Next, we use a series of informative priors that permit departures from MAR. Rather than use point mass priors, we employ normal priors centered away from zero and having unit variance. Specifically,

$$
(\psi_2, \psi_3)^{\mathrm{T}} \mid (d_2, d_3)^{\mathrm{T}} \;\sim\; N\left(\begin{pmatrix} d_2 \\ d_3 \end{pmatrix}, \begin{pmatrix} 1 & 0 \\ 0 & 1 \end{pmatrix} \right). \tag{10.15}
$$

The mean parameters are varied over the set

$$
(d_2, d_3) \;\in\; \{-0.5, 0, 0.5, 1.0\}^{\otimes 2}.
$$

Priors for $(\boldsymbol{\beta}, \psi_0, \psi)$ are specified as independent diffuse normal priors with mean zero and variance 100.

Table 10.3 *OASIS trial: posterior means (SD) for hazard function parameters, treatment-specific smoking rates, and treatment effects from selection model analysis using different priors for (ψ_2, ψ_3). Summaries provided here for point mass MAR prior and for normal priors having mean (d_2, d_3) and variance $\boldsymbol{I}$; see (10.15).*

Trt	Parameter	MAR	Prior means (d_2, d_3)		
			$(-1,-1)$	$(0,0)$	$(1,1)$
ET	ψ_0	−2.4 (.50)	−2.5 (.64)	−2.8 (.69)	−3.2 (.80)
	ψ_1	1.6 (.50)	1.8 (1.0)	.88 (.99)	−.02 (1.0)
	ψ_2	—	−.15 (.96)	.50 (.86)	1.2 (.90)
	ψ_3	—	−.16 (.99)	.56 (.91)	1.2 (.91)
ST	ψ_0	−2.0 (.41)	−1.7 (.44)	−1.8 (.54)	−2.6 (.76)
	ψ_1	.47 (.43)	2.6 (.60)	1.7 (1.2)	−.71 (1.3)
	ψ_2	—	−2.0 (.86)	−1.1 (1.2)	.86 (1.1)
	ψ_3	—	−1.9 (.87)	−.95 (1.1)	.87 (1.1)
	$E(Y_4 \mid \text{ET})$	.78 (.05)	.76 (.07)	.83 (.04)	.82 (.04)
	$E(Y_4 \mid \text{ST})$	.88 (.03)	.77 (.06)	.82 (.07)	.89 (.04)
	Trt effect φ	.51	1.09	.97	.61
	95% interval	(.19, 1.06)	(.32, 2.53)	(.26, 2.38)	(.22, 1.38)

10.3.3 Selection model analyses under MAR and MNAR

Results are shown in Table 10.3. Under MAR ($\psi_2 = \psi_3 = 0$), the treatment odds ratio is .5 with a credible interval of $(.2, 1.1)$, indicating a marginally positive effect of the ET treatment in lowering the rate of smoking.

For the ET arm, subjects smoking the previous week were more likely to drop out at the current week than those not smoking, with an odds ratio of about five. A similar relation was seen in the ST arm, but with a smaller odds ratio (1.6) and a credible interval covering one.

Under MNAR, the posteriors for ψ_2 and ψ_3 were relatively diffuse and informative priors were needed to stabilize the posterior. In fact, the priors and posteriors for (ψ_2, ψ_3) are very different for all three informative priors, indicating that observed data are contributing information.

We therefore do not recommend the MNAR analyses with informative priors here because the prior and the posterior for (ψ_2, ψ_3) are very different and therefore do not satisfy the condition (9.12) as recommended in Chapter 9.

Semiparametric selection models offer a possible alternative; this is an area of ongoing work.

10.3.4 Pattern mixture model specification

For the mixture model specification, the full-data response distribution is a mixture over the follow-up times, factored as

$$p(\boldsymbol{y}_{\mathrm{mis}}, \boldsymbol{y}_{\mathrm{obs}}, s \mid z) = p(\boldsymbol{y}_{\mathrm{mis}} \mid \boldsymbol{y}_{\mathrm{obs}}, s, z)\, p(\boldsymbol{y}_{\mathrm{obs}} \mid s, z)\, p(s \mid z).$$

We group the patterns $S = 2$ and $S = 3$ into a single pattern and label it $S = (2, 3)$. Hence for purposes of the pattern mixture model, the realizations of S are $s \in \{1, (2, 3), 4\}$. The component distributions follow

$$Y_1 \mid S = s, Z = z \quad \sim \quad \mathrm{Ber}(\mu_{1z}^{(s)}),$$

$$Y_j \mid Y_{j-1}, S = s, Z = z \quad \sim \quad \mathrm{Ber}(\phi_{jz}^{(s)}), \quad j = 2, 3, 4, \qquad (10.16)$$

where

$$\mathrm{logit}(\phi_{jz}^{(s)}) = \gamma_{jz}^{(s)} + \theta_{jz}^{(k)} Y_{j-1}. \qquad (10.17)$$

For the dropout distribution, we assume

$$S \mid Z = z \quad \sim \quad \mathrm{Mult}(\psi_{1z}, \psi_{(2,3)z}, \psi_{4z}).$$

This model assumes that within dropout pattern, the joint distribution of $\boldsymbol{Y}$ can be captured using a first-order serial dependence structure. Table 10.4 shows the structure of the model in the logit scale; empty cells represent quantities that cannot be identified by observed data.

10.3.5 MAR and MNAR parameterizations

Identifying the model under MAR

As with the Growth Hormone example, model identification under MAR is based on condition (8.17) in Theorem 8.1. To satisfy MAR under the first-order dependence structure given by (10.17), we require

$$p_s(y_j \mid y_{j-1}, z) = p_{\geq j}(y_j \mid y_{j-1}, z)$$

for each $s < j$ and for $j = 2, 3, 4$. * Again, this implies that the distribution $(Y_j \mid Y_{j-1}, Z)$ is equivalent for those dropping out prior to j (i.e., $s < j$) and those still in follow-up at j (i.e., $s \geq j$).

For those still in follow-up at j, define

$$\phi_{jz}^{(\geq j)} \quad = \quad E(Y_j \mid Y_{j-1}, Z = z, S \geq j),$$

___
* With the collapsed patterns, if $s = (2, 3)$, then $s < j$ only when $j = 4$.

Table 10.4 *OASIS trial: representation, on the logit scale, of full-data component distributions model given by (10.17). Cell entries are logits of $\phi_j^{(s)}(y) = E(Y_j \mid Y_{j-1} = y, S = s)$. Empty cells cannot be identified by observed data. Treatment subscripts suppressed for clarity.*

	y	$\phi_2^{(s)}(y)$	$\phi_3^{(s)}(y)$	$\phi_4^{(s)}(y)$
$S = 1$	0			
	1			
$S = (2,3)$	0	$\gamma_2^{(2,3)}$	$\gamma_3^{(2,3)}$	
	1	$\gamma_2^{(2,3)} + \theta_2^{(2,3)}$	$\gamma_3^{(2,3)} + \theta_3^{(2,3)}$	
$S = 4$	0	$\gamma_2^{(4)}$	$\gamma_3^{(4)}$	$\gamma_4^{(4)}$
	1	$\gamma_2^{(4)} + \theta_2^{(4)}$	$\gamma_3^{(4)} + \theta_3^{(4)}$	$\gamma_4^{(4)} + \theta_4^{(4)}$

and let

$$\operatorname{logit}(\phi_{jz}^{(\geq j)}) = \gamma_{jz}^{(\geq j)} + \theta_{jz}^{(\geq j)} Y_{j-1}. \tag{10.18}$$

It can be shown that, in this setting, (10.18) is compatible with the full-data model (10.17); in fact, the regression parameters in (10.18) are determined by those in (10.17).

From Theorem 8.1, the MAR constraint is satisfied when, for each $j > s$,

$$\gamma_{jz}^{(s)} = \gamma_{jz}^{(\geq j)}$$
$$\theta_{jz}^{(s)} = \theta_{jz}^{(\geq j)}.$$

The MAR identification scheme is shown in Table 10.5.

Parameterizing the full-data model under MNAR

The full-data model can be reparameterized in terms of sensitivity parameters. Let $\boldsymbol{\xi}(\boldsymbol{\omega}) = (\boldsymbol{\xi}_{\mathrm{S}}, \boldsymbol{\xi}_{\mathrm{M}})$, where

$$\boldsymbol{\xi}_{\mathrm{S}} = \begin{cases} \theta_j^{(1)}, \gamma_j^{(1)} & j = 2, 3, 4 \\ \theta_4^{(2,3)}, \gamma_4^{(2,3)} & \end{cases}$$

$$\boldsymbol{\xi}_{\mathrm{M}} = \begin{cases} \theta_j^{(2,3)}, \gamma_j^{(2,3)} & j = 2, 3 \\ \theta_j^{(4)}, \gamma_j^{(4)} & j = 2, 3, 4 \\ \mu_1^{(1)}, \mu_1^{(2,3)}, \mu_1^{(4)} & \\ \psi^{(1)}, \psi^{(2,3)}, \psi^{(4)} & \end{cases}$$

Table 10.5 *OASIS trial: identification of full-data model (10.17) under MAR, on the logit scale. Cell entries are (logits of)* $\phi_j^{(s)}(y) = E(Y_j \mid Y_{j-1} = y, S = s)$. *Treatment subscripts suppressed for clarity.*

	y	$\phi_2^{(s)}(y)$	$\phi_3^{(s)}(y)$	$\phi_4^{(s)}(y)$
$S = 1$	0	$\gamma_2^{(\geq 2)}$	$\gamma_3^{(\geq 3)}$	$\gamma_4^{(4)}$
	1	$\gamma_2^{(\geq 2)} + \theta_2^{(\geq 2)}$	$\gamma_3^{(\geq 3)} + \theta_3^{(\geq 3)}$	$\gamma_4^{(4)} + \theta_4^{(4)}$
$S = (2, 3)$	0	$\gamma_2^{(2,3)}$	$\gamma_3^{(2,3)}$	$\gamma_4^{(4)}$
	1	$\gamma_2^{(2,3)} + \theta_2^{(2,3)}$	$\gamma_3^{(2,3)} + \theta_3^{(2,3)}$	$\gamma_4^{(4)} + \theta_4^{(4)}$
$S = 4$	0	$\gamma_2^{(4)}$	$\gamma_3^{(4)}$	$\gamma_4^{(4)}$
	1	$\gamma_2^{(4)} + \theta_2^{(4)}$	$\gamma_3^{(4)} + \theta_3^{(4)}$	$\gamma_4^{(4)} + \theta_4^{(4)}$

(see Table 10.4), where we have suppressed the dependence of these parameters on treatment. We also can characterize departures from MAR via the mapping $\boldsymbol{\xi}_S = \boldsymbol{h}(\boldsymbol{\xi}_M, \boldsymbol{\Delta})$, defined as follows.

Among those in patterns $s = 1$ and $s = (2, 3)$, define, for each $j > s$ and for $z = 0, 1$, a pair of parameters $(\Delta_{0jz}^{(s)}, \Delta_{1jz}^{(s)})$ that satisfy

$$
\begin{aligned}
\gamma_{jz}^{(s)} &= h_1(\boldsymbol{\xi}_M) = \gamma_{jz}^{(\geq j)} + \Delta_{0jz}^{(s)} \\
\theta_{jz}^{(s)} &= h_2(\boldsymbol{\xi}_M) = \theta_{jz}^{(\geq j)} + (\Delta_{1jz}^{(s)} - \Delta_{0jz}^{(s)}).
\end{aligned}
\tag{10.19}
$$

Each $\boldsymbol{\Delta}$ parameter represents a log odds ratio comparing odds of smoking at time j, conditional on Y_{j-1}, between those with $s < j$ (dropped out) and those with $s \geq j$ (still in follow-up). Specifically,

$$
\Delta_{yjz}^{(s)} = \log \left\{ \frac{\text{odds}(Y_j = 1 \mid Y_{j-1} = y, Z = z, S = s)}{\text{odds}(Y_j = 1 \mid Y_{j-1} = y, Z = z, S \geq j)} \right\}.
$$

Setting each $\Delta_{yjz}^{(s)} = 0$ implies MAR.

The dimension of $\boldsymbol{\Delta}$ is 16. To reduce the parameter space, we make the simplifying assumption that $\Delta_{yjz}^{(s)} = \Delta_y$, leaving the two-dimensional parameter $\boldsymbol{\Delta} = (\Delta_0, \Delta_1)$. The simplification implies that departures from MAR at time j may differ depending on Y_{j-1}, but conditionally on Y_{j-1} are constant across measurement time, treatment group, and dropout time. The MNAR model is shown in Table 10.6. Naturally, other simplifications can be considered, depending on the application.

Prior distributions for parameters can still be specified as

$$
p(\boldsymbol{\xi}_S, \boldsymbol{\xi}_M, \boldsymbol{\Delta}) = p(\boldsymbol{\xi}_S \mid \boldsymbol{\xi}_M, \boldsymbol{\Delta}) \, p(\boldsymbol{\Delta} \mid \boldsymbol{\xi}_M) \, p(\boldsymbol{\xi}_M),
$$

Table 10.6 *OASIS trial: full-data model (10.17) under MNAR, with parameters* (Δ_0, Δ_1) *capturing departures from MAR on the logit scale. Cell entries are logits of* $\phi_j^{(s)}(y) = E(Y_j \mid Y_{j-1} = y, S = s)$. *Treatment subscripts suppressed for clarity.*

	y	$\phi_2^{(s)}(y)$	$\phi_3^{(s)}(y)$	$\phi_4^{(s)}(y)$
$S = 1$	0	$\gamma_2^{(\geq 2)} + \Delta_0$	$\gamma_3^{(\geq 3)} + \Delta_0$	$\gamma_4^{(4)} + \Delta_0$
	1	$\gamma_2^{(\geq 2)} + \theta_2^{(\geq 2)} + \Delta_1$	$\gamma_3^{(\geq 3)} + \theta_3^{(\geq 3)} + \Delta_1$	$\gamma_4^{(4)} + \theta_4^{(4)} + \Delta_1$
$S = (2,3)$	0	$\gamma_2^{(2,3)}$	$\gamma_3^{(2,3)}$	$\gamma_4^{(4)} + \Delta_0$
	1	$\gamma_2^{(2,3)} + \theta_2^{(2,3)}$	$\gamma_3^{(2,3)} + \theta_3^{(2,3)}$	$\gamma_4^{(4)} + \theta_4^{(4)} + \Delta_1$
$S = 4$	0	$\gamma_2^{(4)}$	$\gamma_3^{(4)}$	$\gamma_4^{(4)}$
	1	$\gamma_2^{(4)} + \theta_2^{(4)}$	$\gamma_3^{(4)} + \theta_3^{(4)}$	$\gamma_4^{(4)} + \theta_4^{(4)}$

where, as usual,

$$p(\boldsymbol{\xi}_{\mathrm{S}} \mid \boldsymbol{\xi}_{\mathrm{M}}, \boldsymbol{\Delta}) \;\;=\;\; I\{\boldsymbol{\xi}_{\mathrm{S}} = \boldsymbol{h}(\boldsymbol{\xi}_{\mathrm{M}}, \boldsymbol{\Delta})\}.$$

10.3.6 Pattern mixture analysis under MAR

As a starting point, we present analysis under MAR assumption. First we assign point mass at zero to (Δ_0, Δ_1) separately by both treatment arms via

$$p(\Delta_0, \Delta_1 \mid \boldsymbol{\xi}_{\mathrm{M}}) \;\;=\;\; I\{(\Delta_0, \Delta_1) = (0,0)\}.$$

This prior assumes MAR with 100% certainty. Priors for components of $\boldsymbol{\xi}_{\mathrm{M}}$ are diffuse but proper. Results are shown in Table 10.7, compared side by side with summaries using available data separately at each time point. The table suggests that the MAR analysis varies very little from simply computing smoking rates from available data.

10.3.7 Pattern mixture analysis under MNAR using elicited priors

Under MAR, the probability of smoking at time j, conditional on observed smoking history up to but not including j, is equivalent between dropouts and those who continue the study. Although research on smoking cessation has suggested that dropouts are more likely to be smokers after leaving a study (Lichtenstein and Glasgow, 1992), the data offer no evidence about this.

The mixture model factorization allows assumptions about the missing data mechanism to be completely encoded in the prior $p(\boldsymbol{\Delta} \mid \boldsymbol{\xi}_{\mathrm{M}})$. In this section,

Table 10.7 *OASIS trial: inference about time-specific smoking rates using available data at each time point and using a pattern mixture model with an MAR point mass prior.*

Treatment	Month	Available Data	MAR Prior
ET	1	.83	.83
	3	.84	.87
	6	.79	.82
	12	.76	.78
ST	1	.85	.85
	3	.85	.85
	6	.84	.84
	12	.88	.88
Trt effect φ		.43	.51
95% interval			(.19, 1.06)

we summarize the use of priors elicited from a panel of experts. Complete details appear in Lee et al. (2007) and Lee (2006); the process and analyses are briefly summarized here.

Eliciting information and constructing prior distributions

Lee et al. (2007) convened a panel of four faculty investigators who are experts in smoking cessation in order to elicit information about $\mathbf{\Delta} = (\Delta_0, \Delta_1)$. Elicitation was conducted using a written questionnaire that provided detailed information about the study and its target population. The sensitivity parameters Δ_0 and Δ_1 are measured on the odds ratio scale, but prior information was elicited on the probability scale for the sake of transparency.

The main part of the questionnaire asked each expert to provide a *best guess* and *90% interval* for the probability of smoking at t_j among those who dropped out at t_{j-1}. To help calibrate the response, experts were asked to provide their answer conditionally on hypothetical rates for those who did not drop out. Formally, each expert was provided with a range of values for

$$E(Y_j \mid Y_{j-1} = y, S \geq j), \quad y = 0, 1$$

and asked to provide a best guess and 90% interval for

$$E(Y_j \mid Y_{j-1} = y, S = j - 1), \quad y = 0, 1.$$

Next, each expert's best guess and lower and upper bounds of their 90% interval for smoking rates among dropouts at each time j were converted to the log odds scale, based on the observed smoking rates at each time among non-dropouts; i.e., based on observed values of $E(Y_j \mid Y_{j-1} = y, S \geq j)$. In that sense we are eliciting priors for $\Delta \mid \boldsymbol{\xi}_M$ because $E(Y_j \mid Y_{j-1}, S \geq j)$ is a function of $\boldsymbol{\xi}_M$.

The log odds ratios corresponding to the best guess and interval boundaries were then averaged over time, weighting by sample size, to obtain a summary guess and summary interval for each expert. This yields, for each expert, a 90% interval and a modal value for both Δ_0 and Δ_1.

Not surprisingly, the expert-specific intervals for the Δ's were asymmetric. We therefore assumed a skew-normal prior for each expert (Azzalini and Valle, 1996). The skew-normal distribution $p(\cdot \mid \mu, \eta, \nu)$ has three parameters: location μ, scale η, and shape (or skewness) ν. Two percentiles and a mode are sufficient to uniquely determine the parameters. Specifically, using summaries of best guess (mode), 5th percentile, and 95th percentile as inputs, the following system of equations was solved for (μ, η, ν), separately for each expert and each of Δ_0, Δ_1:

$$\arg \max_{\Delta} p(\Delta \mid \mu, \eta, \nu) = \text{mode},$$

$$\int_{-\infty}^{L} p(\Delta \mid \mu, \eta, \nu) \, d\Delta = 0.05,$$

$$\int_{-\infty}^{U} p(\Delta \mid \mu, \eta, \nu) \, d\Delta = 0.95.$$

Here, L and U are, respectively, the lower and upper bounds of the 90% interval for Δ, elicited from the experts.

Finally, a four-component mixture (one for each expert) of skew normal distributions was used for each Δ parameter, with each expert's prior contributing equally to the mixture.

Summary of analysis

Figure 10.3 shows the effect of the informative priors relative to MAR for the ET arm by examining the conditional smoking probabilities $E(Y_j \mid Y_{j-1} = 0, S = 1)$, for $j = 2, 3, 4$. In each case, the expert-specific prior shifts the smoking probabilities toward one. This figure shows that even for participants who were not smoking at t_{j-1}, those who drop out after time 1 ($S = 1$) are believed to have substantially higher smoking probability compared to those who continue ($S \geq 2$). Under MAR, the probabilities would be equal between dropouts and non-dropouts.

Figure 10.4 shows the distribution of $E(Y_4)$ for each treatment arm; it separates out posteriors derived by using individual expert-specific priors, the mixture (of experts) prior, and the MAR prior. Again, relative to MAR, we see $E(Y_4)$ shifted toward one.

The full summary of time-specific smoking rates and the treatment odds ratio is given in Table 10.8, and the posterior distribution of the treatment odds ratio is given in Figure 10.5. Using the informative prior, the odds ratio is attenuated toward 1 relative to MAR, and shows more variability (posterior mean .5 with 95% interval (.2, 1.1) under MAR; posterior mean .7 with 95% interval (.2, 1.4) using informative priors). The additional variability can also be seen in Figure 10.5 as the right tail under MNAR is heavier than under

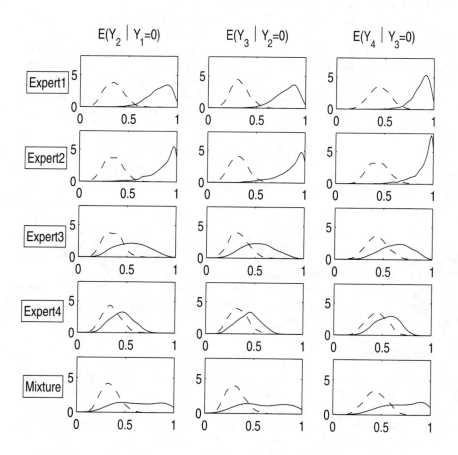

Figure 10.3 *OASIS trial: for the ET arm, comparison of conditional means $E(Y_j \mid Y_{j-1} = 0, S = 1)$ for $j = 2, 3, 4$ under MNAR using expert-specific and mixture-of-experts priors (solid lines), and under MAR (dashed line).*

MAR. Whereas the MAR analysis suggests that ET is moderately superior, the MNAR analysis incorporating expert opinion — along with uncertainty stemming from within-expert and between-expert variation — is more equivocal. Because dropout rate is greater on the ET arm, the relative adjustment in smoking rates under MNAR also is greater.

10.3.8 Summary: selection vs. pattern mixture approaches

We fit both parametric selection models and mixture models under MAR and by using informative priors. This analysis has brought out several key issues. First, the identifiability of all parameters in parametric selection models makes them ill-suited for sensitivity analysis and constructing informative priors. Semiparametric selection models would be a possible alternative. On the other

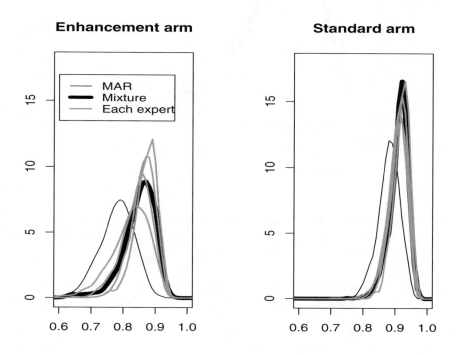

Figure 10.4 *OASIS trial: posterior distribution of* $E(Y_4)$*, smoking rate at month 12, by treatment. Posteriors shown correspond to the following priors for* Δ*: point mass at zero (MAR), expert-specific priors, and the mixture of expert-specific priors (MNAR).*

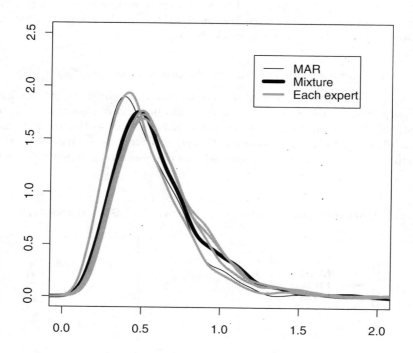

Figure 10.5 *OASIS trial: posterior distribution of the treatment odds ratio φ under priors for Δ corresponding to MAR, expert-specific opinion, and a mixture of expert opinion. The MAR prior places point mass at $(\Delta_0, \Delta_1) = (0,0)$.*

hand, mixture models are well suited because the parameter space is easier to separate. The various pattern mixture models fit here all have the same observed-data likelihood.

Second, eliciting informative priors can be complex. A key component of the elicitation is to develop tools for which expert opinion can be accurately acquired. Here, experts were asked to condition their opinion about smoking rate at time j among dropouts using hypothetical smoking rates for non-dropouts, conditional on smoking status Y_j at time $j - 1$. This helped the experts calibrate their priors.

Finally, we saw that the (ignorable) selection model MAR analysis and the MAR mixture model analysis were identical. This is not a general phe-

Table 10.8 *OASIS trial: posterior inferences about time-specific smoking probability, derived using informative priors elicited from experts. The MAR inferences are shown for comparison.*

				Expert-Specific Posteriors			
Treatment	Month	MAR	MNAR	1	2	3	4
	1	.83	.83	.83	.83	.83	.83
ET	3	.87	.88	.87	.89	.89	.88
	6	.82	.85	.83	.87	.87	.85
	12	.78	.84	.83	.87	.87	.83
	1	.85	.85	.85	.85	.85	.85
ST	3	.85	.87	.86	.87	.87	.86
	6	.84	.86	.86	.87	.86	.86
	12	.88	.90	.90	.91	.90	.90
Trt effect φ		.51	.65	.65	.62	.71	.68
		(.2, 1.1)	(.2, 1.4)				

nomenon. The reason for their equivalence here is that with longitudinal binary data, a mixture of first-order serial dependence models can also be represented as a single first-order serial dependence model. Hence inferences about the full data will be the same under MAR.

10.4 Pediatric AIDS trial: Mixture of varying coefficient models for continuous dropout

10.4.1 Overview

In this example we use a mixture of varying coefficient models to analyze data from the pediatric clinical trial described in Section 1.7. Individuals were scheduled for follow-up every 3 months, but the actual measurement times varied considerably from individual to individual. Dropout time was fixed at the time of the last observed measurement. The observed CD4 data, in the log scale, are shown in Figure 1.5, with several individual trajectories highlighted. From the small random sample of individuals, we see that those with longer follow-up tend to have higher overall CD4 counts and less steep declines over time.

Our objective is to characterize the average CD4 count over time in both treatment arms, and to draw inference about the difference in mean CD4 at the end of the study period. We assume that the average (log) CD4 is linear in time with slope and intercept that depend on dropout time. The conditional distributions (given dropout time) are averaged across the distribution of dropout times separately by treatment, using a mixture of varying coefficient models (Section 8.4.4). The intercept and slope of CD4 will be given by smooth but unknown functions $\beta_0(u)$ and $\beta_1(u)$ of dropout time.

Figure 10.6 is an exploratory plot that can be used to motivate the mixture model. It shows individual-specific OLS slope estimates plotted against dropout time, and suggests a possibly nonlinear association between slope and dropout time. However, it should be kept in mind that the slope estimates for earlier dropouts are based on fewer observations and tend to have higher variance.

Inference from the mixture of VCM will be compared to inference under a standard regression model that assumes MAR. Hogan et al. (2004a) analyze these data using a similar model; the Bayesian formulation given here derives mainly from Su and Hogan (2007), who provide a more extensive treatment, including detailed sensitivity analysis.

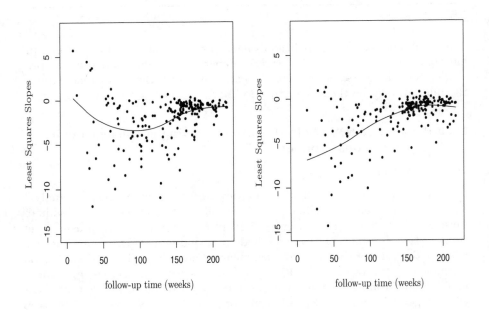

Figure 10.6 *Pediatric AIDS study: OLS slopes vs. follow-up time for each treatment, with unweighted lowess smoother fit to each set of points.*

10.4.2 Model specification: CD4 counts

Centering and rescaling the time axis leads to a more interpretable treatment effect and more stable implementation of posterior sampling. Let t^0 and U^0, respectively, denote measurement time and dropout time in the original scale. We use the shifted and rescaled time variables $t = (t^0 - C)/L$ and $U = (U^0 - C)/L$, where $C = \overline{U}^0 = 139$ and $L = \max_i U_i^0$. The length of the rescaled time axis is equal to one and it is centered at the (sample) average follow-up time.

The model used here is described more generally in Section 8.4.4; the specific formulation is similar to Example 8.8. Treatment assignment is denoted by Z (1 = high dose AZT, 0 = low dose AZT). Let t denote (rescaled) time elapsed from beginning of follow-up, let $Y(t) = \log\{CD4(t) + 1\}$, and let U denote (rescaled) length of follow-up (i.e., dropout time). Our objective is to characterize the distribution of $Y(t)$ over the time interval.

Figure 1.5 shows, in the original time axis, a scatterplot of log CD4 counts vs. time, with individual profiles highlighted. Even in a small sample of individuals, we see that longer follow-up time tends to be associated with greater intercept and slope.

As with Example 8.8, we assume the process $Y_i(t)$ is potentially observed at a set of points $\mathscr{T}_i = \{t_{i1}, \ldots, t_{i,n_i}\}$, giving rise to the full-data vector

$$\boldsymbol{Y}_i = (Y_i(t_{i1}), \ldots, Y_i(t_{i,n_i}))^{\mathrm{T}} = (Y_{i1}, \ldots, Y_{i,n_i})^{\mathrm{T}}.$$

Conditional on dropout, we assume the log CD4 trajectory is linear in time. Let $\boldsymbol{x}_{ij} = (1, t_{ij})$, so that $\boldsymbol{X}_i = (\boldsymbol{x}_{i1}^{\mathrm{T}}, \ldots, \boldsymbol{x}_{i,n_i}^{\mathrm{T}})^{\mathrm{T}}$ is the $n_i \times 2$ design matrix.

The full-data distribution $p(\boldsymbol{y}, u \mid z)$ is factored as $p(\boldsymbol{y} \mid u, z) \, p(u \mid z)$; the first factor follows

$$\boldsymbol{Y}_i \mid U_i = u, Z_i = z \quad \sim \quad N(\boldsymbol{\mu}_{iz}(u), \boldsymbol{\Sigma}_{iz}). \tag{10.20}$$

The conditional mean $\boldsymbol{\mu}_{iz}(u)$ is an $n_i \times 1$ vector having elements

$$\mu_{ijz}(u) = E(Y_{ij} \mid u, z) = \beta_{0z}(u) + \beta_{1z}(u)t_{ij}, \tag{10.21}$$

where $\beta_{0z}(u)$ and $\beta_{1z}(u)$ are smooth but unspecified functions of u. When using the rescaled time axis, the intercept term $\beta_{0z}(u)$ is the full-data mean of log CD4 count at time $t = \overline{U}^0 = 139$ among those who drop out at time u; the slope $\beta_{1z}(u)$ is the full-data mean change in log CD4 from baseline to end of the study among those dropping out at u. When both $\beta_0(u)$ and $\beta_1(u)$ are constant in u, we have MAR.

The variance $\boldsymbol{\Sigma}_{iz}$ is parameterized using a random effects structure

$$\boldsymbol{\Sigma}_{iz} = \boldsymbol{\Sigma}_{iz}(\boldsymbol{\Omega}, \sigma) = \boldsymbol{x}_i \boldsymbol{\Omega} \boldsymbol{x}_i^{\mathrm{T}} + \sigma^2 \boldsymbol{I}_{n_i},$$

where $\boldsymbol{\Omega}$ is a 2×2 positive definite matrix and $\boldsymbol{I}_{n_i}$ is an identity matrix having dimension n_i.

The smooth functions $\beta_{0z}(u)$ and $\beta_{1z}(u)$ are represented in terms of penalized splines. We use a low-rank thin-plate spline basis (Crainiceanu et al., 2005). Suppressing individual-level subscripts for now, we have

$$\beta(u; \boldsymbol{\alpha}, \boldsymbol{\psi}) = \alpha_0 + \alpha_1 u + \sum_{l=1}^{k} \psi_l |u - s_l|^3, \qquad (10.22)$$

where s_l are fixed knot points, $\boldsymbol{\alpha} = (\alpha_0, \alpha_1)^{\mathrm{T}}$, and $\boldsymbol{\psi} = (\psi_1, \ldots, \psi_k)^{\mathrm{T}}$. In the full-data loglikelihood, we introduce a penalty term that shrinks the coefficients of (10.22) to induce smoothness; it takes the form $\lambda \boldsymbol{\phi}^{\mathrm{T}} \boldsymbol{D} \boldsymbol{\phi}$, where $\lambda > 0$ is a smoothing parameter, $\boldsymbol{\phi}^{\mathrm{T}} = (\boldsymbol{\alpha}^{\mathrm{T}}, \boldsymbol{\psi}^{\mathrm{T}})$, and $\boldsymbol{D}$ is a symmetric matrix with dimension $2 + k$. Following Crainiceanu et al. (2005), we only penalize higher-order coefficients $\boldsymbol{\psi}$, so that

$$\boldsymbol{D} = \begin{bmatrix} \boldsymbol{0}_{2\times 2} & \boldsymbol{0}_{2\times k} \\ \boldsymbol{0}_{k\times 2} & \boldsymbol{Q}_{k\times k} \end{bmatrix},$$

where the (l, l') entry of $\boldsymbol{Q}$ is $|s_l - s_{l'}|^3$. Now write $\boldsymbol{v}(u) = (1, u)$ and $\boldsymbol{w}(u) = (|u - s_1|^3, \ldots, |u - s_k|^3)$, and let $\widetilde{\boldsymbol{\psi}} = \boldsymbol{Q}^{\frac{1}{2}} \boldsymbol{\psi}$ and $\widetilde{\boldsymbol{w}}(u) = \boldsymbol{w}(u) \boldsymbol{Q}^{-\frac{1}{2}}$.

The penalized splines have a linear model representation as

$$\beta(u, \boldsymbol{\phi}) = \boldsymbol{v}(u) \boldsymbol{\alpha} + \widetilde{\boldsymbol{w}}(u) \widetilde{\boldsymbol{\psi}},$$

where the $\widetilde{\boldsymbol{\psi}} \sim N(\boldsymbol{0}, \tau^2 \boldsymbol{I}_k)$ are viewed as random effects with variance $\tau^2 = 1/(2\lambda)$. Applying this separately to each of the four functions (intercept and slope function for each treatment group), we have, for $l = 0, 1$ and $z = 0, 1$,

$$\beta_{lz}(u, \boldsymbol{\phi}) = \beta(u, \boldsymbol{\phi}_{lz}) = \boldsymbol{v}(u) \boldsymbol{\alpha}_{lz} + \widetilde{\boldsymbol{w}}(u) \widetilde{\boldsymbol{\psi}}_{lz},$$

with $\widetilde{\boldsymbol{\psi}}_{lz} \sim N(\boldsymbol{0}, \tau_{lz}^2 \boldsymbol{I}_k)$.

The conditional distribution of CD4 counts conditional on dropout time (10.20) can be rewritten as

$$(\boldsymbol{Y}_i \mid u, z) \quad \sim \quad N(\ \boldsymbol{X}_i \boldsymbol{\beta}(u; \boldsymbol{\phi}_z, \boldsymbol{\lambda}_z),\ \boldsymbol{\Sigma}(\boldsymbol{\Omega}_z, \sigma_z)\ ),$$

where $\boldsymbol{\phi}_z^{\mathrm{T}} = (\boldsymbol{\alpha}_{0z}^{\mathrm{T}}, \boldsymbol{\alpha}_{1z}^{\mathrm{T}}, \widetilde{\boldsymbol{\psi}}_{0z}^{\mathrm{T}}, \widetilde{\boldsymbol{\psi}}_{1z}^{\mathrm{T}})$ and $\boldsymbol{\tau}_z^2 = (\tau_{0z}^2, \tau_{1z}^2)^{\mathrm{T}}$. Priors for this part of the model are

$$\begin{aligned} \boldsymbol{\alpha}_{lz} &\sim N(\boldsymbol{0}, 10^4 \boldsymbol{I}_2), \\ \widetilde{\boldsymbol{\psi}}_{lz} \mid \tau_{lz} &\sim N(\boldsymbol{0}, \tau_{lz}^2 \boldsymbol{I}_k), \\ \tau_{lz}^{-2} &\sim \text{Gamma}(.01, 1/.01), \\ \boldsymbol{\Omega}_z^{-1} &\sim \text{Wishart}(3, 3\boldsymbol{I}), \\ \sigma_z^{-2} &\sim \text{Gamma}(.01, 1/.01), \end{aligned}$$

for $l = 0, 1$ and $z = 0, 1$.

10.4.3 Model specification: dropout times

For treatment groups $z = 0, 1$, the distributions $p(u \mid z)$ are left essentially unspecified, and we use Bayesian nonparametric approaches to characterize them. In particular, we use a Dirichlet mixture of normal distributions (Section 3.6), where

$$
\begin{aligned}
U_i \mid z &\sim N(\alpha_{iz}, \nu_z), \\
\alpha_{iz} &\sim G_z, \\
G_z &\sim \mathrm{DP}(F_0, a),
\end{aligned}
$$

with F_0 a normal distribution. Technical details on sampling can be found in MacEachern and Müller (1998).

10.4.4 Summary of analyses under MAR and MNAR

The primary objective is to draw inference about the difference in mean change in log CD4 cell count at the end of follow-up; denote this by θ. In the mixture model, this corresponds to the difference in integrated slope functions $\beta_{1z}(u)$ between treatment groups,

$$
\theta = \int \beta_{11}(u)\, p(u \mid z = 1)\, du - \int \beta_{10}(u)\, p(u \mid z = 0)\, du.
$$

Treatment effect is summarized using the posterior mean and standard deviation for θ and the posterior probability that $\theta > 0$.

In addition to the mixture model, we fit a standard random effects model that assumes MAR; the latter corresponds to a special case of the mixture model where the $\beta_{lz}(u)$ are constant functions of u.

Figure 10.7 shows posterior means for the smooth functions $\beta_{lz}(u)$ (for $l = 0, 1; z = 0, 1$), together with posterior estimates of individual-specific intercepts and slopes. The plots of intercept vs. dropout time indicate that mean CD4 is higher among those with longer follow-up, with a more pronounced effect in the low dose arm; likewise, CD4 slopes are higher among those with longer follow-up times.

Inferences about treatment-specific intercepts and slopes, and treatment effect, are summarized in Table 10.9. In general, inferences under the mixture model indicate that overall mean (intercept term) and slope over time are lower under the MNAR assumption. The effect of the MNAR assumption (relative to MAR) is most pronounced in the low dose arm, where the association between CD4 count and dropout time is stronger (Figure 10.7). In fact, the shift in intercepts and slopes from MAR to MNAR exceeds two posterior standard deviations (PSD) for all parameters except the intercept on high dose. The PSD is generally greater for the mixture model.

The MNAR assumption also influences the treatment effect; under MAR, the posterior mean of θ is $-.37$ (PSD $= .21$) with $P(\theta > 0) = .96$, suggesting superiority of high dose; under MNAR we have posterior mean .01 (PSD = .25) and $P(\theta > 0) = .49$, which is far more equivocal.

10.4.5 Summary

Studies like the Pediatric AIDS trial require methods that can handle continuous time dropout. The mixture model given here induces an MNAR dropout process by assuming the full-data response distribution, conditional on dropout time u, has mean that is linear in time; the intercept and slope parameters are smooth but unspecified functions of u.

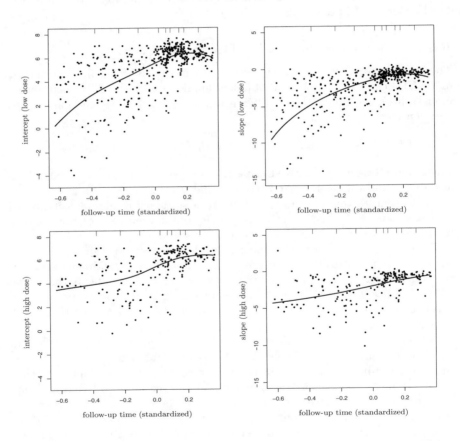

Figure 10.7 *Pediatric AIDS study: posterior (pointwise) mean of intercept function $\beta_{0z}(u)$ and slope function $\beta_{1z}(u)$, separately by treatment ($z = 0, 1$). Points are individual-specific posterior mean intercepts and slopes. Tick marks on top edge of each plot are deciles of the dropout distribution.*

Table 10.9 *Pediatric AIDS trial: posterior means and standard deviations for key parameters from random effects model (REM) fit under MAR, and mixture of varying coefficient model (VCM), which assumes MNAR.*

		Model	
Dose	Parameter	REM (MAR)	VCM (MNAR)
Low	Intercept	5.6 (.13)	5.2 (.11)
	Slope	−1.6 (.14)	−2.1 (.13)
High	Intercept	5.4 (.14)	5.3 (.14)
	Slope	−1.9 (.16)	−2.1 (.19)
	Difference (θ)	−.37 (.21)	.01 (.25)
	$P(\theta > 0)$	.96	.49

An advantage of the model is that the functional form of the missingness mechanism does not have to be known; it is characterized by functions that can be left unspecified. However, it *is* necessary to specify the form of $E\{Y(t) \mid u\}$ as a function of t; here we assume it is linear with intercept $\beta_0(u)$ and intercept $\beta_1(u)$. A key consequence is that conditionally on $U = u$, the mean of $Y(t)$ has the same functional form before and after dropout; for the linear case, the implication is that the slope is the same prior to and after u.

This property can be relaxed by expanding the model to accommodate sensitivity parameters that allow (for example) a different slope following dropout. For example, we can expand (10.21) by introducing a sensitivity parameter $\boldsymbol{\xi} = (\xi_0, \xi_1)$, so that

$$E(Y_{ij} \mid u, z) = \beta_{0z}(u) + \beta_{1z}(u)t_{ij} + \xi_z(t_{ij} - u)_+,$$

where $a_+ = aI\{a > 0\}$ is the positive part of a. In this formulation, $\beta_{1z}(u)$ is the slope prior to dropout time u and $\beta_{1z}(u) + \xi_z$ is the slope after u. More generally, ξ_z can depend on u. Another generalization of the model allows the variance-covariance structure to vary smoothly with u; this can be accomplished using an appropriate parameterization of $\boldsymbol{\Sigma}$. Details can be found in Su and Hogan (2007).

Distributions

Discrete Distributions

Binomial $\hspace{8cm} Y \mid n, \theta \sim Bin(n, \theta)$

The binomial model characterizes the number of successes in n independent trials, where success probability on each trial is θ. It has mass function

$$p(y \mid \theta, n) = \binom{n}{y} \theta^y (1-\theta)^{n-y},$$

where $y \in \{0, 1, \ldots, n\}$ and $0 < \theta < 1$.

Multinomial with k categories $\hspace{5cm} \boldsymbol{Y} \mid n, \boldsymbol{\theta} \sim Mult(n, \boldsymbol{\theta})$

The multinomial model characterizes the distribution of n events across k mutually exclusive categories labeled $j = 1, \ldots, k$, where Y_j is the count and θ_j is the probability for category j. The multinomial random variable has mass function

$$p(\boldsymbol{y} \mid n, \boldsymbol{\theta}) = \binom{n}{y_1 \, y_2 \cdots y_k} \theta_1^{y_1} \theta_2^{y_2} \cdots \theta_k^{y_k},$$

where $\boldsymbol{y} = (y_1, \ldots, y_k)^{\mathrm{T}}$, $y_j \geq 0$, $\sum_{j=1}^{k} y_j = n$, $0 < \theta_j < 1$, and $\sum_{j=1}^{k} \theta_j = 1$.

Continuous distributions

Dirichlet with k categories $\hspace{6cm} \boldsymbol{\theta} \sim Dir(\boldsymbol{\beta})$

The Dirichlet distribution characterizes the joint distribution of a vector of random variables $\boldsymbol{\theta} = (\theta_1, \ldots, \theta_k)$ having $\theta_j > 0$ and $\sum_j \theta_j = 1$. It is frequently used as a prior for multinomial probabilities. It density function is

$$p(\boldsymbol{\theta} \mid \boldsymbol{\beta}) = \frac{\Gamma(\beta_1 + \cdots + \beta_k)}{\prod_{j=1}^{k} \Gamma(\beta_j)} \theta_1^{\beta_1 - 1} \cdots \theta_k^{\beta_k - 1},$$

where $\beta_j > 0$ and $\Gamma(\cdot)$ is the gamma function.

Gamma $\hspace{10cm} Y \sim Gamma(\alpha, \beta)$

A random variable Y having the gamma distribution has density function

$$p(y \mid \alpha, \beta) = \frac{1}{\Gamma(\alpha)\beta^{\alpha}} \exp(-y/\beta) y^{\alpha-1},$$

with $y > 0$, $\alpha > 0$, and $\beta > 0$. Its mean is $E(Y) = \alpha\beta$.

Normal $\hspace{10cm} Y \sim N(\mu, \sigma^2)$

A random variable Y having the normal (Gaussian) distribution with mean μ and variance σ^2 has density function

$$p(y \mid \mu, \sigma) = (2\pi)^{-1/2} \sigma^{-1} \exp\left\{-(x - \mu)^2/2\sigma^2\right\}$$

with $-\infty < y < \infty$, $-\infty < \mu < \infty$, and $\sigma > 0$.

Multivariate normal (q-dimensional) $\hspace{5cm} \boldsymbol{Y} \sim N(\boldsymbol{\mu}, \boldsymbol{\Sigma})$

A q-dimensional normal random variable $\boldsymbol{Y} = (Y_1, \ldots, Y_q)^{\mathrm{T}}$ with mean $\boldsymbol{\mu} = (\mu_1, \ldots, \mu_q)^{\mathrm{T}}$ and $q \times q$ variance matrix $\boldsymbol{\Sigma}$ has density function

$$p(\boldsymbol{y} \mid \boldsymbol{\mu}, \boldsymbol{\Sigma}) = (2\pi)^{-q/2} |\boldsymbol{\Sigma}|^{-1/2} \exp\left\{-\tfrac{1}{2}(\boldsymbol{y} - \boldsymbol{\mu})^{\mathrm{T}} \boldsymbol{\Sigma}^{-1}(\boldsymbol{y} - \boldsymbol{\mu})\right\},$$

where $-\infty < y_j < \infty$, $-\infty < \mu_j < \infty$, and $\boldsymbol{\Sigma}$ is positive definite.

t distribution $\hspace{9cm} Y \sim \mathcal{T}_{\nu}(\mu, \sigma^2)$

A random variable Y having the t distribution with mean μ, scale σ^2 and ν degrees of freedom has density

$$p(y \mid \mu, \sigma^2, \nu) = \frac{\Gamma((\nu + 1)/2)}{\Gamma(\nu/2)(\nu\sigma\pi)^{1/2}} \left\{1 + (y - \mu)^2/\nu\sigma^2\right\}^{-(\nu+1)/2},$$

where $-\infty < y < \infty$, $-\infty < \mu < \infty$, $\sigma^2 > 0$, and $\nu > 0$.

Multivariate t (q-dimensional) $\hspace{5cm} \boldsymbol{Y} \sim \mathcal{T}_{\nu}(\boldsymbol{\mu}, \boldsymbol{\Sigma})$

A q dimensional vector $\boldsymbol{Y}$ having the multivariate t distribution has density function

$$p(\boldsymbol{y} \mid \boldsymbol{\mu}, \boldsymbol{\Sigma}, \nu) = \frac{\Gamma((\nu + p)/2)}{\Gamma(\nu/2)(\nu\pi)^{p/2}} |\boldsymbol{\Sigma}|^{-1/2} \left\{1 + (\boldsymbol{y} - \boldsymbol{\mu})^T \boldsymbol{\Sigma}^{-1}(\boldsymbol{y} - \boldsymbol{\mu})/\nu\right\}^{-(\nu+p)/2}$$

with $-\infty < Y_j < \infty$, $-\infty < \mu_j < \infty$, $\nu > 0$, and $\boldsymbol{\Sigma}$ positive definite.

Wishart (q-dimensional covariance matrix) $\hspace{3.5cm} \boldsymbol{S} \sim Wishart(\nu, \boldsymbol{A})$

A q dimensional covariance matrix from the Wishart distribution with scale matrix $\boldsymbol{A}$ and degrees of freedom ν has density function

$$p(\boldsymbol{S} \mid \boldsymbol{A}, \nu) = c(\nu, q) |\boldsymbol{A}|^{-\nu/2} |\boldsymbol{S}|^{(\nu-q-1)/2} \exp\left\{-\mathrm{tr}(\boldsymbol{A}^{-1}\boldsymbol{S})\right\},$$

where

$$c(\nu, q) = \left\{ 2^{\nu q/2} \, \pi^{q(q-1)/4} \prod_{i=1}^{q} \Gamma\left(\frac{q+1-i}{2}\right) \right\}^{-1},$$

$\boldsymbol{S}$ and $\boldsymbol{A}$ are positive definite, and $\nu \geq q$. The Wishart distribution is frequently used as a prior for an inverse covariance matrix. It has mean $E(\boldsymbol{S}) = \nu \boldsymbol{A}$.

Uniform $Y \sim \mathit{Unif}(a, b)$

The uniform distribution has density

$$p(y \mid a, b) = \frac{1}{b - a}$$

where $a \leq y \leq b$ and $-\infty < a < b < \infty$.

Bibliography

AKAIKE, H. (1973). Information theory and an extension of the maximum likelihood principle, in *Proceedings of 2nd International Symposium on Information theory.* PETROV, B. & CSAKI, F., eds. Akademiai Kiado [Budapest], 267–281.

ALBERT, J. H. & CHIB, S. (1993). Bayesian analysis of binary and polychotomous response data. *Journal of the American Statistical Association* 88 669–679.

ALBERT, P. S. (2000). A transitional model for longitudinal binary data subject to nonignorable missing data. *Biometrics* 56 602–608.

ALBERT, P. S., FOLLMANN, D. A., WANG, S. A. & SUH, E. B. (2002). A latent autoregressive model for longitudinal binary data subject to informative missingness. *Biometrics* 58 631–642.

ANDERSON, T. (1984). *An Introduction to Multivariate Statistical Analysis, 2nd Edition.* Wiley.

ANGRIST, J. D., IMBENS, G. W. & RUBIN, D. B. (1996). Identification of causal effects using instrumental variables (with discussion). *Journal of the American Statistical Association* 91 444–472.

AZZALINI, A. & VALLE, A. D. (1996). The multivariate skew-normal distribution. *Biometrika* 83 715–726.

BAKER, S. G. (1995). Marginal regression for repeated binary data with outcome subject to non-ignorable non-response. *Biometrics* 51 1042–1052.

BAKER, S. G., KO, C. W. & GRAUBARD, B. I. (2003). A sensitivity analysis for nonrandomly missing categorical data arising from a national health disability survey. *Biostatistics* 4 41–56.

BAKER, S. G. & LAIRD, N. M. (1988). Regression analysis for categorical variables with outcome subject to nonignorable nonresponse (Corr: V83 p1232). *Journal of the American Statistical Association* 83 62–69.

BANDEEN-ROCHE, K., MIGLIORETTI, D. L., ZEGER, S. L. & RATHOUZ, P. J. (1997). Latent variable regression for multiple discrete outcomes. *Journal of the American Statistical Association* 92 1375–1386.

BARNARD, J., McCULLOCH, R. & MENG, X.-L. (2000). Modeling covariance matrices in terms of standard deviations and correlations, with application to shrinkage. *Statistica Sinica* 10 1281–1311.

BARTHOLOMEW, D. J. & KNOTT, M. (1999). *Latent Variable Models and Factor Analysis*. Edward Arnold Publishers Ltd.

BERGER, J. O. (1985). *Statistical Decision Theory and Bayesian Analysis (Second Edition)*. Springer-Verlag Inc.

BERGER, J. O. & BERNARDO, J. M. (1992). On the development of reference priors (Disc: P49-60), in *Bayesian Statistics 4. Proceedings of the Fourth Valencia International Meeting*. BERNARDO, J. M., BERGER, J. O., DAWID, A. P. & SMITH, A. F. M., eds. Clarendon Press [Oxford University Press], 35–49.

BERGER, J. O. & O'HAGAN, A. (1988). Ranges of posterior probabilities for unimodal priors with specified quantiles, in *Bayesian Statistics 3*. BERNARDO, J. M., DEGROOT, M. H., LINDLEY, D. V. & SMITH, A. F. M., eds. Clarendon Press [Oxford University Press], 45–65.

BERGER, J. O. & PERICCHI, L. R. (1996). The intrinsic Bayes factor for linear models, in *Bayesian Statistics 5 – Proceedings of the Fifth Valencia International Meeting*. BERNARDO, J., BERGER, J., DAWID, A. & SMITH, A., eds. 25–44.

BERGER, J. O., STRAWDERMAN, W. & TANG, D. (2005). Posterior propriety and admissibility of hyperpriors in normal hierarchical models. *The Annals of Statistics* 33 606–646.

BERNARDO, J. M. (2006). Reference analysis, in *Handbook of Statistics*. DEY, D. & RAO, C., eds. Elsevier, 17–90.

BERRY, S. M., CARROLL, R. J. & RUPPERT, D. (2002). Bayesian smoothing and regression splines for measurement error problems. *Journal of the American Statistical Association* 97 160–169.

BIRMINGHAM, J. & FITZMAURICE, G. M. (2002). A pattern-mixture model for longitudinal binary responses with nonignorable nonresponse. *Biometrics* 58 989–996.

BOTTS, C. & DANIELS, M. J. (2007). A flexible approach to Bayesian multiple curve fitting. Tech. rep., Department of Statistics, University of Florida.

BRADY, M., MCGRATH, N., BROUWERS, P., GELBER, R., FOWLER, M. G., YOGEV, R., HUTTON, N., BRYSON, Y. J., MITCHELL, C. D., FIKRIG, S., BORKOWSKY, W., JIMENEZ, E., MCSHERRY, G., RUBINSTEIN, A., WILFERT, C. M., MCINTOSH, K., ELKINS, M. M. & WEINTRUB, P. S. (1996). Randomized study of the tolerance and efficacy of high- versus low-dose zidovudine in human immunodeficiency virus-infected children with mild to moderate symptoms (AIDS Clinical Trials Group 128). *Journal of Infectious Diseases* 173 1097–1106.

BRESLOW, N. E. & CLAYTON, D. G. (1993). Approximate inference in generalized linear mixed models. *Journal of the American Statistical Association* 88 9–25.

BROOKS, S. P. & GELMAN, A. (1998). General methods for monitoring convergence of iterative simulations. *Journal of Computational and Graphical Statistics* 7 434–455.

BROWN, P. J., LE, N. D. & ZIDEK, J. V. (1994). Inference for a covariance matrix, in *Aspects of Uncertainty. A Tribute to D. V. Lindley.* FREEMAN, P. R. & SMITH, A. F. M., eds. John Wiley & Sons, 77–92.

BURNHAM, K. P. & ANDERSON, D. R. (1998). *Model Selection and Inference: A Practical Information-theoretic Approach.* Springer-Verlag Inc.

CAREY, V. J. & ROSNER, B. A. (2001). Analysis of longitudinally observed irregularly timed multivariate outcomes: regression with focus on cross-component correlation. *Statistics in Medicine* 20 21–31.

CARLIN, B. P. & LOUIS, T. A. (2000). *Bayes and Empirical Bayes Methods for Data Analysis.* Chapman & Hall Ltd.

CARROLL, R. J., RUPPERT, D. & STEFANSKI, L. A. (1995). *Measurement Error in Nonlinear Models.* Chapman and Hall.

CASELLA, G. & BERGER, R. L. (2001). *Statistical Inference.* Duxbury Press.

CATALANO, P. J. & RYAN, L. M. (1992). Bivariate latent variable models for clustered discrete and continuous outcomes. *Journal of the American Statistical Association* 87 651–658.

CELEUX, G., FORBES, F., ROBERT, C. & TITTERINGTON, D. (2006). Deviance information criteria for missing data models. *Bayesian Analysis* 1 651–674.

CHALONER, K. (1996). Elicitation of prior distributions, in *Bayesian Biostatistics.* BERRY, D. A. & STANGL, D. K., eds. Marcel Dekker Inc., 141–156.

CHEN, H. Y. & LITTLE, R. J. A. (1999). A test of missing completely at random for generalised estimating equations with missing data. *Biometrika* 86 1–13.

CHEN, M.-H., IBRAHIM, J. G., SHAO, Q.-M. & WEISS, R. E. (2003). Prior elicitation for model selection and estimation in generalized linear mixed models. *Journal of Statistical Planning and Inference* 111 57–76.

CHEN, Z. & DUNSON, D. B. (2003). Random effects selection in linear mixed models. *Biometrics* 59 762–769.

CHIANG, C.-T., RICE, J. A. & WU, C. O. (2001). Smoothing spline estimation for varying coefficient models with repeatedly measured dependent variables. *Journal of the American Statistical Association* 96 605–619.

CHIB, S. & GREENBERG, E. (1998). Analysis of multivariate probit models. *Biometrika* 85 347–361.

CHIU, T. Y. M., LEONARD, T. & TSUI, K.-W. (1996). The matrix-logarithmic covariance model. *Journal of the American Statistical Association* 91 198–210.

CHRISTIANSEN, C. L. & MORRIS, C. N. (1997). Hierarchical Poisson regression modeling. *Journal of the American Statistical Association* 92 618–632.

CNAAN, A., LAIRD, N. M. & SLASOR, P. (1997). Using the general linear mixed model to analyse unbalanced repeated measures and longitudinal data. *Statistics in Medicine* 16 2349–2380.

COPAS, J. & EGUCHI, S. (2005). Local model uncertainty and incomplete-data bias (with discussion). *Journal of the Royal Statistical Society, Series B: Statistical Methodology* 67 459–513.

COWLES, M. K. & CARLIN, B. P. (1996). Markov chain Monte Carlo convergence diagnostics: a comparative review. *Journal of the American Statistical Association* 91 883–904.

COX, D. R. & REID, N. (1987). Parameter orthogonality and approximate conditional inference (C/R: P18-39). *Journal of the Royal Statistical Society, Series B: Methodological* 49 1–18.

CRAINICEANU, C. M., RUPPERT, D. & CARROLL, R. J. (2007). Spatially adaptive penalized splines with heteroscedastic errors. *Journal of Computational and Graphical Statistics (in press)* .

CRAINICEANU, C. M., RUPPERT, D. & WAND, M. P. (2005). Bayesian analysis for penalized spline regression using WinBUGS. *Journal of Statistical Software* 14.

CZADO, C. (2000). Multivariate regression analysis of panel data with binary outcomes applied to unemployment data. *Statistical Papers* 41 281–304.

DAMIEN, P., WAKEFIELD, J. & WALKER, S. W. (1999). Gibbs sampling for Bayesian non-conjugate and hierarchical models by using auxiliary variables. *Journal of the Royal Statistical Society, Series B: Statistical Methodology* 61 331–344.

DANIELS, M. J. (1998). Computing posterior distributions for covariance matrices, in *Computing Science and Statistics. Dimension Reduction, Computational Complexity and Information. Proceedings of the 30th Symposium on the Interface.* WEISBERG, S., ed. Interface Foundation of North America, 192–196.

DANIELS, M. J. (1999). A prior for the variance in hierarchical models. *The Canadian Journal of Statistics / La Revue Canadienne de Statistique* 27 567–578.

DANIELS, M. J. (2005). Shrinkage priors for the dependence structure in longitudinal data. *Journal of Statistical Planning and Inference* 127 119–130.

DANIELS, M. J. (2006). Bayesian modelling of several covariance matrices and some results on the propriety of the posterior for linear regression with correlated and/or heterogeneous errors. *Journal of Multivariate Analysis* 97 1185–1207.

DANIELS, M. J. & GATSONIS, C. (1999). Hierarchical generalized linear models in the analysis of variations in health care utilization. *Journal of the American Statistical Association* 94 29–42.

DANIELS, M. J. & KASS, R. E. (1999). Nonconjugate Bayesian estimation of covariance matrices and its use in hierarchical models. *Journal of the American Statistical Association* 94 1254–1263.

DANIELS, M. J. & KASS, R. E. (2001). Shrinkage estimators for covariance matrices. *Biometrics* 57 1173–1184.

DANIELS, M. J. & POURAHMADI, M. (2002). Bayesian analysis of covariance matrices and dynamic models for longitudinal data. *Biometrika* 89 553–566.

DANIELS, M. J. & POURAHMADI, M. (2007). Modeling correlation matrices via ordered partial correlations. Tech. rep., Department of Statistics, University of Florida.

DANIELS, M. J. & SCHARFSTEIN, D. O. (2007). A Bayesian analysis of informative missing data with competing causes of dropout. Tech. rep., Department of Statistics, University of Florida.

DANIELS, M. J. & ZHAO, Y. (2003). Modelling the random effects covariance matrix in longitudinal data. *Statistics in Medicine* 22 1631–1647.

DATTA, G. S. & GHOSH, J. K. (1995). On priors providing frequentist validity for Bayesian inference. *Biometrika* 82 37–45.

DAVIDIAN, M. & GILTINAN, D. M. (1998). *Nonlinear Models for Repeated Measurement Data*. Chapman & Hall Ltd.

DAVIDIAN, M. & GILTINAN, D. M. (2003). Nonlinear models for repeated measurement data: an overview and update. *Journal of Agricultural, Biological, and Environmental Statistics* 8 387–419.

DAWID, A. P. (1979). Conditional independence in statistical theory. *Journal of the Royal Statistical Society, Series B* 41 1–15.

DE GRUTTOLA, V. & TU, X. M. (1994). Modelling progression of CD4-lymphocyte count and its relationship to survival time. *Biometrics* 50 1003–1014.

DEMPSTER, A. P., LAIRD, N. M. & RUBIN, D. B. (1977). Maximum likelihood from incomplete data via the EM algorithm (C/R: P22-37). *Journal of the Royal Statistical Society, Series B: Methodological* 39 1–22.

DENISON, D. G. T., MALLICK, B. K. & SMITH, A. F. M. (1998). Automatic Bayesian curve fitting. *Journal of the Royal Statistical Society, Series B: Statistical Methodology* 60 333–350.

DIGGLE, P. J. (1999). Comment on "Adjusting for nonignorable drop-out using semiparametric nonresponse models". *Journal of the American Statistical Association* 94 1128–1129.

DIGGLE, P. J., HEAGERTY, P. J., LIANG, K.-Y. & ZEGER, S. L. (2002). *Analysis of Longitudinal Data, Second Edition.* Oxford University Press.

DIGGLE, P. J. & KENWARD, M. G. (1994). Informative drop-out in longitudinal data analysis (Disc: P73-93). *Applied Statistics* 43 49–73.

DIMATEO, I., KASS, R. E. & GENOVESE, C. (2001). Bayesian curve fitting with free-knot splines. *Biometrika* 88 1055–1071.

DOBSON, A. & HENDERSON, R. (2003). Diagnostics for joint longitudinal and dropout time modeling. *Biometrics* 59 741–751.

DRAPER, D. (1995). Assessment and propagation of model uncertainty. *Journal of the Royal Statistical Society. Series B (Methodological)* 57 45–97.

DUNSON, D. B. (2003). Dynamic latent trait models for multidimensional longitudinal data. *Journal of the American Statistical Association* 98 555–563.

EBERLY, L. E. & CASELLA, G. (2003). Estimating Bayesian credible intervals. *Journal of Statistical Planning and Inference* 112 115–132.

EILERS, P. H. C. & MARX, B. D. (1996). Flexible smoothing with B-splines and penalties (Disc: P102-121). *Statistical Science* 11 89–102.

EKHOLM, A. & SKINNER, C. (1998). The Muscatine children's obesity data reanalysed using pattern mixture models. *Journal of the Royal Statistical Society, Series C: Applied Statistics* 47 251–263.

ESCOBAR, M. D. (1994). Estimating normal means with a Dirichlet process prior. *Journal of the American Statistical Association* 89 268–277.

ESCOBAR, M. D. (1995). Nonparametric Bayesian methods in hierarchical models. *Journal of Statistical Planning and Inference* 43 97–106.

FAUCETT, C. L. & THOMAS, D. C. (1996). Simultaneously modelling censored survival data and repeatedly measured covariates: a Gibbs sampling approach. *Statistics in Medicine* 15 1663–1685.

FERGUSON, T. S. (1982). Sequential estimation with Dirichlet process priors, in *Statistical Decision Theory and Related Topics III, in two volumes, Volume 1.* GUPTA, S. S. & BERGER, J. O., eds. Academic Press, 385–401.

FERRER, E. & MCARDLE, J. J. (2003). Alternative structural models for multivariate longitudinal data analysis. *Structural Equation Modeling* 10 493–524.

FIRTH, D. (1987). Discussion of parameter orthogonality and approximate conditional inference. *Journal of the Royal Statistical Society, Series B: Methodological* 49 22–23.

FITZMAURICE, G. M. (2003). Methods for handling dropouts in longitudinal clinical trials. *Statistica Neerlandica* 57 75–99.

FITZMAURICE, G. M. & LAIRD, N. M. (1993). A likelihood-based method for analysing longitudinal binary responses. *Biometrika* 80 141–151.

FITZMAURICE, G. M. & LAIRD, N. M. (1997). Regression models for mixed discrete and continuous responses with potentially missing values. *Biometrics* 53 110–122.

FITZMAURICE, G. M. & LAIRD, N. M. (2000). Generalized linear mixture models for handling nonignorable dropouts in longitudinal studies. *Biostatistics* 1 141–156.

FITZMAURICE, G. M., LAIRD, N. M. & LIPSITZ, S. R. (1994). Analysing incomplete longitudinal binary responses: a likelihood-based approach. *Biometrics* 50 601–612.

FITZMAURICE, G. M., LAIRD, N. M. & WARE, J. H. (2004). *Applied Longitudinal Analysis.* Wiley Interscience.

FITZMAURICE, G. M., LAIRD, N. M. & ZAHNER, G. E. P. (1996). Multivariate logistic models for incomplete binary responses. *Journal of the American Statistical Association* 91 99–108.

FITZMAURICE, G. M., MOLENBERGHS, G. & LIPSITZ, S. R. (1995). Regression models for longitudinal binary responses with informative drop-outs. *Journal of the Royal Statistical Society, Series B: Methodological* 57 691–704.

FOLLMANN, D. & WU, M. C. (1995). An approximate generalized linear model with random effects for informative missing data (Corr: 97V53 p384). *Biometrics* 51 151–168.

FORSTER, J. J. & SMITH, P. W. F. (1998). Model-based inference for categorical survey data subject to non-ignorable non-response. *Journal of the Royal Statistical Society: Series B (Statistical Methodology)* 60 57–70.

FRANGAKIS, C. E. & RUBIN, D. B. (2002). Principal stratification in causal inference. *Biometrics* 58 21–29.

FULLER, W. A. (1987). *Measurement Error Models.* Wiley.

GARTHWAITE, P. H. & AL-AWADHI, S. A. (2001). Non-conjugate prior distribution assessment for multivariate normal sampling. *Journal of the Royal Statistical Society, Series B: Statistical Methodology* 63 95–110.

GARTHWAITE, P. H., KADANE, J. B. & O'HAGAN, A. (2005). Statistical methods for eliciting probability distributions. *Journal of the American Statistical Association* 100 680–700.

GELFAND, A. E., DEY, D. K. & CHANG, H. (1992). Model determination using predictive distributions, with implementation via sampling-based methods (Disc: P160–167), in *Bayesian Statistics 4. Proceedings of the*

Fourth Valencia International Meeting. BERNARDO, J. M., BERGER, J. O., DAWID, A. P. & SMITH, A. F. M., eds. Clarendon Press [Oxford University Press], 147–159.

GELFAND, A. E. & GHOSH, S. K. (1998). Model choice: a minimum posterior predictive loss approach. *Biometrika* 85 1–11.

GELFAND, A. E. & SMITH, A. F. M. (1990). Sampling-based approaches to calculating marginal densities. *Journal of the American Statistical Association* 85 398–409.

GELMAN, A. (2006). Prior distributions for variance parameters in hierarchical models. *Bayesian Analysis* 1 515–534.

GELMAN, A., MENG, X.-L. & STERN, H. (1996). Posterior predictive assessment of model fitness via realized discrepancies (Disc: P760–807). *Statistica Sinica* 6 733–760.

GELMAN, A. & RUBIN, D. B. (1992). Inference from iterative simulation using multiple sequences (Disc: P483–501, 503–511). *Statistical Science* 7 457–472.

GELMAN, A., VAN MECHELEN, I., VERBEKE, G., HEITJAN, D. F. & MEULDERS, M. (2005). Multiple imputation for model checking: completed-data plots with missing latent data. *Biometrics* 61 74–85.

GEMAN, S. & GEMAN, D. (1984). Stochastic relaxation, Gibbs distributions, and the Bayesian restoration of images. *IEEE Transactions on Pattern Analysis and Machine Intelligence* 6 721–741.

GHOSH, M. & HEO, J. (2003). Default Bayesian priors for regression models with first-order autoregressive residuals. *Journal of Time Series Analysis* 24 269–282.

GILKS, W. R. & WILD, P. (1992). Adaptive rejection sampling for Gibbs sampling. *Applied Statistics* 41 337–348.

GILL, R. D. & ROBINS, J. M. (1997). Sequential models for coarsening and inference, in *Proceedings of the First Seattle Symposium in Biostatistics: Survival Analysis*. LIN, D. & FLEMING, T. R., eds. Springer-Verlag, 295–305.

GILL, R. D. & ROBINS, J. M. (2001). Causal inference for complex longitudinal data: the continuous case. *The Annals of Statistics* 29 1785–1811.

GREEN, P. J. (1995). Reversible jump Markov chain Monte Carlo computation and Bayesian model determination. *Biometrika* 82 711–732.

GUEORGUIEVA, R. V. & AGRESTI, A. (2001). A correlated probit model for joint modeling of clustered binary and continuous responses. *Journal of the American Statistical Association* 96 1102–1112.

GUO, W. (2004). Functional data analysis in longitudinal settings using smoothing splines. *Statistical Methods in Medical Research* 13 49–62.

GUSTAFSON, P. (1997). Large hierarchical Bayesian analysis of multivariate survival data. *Biometrics* 53 230–242.

GUSTAFSON, P. (2006). On model expansion, model contraction, identifiability, and prior information: two illustrative scenarios involving mismeasured variables. *Statistical Science* 20 111–140.

GUSTAFSON, P., MACNAB, Y. & WEN, S. (2004). On the value of derivative evaluations and random walk suppression in Markov chain Monte Carlo algorithms. *Statistics and Computing* 14 23–38.

HASTIE, T. J. & TIBSHIRANI, R. J. (1990). *Generalized Additive Models.* Chapman & Hall/CRC.

HASTINGS, W. K. (1970). Monte Carlo sampling methods using Markov chains and their applications. *Biometrika* 57 97–109.

HEAGERTY, P. J. (1999). Marginally specified logistic-normal models for longitudinal binary data. *Biometrics* 55 688–698.

HEAGERTY, P. J. (2002). Marginalized transition models and likelihood inference for longitudinal categorical data. *Biometrics* 58 342–351.

HEAGERTY, P. J. & KURLAND, B. F. (2001). Misspecified maximum likelihood estimates and generalised linear mixed models. *Biometrika* 88 973–985.

HECKMAN, J. J. (1976). The common structure of statistical models of truncation, sample selection and limited dependent variables and a simple estimator for such models. *Annals of Economic and Social Measurement* 5 475–492.

HECKMAN, J. J. (1979). Sample selection bias as a specification error. *Econometrica* 47 153–161.

HEDEKER, D. & GIBBONS, R. D. (2006). *Longitudinal Data Analysis.* John Wiley & Sons.

HEITJAN, D. F. & RUBIN, D. B. (1991). Ignorability and coarse data. *The Annals of Statistics* 19 2244–2253.

HENDERSON, R., DIGGLE, P. J. & DOBSON, A. (2000). Joint modelling of longitudinal measurements and event time data. *Biostatistics (Oxford)* 1 465–480.

HERNÁN, M. A., BRUMBACK, B. & ROBINS, J. M. (2001). Marginal structural models to estimate the joint causal effect of nonrandomized treatments. *Journal of the American Statistical Association* 96 440–448.

HJORT, N. L., DAHL, F. A. & STEINBAKK, G. H. (2006). Post-processing posterior predictive P values. *Journal of the American Statistical Association* 101 1157–1174.

HOBERT, J. P. & CASELLA, G. (1996). The effect of improper priors on Gibbs sampling in hierarchical linear mixed models. *Journal of the American Statistical Association* 91 1461–1473.

HODGES, J. S. (1998). Some algebra and geometry for hierarchical models, applied to diagnostics (Disc: P521–536). *Journal of the Royal Statistical Society, Series B: Statistical Methodology* 60 497–521.

HOGAN, J. W. & DANIELS, M. J. (2002). A hierarchical modelling approach to analysing longitudinal data with drop-out and non-compliance, with application to an equivalence trial in paediatric acquired immune deficiency syndrome. *Journal of the Royal Statistical Society, Series C: Applied Statistics* 51 1–21.

HOGAN, J. W. & LAIRD, N. M. (1996). Intention-to-treat analyses for incomplete repeated measures data. *Biometrics* 52 1002–1017.

HOGAN, J. W. & LAIRD, N. M. (1997a). Mixture models for the joint distribution of repeated measures and event times. *Statistics in Medicine* 16 239–257.

HOGAN, J. W. & LAIRD, N. M. (1997b). Model-based approaches to analysing incomplete longitudinal and failure time data. *Statistics in Medicine* 16 259–272.

HOGAN, J. W., LIN, X. & HERMAN, B. (2004a). Mixtures of varying coefficient models for longitudinal data with discrete or continuous nonignorable dropout. *Biometrics* 60 854–864.

HOGAN, J. W., ROY, J. & KORKONTZELOU, C. (2004b). Handling dropout in longitudinal studies. *Statistics in Medicine* 23 1455–1497.

HOGAN, J. W. & WANG, F. (2001). Model selection for longitudinal repeated measures subject to informative dropout. Tech. rep., Center for Statistical Science, Brown University.

HOOVER, D. R., RICE, J. A., WU, C. O. & YANG, L.-P. (1998). Nonparametric smoothing estimates of time-varying coefficient models with longitudinal data. *Biometrika* 85 809–822.

IBRAHIM, J. G. & CHEN, M.-H. (2000). Power prior distributions for regression models. *Statistical Science* 15 46–60.

ILK, O. & DANIELS, M. J. (2007). Marginalized transition random effects models for multivariate longitudinal binary data. *Canadian Journal of Statistics* 35 105–123.

JACOBSEN, M. & KEIDING, N. (1995). Coarsening at random in general sample spaces and random censoring in continuous time. *The Annals of Statistics* 23 774–786.

JEFFREYS, H. (1961). *Theory of Probability, 3rd Edition*. Oxford University Press.

JENNRICH, R. I. & SCHLUCHTER, M. D. (1986). Unbalanced repeated-measures models with structured covariance matrices. *Biometrics* 42 805–820.

JONES, C. Y., HOGAN, J. W., SNYDER, B., KLEIN, R. S., ROMPALO, A., SCHUMAN, P. & CARPENTER, C. C. J. (2003). Overweight and human immunodeficiency virus (HIV) in women: associations between HIV progression and changes in body mass index in women in the HIV Epidemiology Research Study cohort. *Cinical Infectious Diseases* 37 S69–S80.

KADANE, J. B., DICKEY, J. M., WINKLER, R. L., SMITH, W. S. & PETERS, S. C. (1980). Interactive elicitation of opinion for a normal linear model. *Journal of the American Statistical Association* 75 845–854.

KADANE, J. B. & WOLFSON, L. J. (1998). Experiences in elicitation (Disc: P55–68). *Journal of the Royal Statistical Society, Series D: The Statistician* 47 3–19.

KASS, R. E. & RAFTERY, A. E. (1995). Bayes factors. *Journal of the American Statistical Association* 90 773–795.

KASS, R. E. & WASSERMAN, L. (1995). A reference Bayesian test for nested hypotheses and its relationship to the Schwarz criterion. *Journal of the American Statistical Association* 90 928–934.

KASS, R. E. & WASSERMAN, L. (1996). The selection of prior distributions by formal rules (Corr: 1998V93 p412). *Journal of the American Statistical Association* 91 1343–1370.

KENWARD, M. G. (1998). Selection models for repeated measurements with non-random dropout: an illustration of sensitivity. *Statistics in Medicine* 17 2723–2732.

KENWARD, M. G. & MOLENBERGHS, G. (1999). Parametric models for incomplete continuous and categorical longitudinal data. *Statistical Methods in Medical Research* 8 51–83.

KENWARD, M. G., MOLENBERGHS, G. & THIJS, H. (2003). Pattern-mixture models with proper time dependence. *Biometrika* 90 53–71.

KIEL, D., PUHL, J., ROSEN, C., BERG, K., MURPHY, J. & MACLEAN, D. (1998). Lack of association between insulin-like growth factor I and body composition, muscle strength, physical performance or self reported mobility among older persons with functional limitations. *Journal of the American Geriatrics Society* 46 822–828.

KLEINMAN, K. P. & IBRAHIM, J. G. (1998a). A semi-parametric Bayesian approach to generalized linear mixed models. *Statistics in Medicine* 17 2579–2596.

KLEINMAN, K. P. & IBRAHIM, J. G. (1998b). A semiparametric Bayesian approach to the random effects model. *Biometrics* 54 921–938.

KULLBACK, S. & LEIBLER, R. (1951). On information and sufficiency. *Annals of Mathematical Statistics* 22 79–86.

KURLAND, B. F. & HEAGERTY, P. J. (2004). Marginalized transition models for longitudinal binary data with ignorable and non-ignorable drop-out. *Statistics in Medicine* 23 2673–2695.

KURLAND, B. F. & HEAGERTY, P. J. (2005). Directly parameterized regression conditioning on being alive: analysis of longitudinal data truncated by deaths. *Biostatistics* 6 241–258.

KUTNER, M. H., NACHTSHEIM, C. J., WASSERMAN, W. & NETER, J. (2003). *Applied linear regression models*. McGraw-Hill/Irwin.

LAIRD, N. M. (1988). Missing data in longitudinal studies. *Statistics in Medicine* 7 305–315.

LAIRD, N. M. (2004). *Analysis of Longitudinal and Cluster-Correlated Data*. Institute of Mathematical Statistics.

LAMBERT, P. & VANDENHENDE, F. (2002). A copula-based model for multivariate non-normal longitudinal data: analysis of a dose titration safety study on a new antidepressant. *Statistics in Medicine* 21 3197–3217.

LANCASTER, T. (2004). *An Introduction to Modern Bayesian Econometrics*. Blackwell.

LANGE, K. L., LITTLE, R. J. A. & TAYLOR, J. M. G. (1989). Robust statistical modeling using the *t*-distribution. *Journal of the American Statistical Association* 84 881–896.

LAPIERRE, Y. D., NAI, N. V., CHAUINARD, G., AWAD, A. G., SAXENA, B., JAMES, B., MCCLURE, D. J., BAKISH, D., MAX, P., MANCHANDA, R., BEAUDRY, P., BLOOM, D., ROTSTEIN, E., ANCILL, R., SANDOR, P., SLADEN-DEW, N., DURAND, C., CHANDRASENA, R., HORN, E., ELLIOT, D., DAS, M., RAVINDRA, A. & MATSOS, G. (1990). A controlled dose-ranging study of remoxipride and haloperidol in schizophrenia: a Canadian multicentre trial. *Acta Psychiatric Scandinavica* 82 72–76.

LAUD, P. W. & IBRAHIM, J. G. (1996). Predictive specification of prior model probabilities in variable selection. *Biometrika* 83 267–274.

LAVINE, M. (1992). Some aspects of Polya tree distributions for statistical modelling. *The Annals of Statistics* 20 1222–1235.

LAVINE, M. (1994). More aspects of Polya tree distributions for statistical modelling. *The Annals of Statistics* 22 1161–1176.

LEE, J. (2006). *Sensitivity analysis for longitudinal binary data with dropout*. Ph.D. thesis, Brown University.

LEE, J. & BERGER, J. O. (2001). Semiparametric Bayesian analysis of selection models. *Journal of the American Statistical Association* 96 1397–1409.

LEE, J. Y., HOGAN, J. W. & HITSMAN, B. (2007). Expert opinion, informative priors and sensitivity analysis for longitudinal binary data with informative dropout. *Biostatistics (in revision)* .

LEE, K. & DANIELS, M. J. (2007). A class of Markov models for longitudinal ordinal data. *Biometrics* 63 1060–1067.

LEONARD, T. & HSU, J. S. J. (1992). Bayesian inference for a covariance matrix. *The Annals of Statistics* 20 1669–1696.

LIANG, H., WU, H. & CARROLL, R. J. (2003). The relationship between virologic and immunologic responses in aids clinical research using mixed-effects varying-coefficient models with measurement error. *Biostatistics* 4 297–312.

LIANG, K.-Y. & ZEGER, S. L. (1986). Longitudinal data analysis using generalized linear models. *Biometrika* 73 13–22.

LICHTENSTEIN, E. & GLASGOW, R. E. (1992). Smoking cessation: what have we learned over the past decade? *Journal of Consulting and Clinical Psychology* 60 518–527.

LIECHTY, J. C., LIECHTY, M. W. & MULLER, P. (2004). Bayesian correlation estimation. *Biometrika* 91 1–14.

LIN, D. Y. & YING, Z. (2001). Semiparametric and nonparametric regression analysis of longitudinal data. *Journal of the American Statistical Association* 96 103–126.

LIN, H., MCCULLOCH, C. E. & ROSENHECK, R. A. (2004a). Latent pattern mixture models for informative intermittent missing data in longitudinal studies. *Biometrics* 60 295–305.

LIN, H., SCHARFSTEIN, D. O. & ROSENHECK, R. A. (2004b). Analysis of longitudinal data with irregular, outcome-dependent follow-up. *Journal of the Royal Statistical Society, Series B: Statistical Methodology* 66 791–813.

LIN, X. & CARROLL, R. J. (2001). Semiparametric regression for clustered data. *Biometrika* 88 1179–1185.

LIN, X. & ZHANG, D. (1999). Inference in generalized additive mixed models by using smoothing splines. *Journal of the Royal Statistical Society, Series B: Statistical Methodology* 61 381–400.

LITTLE, R. J. A. (1993). Pattern-mixture models for multivariate incomplete data. *Journal of the American Statistical Association* 88 125–134.

LITTLE, R. J. A. (1994). A class of pattern-mixture models for normal incomplete data. *Biometrika* 81 471–483.

LITTLE, R. J. A. (1995). Modeling the drop-out mechanism in repeated-measures studies. *Journal of the American Statistical Association* 90 1112–1121.

LITTLE, R. J. A. & RUBIN, D. B. (1999). Comment on "Adjusting for nonignorable drop-out using semiparametric nonresponse models". *Journal of the American Statistical Association* 94 1130–1132.

LITTLE, R. J. A. & RUBIN, D. B. (2002). *Statistical Analysis with Missing Data*. John Wiley & Sons.

LIU, C. (2001). Comment on "The art of data augmentation" (Pkg: P1–111). *Journal of Computational and Graphical Statistics* 10 75–81.

LIU, C. & RUBIN, D. B. (1998). Ellipsoidally symmetric extensions of the general location model for mixed categorical and continuous data. *Biometrika* 85 673–688.

LIU, J. S., WONG, W. H. & KONG, A. (1994). Covariance structure of the Gibbs sampler with applications to the comparisons of estimators and augmentation schemes. *Biometrika* 81 27–40.

LIU, X. & DANIELS, M. J. (2006). A new algorithm for simulating a correlation matrix based on parameter expansion and re-parameterization. *Journal of Computational and Graphical Statistics* 15 897–914.

LIU, X. & DANIELS, M. J. (2007). Joint models for the association of a longitudinal binary and continuous process. Tech. rep., Department of Statistics, University of Florida.

LIU, X., WATERNAUX, C. & PETKOVA, E. (1999). Influence of human immunodeficiency virus infection on neurological impairment: an analysis of longitudinal binary data with informative drop-out. *Journal of the Royal Statistical Society, Series C: Applied Statistics* 48 103–115.

MACEACHERN, S. N. (1994). Estimating normal means with a conjugate style Dirichlet process prior. *Communications in Statistics: Simulation and Computation* 23 727–741.

MACEACHERN, S. N. & MÜLLER, P. (1998). Estimating mixture of Dirichlet process models. *Journal of Computational and Graphical Statistics* 7 223–238.

MANLY, B. F. J. & RAYNER, J. C. W. (1987). The comparison of sample covariance matrices using likelihood ratio tests. *Biometrika* 74 841–847.

MARCUS, B. H., ALBRECHT, A. E., KING, T. K., PARISI, A. F., PINTO, B. M., ROBERTS, M., NIAURA, R. S. & ABRAMS, D. B. (1999). The efficacy of exercise as an aid for smoking cessation in women: a randomized controlled trial. *Archives of Internal Medicine* 159 1169–1171.

MARCUS, B. H., LEWIS, B., HOGAN, J. W., KING, T. K., ALBRECHT, A. E., BOCK, B. & PARISI, A. F. (2005). The efficacy of moderate-intensity exercise as an aid for smoking cessation in women: a randomized controlled trial. *Nicotine and Tobacco Research* 7 397–404.

McCULLAGH, P. & NELDER, J. A. (1989). *Generalized Linear Models*. Chapman & Hall Ltd.

MIGLIORETTI, D. L. (2003). Latent transition regression for mixed outcomes. *Biometrics* 59 710–720.

MIGLIORETTI, D. L. & HEAGERTY, P. J. (2004). Marginal modeling of multilevel binary data with time-varying covariates. *Biostatistics* 5 381–398.

MOLENBERGHS, G. & KENWARD, M. G. (2007). *Missing Data in Clinical Studies*. John Wiley & Sons.

MOLENBERGHS, G., KENWARD, M. G. & LESAFFRE, E. (1997). The analysis of longitudinal ordinal data with nonrandom drop-out. *Biometrika* 84 33–44.

MOLENBERGHS, G., MICHIELS, B., KENWARD, M. G. & DIGGLE, P. J. (1998). Monotone missing data and pattern-mixture models. *Statistica Neerlandica* 52 153–161.

MOLENBERGHS, G. & VERBEKE, G. (2006). *Models for Discrete Longitudinal Data*. Springer-Verlag.

MONAHAN, J. F. (1983). Fully Bayesian analysis of ARMA time series models. *Journal of Econometrics* 21 307–331.

MORI, M., WOODWORTH, G. & WOOLSON, R. F. (1992). Application of empirical bayes methodology to estimation of changes in the presence of informative right censoring. *Statistics in Medicine* 11 621–631.

MUKERJEE, R. & GHOSH, M. (1997). Second-order probability matching priors. *Biometrika* 84 970–975.

NANDRAM, B. & CHOI, J. W. (2002a). A Bayesian analysis of a proportion under non-ignorable non-response. *Statistics in Medicine* 21 1189–1212.

NANDRAM, B. & CHOI, J. W. (2002b). Hierarchical Bayesian nonresponse models for binary data from small areas with uncertainty about ignorability. *Journal of the American Statistical Association* 97 381–388.

NATARAJAN, R. (2001). On the propriety of a modified Jeffreys's prior for variance components in binary random effects models. *Statistics & Probability Letters* 51 409–414.

NATARAJAN, R. & KASS, R. E. (2000). Reference Bayesian methods for generalized linear mixed models. *Journal of the American Statistical Association* 95 227–237.

NATARAJAN, R. & McCULLOCH, C. E. (1995). A note on the existence of the posterior distribution for a class of mixed models for binomial responses. *Biometrika* 82 639–643.

NEAL, R. M. (1996). *Bayesian Neural Networks*. Springer-Verlag Inc.

NEAL, R. M. (2003). Slice sampling. *The Annals of Statistics* 31 705–767.

NELSEN, R. B. (1999). *An Introduction to Copulas.* Springer-Verlag Inc.

NÚÑEZ ANTÓN, V. & ZIMMERMANN, D. L. (2000). Modeling nonstationary longitudinal data. *Biometrics* 56 699–705.

OAKLEY, J. & O'HAGAN, A. (2007). Uncertainty in prior elicitations: a nonparametric approach. *Biometrika* 94 427–441.

O'BRIEN, S. M. & DUNSON, D. B. (2004). Bayesian multivariate logistic regression. *Biometrics* 60 739–746.

O'HAGAN, A. (1998). Eliciting expert beliefs in substantial practical applications (Disc: P55–68). *Journal of the Royal Statistical Society, Series D: The Statistician* 47 21–35.

OVERALL, J. E. & GORHAM, D. R. (1988). The Brief Psychiatric Rating Scale (BPRS): recent developments in ascertainment and scaling. *Psychopharmacology Bulletin* 22 97–99.

PERUGGIA, M. (1997). On the variability of case-deletion importance sampling weights in the Bayesian linear model. *Journal of the American Statistical Association* 92 199–207.

POURAHMADI, M. (1999). Joint mean-covariance models with applications to longitudinal data: unconstrained parameterisation. *Biometrika* 86 677–690.

POURAHMADI, M. (2000). Maximum likelihood estimation of generalised linear models for multivariate normal covariance matrix. *Biometrika* 87 425–435.

POURAHMADI, M. (2007). Cholesky decompositions and estimation of a covariance matrix: orthogonality of variance-correlation parameters. Tech. rep., Division of Statistics, Northern Illinois University.

POURAHMADI, M. & DANIELS, M. J. (2002). Dynamic conditionally linear mixed models for longitudinal data. *Biometrics* 58 225–231.

POURAHMADI, M., DANIELS, M. J. & PARK, T. (2007). Simultaneous modelling of several covariance matrices using the modified choleski decomposition with applications. *Journal of Multivariate Analysis* 98 568–587.

PRESS, S. J. (2003). *Subjective and Objective Bayesian Statistics: Principles, Models, and Applications.* John Wiley & Sons.

PULKSTENIS, E. P., TEN HAVE, T. R. & LANDIS, J. R. (1998). Model for the analysis of binary longitudinal pain data subject to informative dropout through remediation. *Journal of the American Statistical Association* 93 438–450.

RIBAUDO, H. J. & THOMPSON, S. G. (2002). The analysis of repeated multivariate binary quality of life data: a hierarchical model approach. *Statistical Methods in Medical Research* 11 69–83.

RICE, J. A. (2004). Functional and longitudinal data analysis: perspectives on smoothing. *Statistica Sinica* 14 631–647.

ROBERTS, G. O. & SAHU, S. K. (1997). Updating schemes, correlation structure, blocking and parameterization for the Gibbs sampler. *Journal of the Royal Statistical Society, Series B: Methodological* 59 291–317.

ROBINS, J. M. (1997). Non-response models for the analysis of non-monotone non-ignorable missing data. *Statistics in Medicine* 16 21–37.

ROBINS, J. M. & RITOV, Y. (1997). Toward a curse of dimensionality appropriate (CODA) asymptotic theory for semi-parametric models. *Statistics in Medicine* 16 285–319.

ROBINS, J. M., ROTNITZKY, A. & ZHAO, L. P. (1995). Analysis of semiparametric regression models for repeated outcomes in the presence of missing data. *Journal of the American Statistical Association* 90 106–121.

ROBINS, J. M., VENTURA, V. & VAN DER VAART, A. (2000). Asymptotic distribution of P values in composite null models (Pkg: P1127–1171). *Journal of the American Statistical Association* 95 1143–1156.

RODRIGUEZ, A., DUNSON, D. B. & GELFAND, A. E. (2007). Nonparametric functional data analysis through Bayesian density estimation. Tech. rep., Duke University.

ROTNITZKY, A., ROBINS, J. M. & SCHARFSTEIN, D. O. (1998). Semiparametric regression for repeated outcomes with nonignorable nonresponse. *Journal of the American Statistical Association* 93 1321–1339.

ROTNITZKY, A., SCHARFSTEIN, D. O., SU, T.-L. & ROBINS, J. M. (2001). Methods for conducting sensitivity analysis of trials with potentially nonignorable competing causes of censoring. *Biometrics* 57 103–113.

ROY, J. (2003). Modeling longitudinal data with nonignorable dropouts using a latent dropout class model. *Biometrics* 59 829–836.

ROY, J. & DANIELS, M. J. (2007). A general class of pattern mixture models for nonignorable dropout with many possible dropout times. *Biometrics (in press)* .

ROY, J. & HOGAN, J. W. (2007). Causal contrasts in randomized trials of two active treatments with noncompliance. Tech. rep., Center for Statistical Sciences, Brown University.

ROY, J. & LIN, X. (2000). Latent variable models for longitudinal data with multiple continuous outcomes. *Biometrics* 56 1047–1054.

ROY, J. & LIN, X. (2002). The analysis of multivariate longitudinal outcomes with nonignorable dropouts and missing covariates: changes in methadone treatment practices. *Journal of the American Statistical Association* 97 40–52.

RUBIN, D. B. (1976). Inference and missing data. *Biometrika* 63 581–590.

RUBIN, D. B. (1977). Formalizing subjective notions about the effect of nonrespondents in sample surveys. *Journal of the American Statistical Association* 72 538–543.

RUBIN, D. B. (1987). *Multiple Imputation for Nonresponse in Surveys*. John Wiley & Sons.

RUBIN, D. B. (2006). Causal inference through potential outcomes and principal stratification: Application to studies with censoring due to death (with discussion). *Statistical Science* 21 299–321.

RUPPERT, D. & CARROLL, R. J. (2000). Spatially-adaptive penalties for spline fitting. *Australian & New Zealand Journal of Statistics* 42 205–223.

RUPPERT, D., WAND, M. P. & CARROLL, R. J. (2003). *Semiparametric Regression*. Cambridge University Press.

SAMMEL, M. D., LIN, X. & RYAN, L. M. (1999). Multivariate linear mixed models for multiple outcomes. *Statistics in Medicine* 18 2479–2492.

SAMMEL, M. D., RYAN, L. M. & LEGLER, J. M. (1997). Latent variable models for mixed discrete and continuous outcomes. *Journal of the Royal Statistical Society, Series B: Methodological* 59 667–678.

SCHAFER, J. L. (1997). *Analysis of Incomplete Multivariate Data*. Chapman & Hall Ltd.

SCHARFSTEIN, D. O., DANIELS, M. J. & ROBINS, J. M. (2003). Incorporating prior beliefs about selection bias into the analysis of randomized trials with missing outcomes. *Biostatistics (Oxford)* 4 495–512.

SCHARFSTEIN, D. O., HALLORAN, M. E., CHU, H. & DANIELS, M. J. (2006). On estimation of vaccine efficacy using validation samples with selection bias. *Biostatistics* 7 615–629.

SCHARFSTEIN, D. O. & IRIZARRY, R. A. (2003). Generalized additive selection models for the analysis of studies with potentially nonignorable missing outcome data. *Biometrics* 59 601–613.

SCHARFSTEIN, D. O., ROTNITZKY, A. & ROBINS, J. M. (1999). Adjusting for nonignorable drop-out using semiparametric nonresponse models (C/R: P1121–1146). *Journal of the American Statistical Association* 94 1096–1120.

SCHILDCROUT, J. & HEAGERTY, P. J. (2007). Marginalized models for long series of binary responses. *Biometrics* 63 322–331.

SETHURAMAN, J. (1994). A constructive definition of Dirichlet priors. *Statistica Sinica* 4 639–650.

SMITH, D., WARREN, D., VLAHOV, D., SCHUMAN, P., STEIN, M., GREENBERG, B. & HOLMBERG, S. (1997). Design and baseline participant characteristics of the Human Immunodeficiency Virus Epidemiology Research Study (HERS). *American Journal of Epidemiology* 146 459–469.

SPIEGELHALTER, D. J., BEST, N. G., CARLIN, B. P. & VAN DER LINDE, A. (2002). Bayesian measures of model complexity and fit (Pkg: P583–639). *Journal of the Royal Statistical Society, Series B: Statistical Methodology* 64 583–616.

SU, L. & HOGAN, J. W. (2007). Likelihood-based semiparametric marginal regression for continuously measured longitudinal binary outcomes. Tech. rep., Center for Statistical Sciences, Brown University.

SY, J. P., TAYLOR, J. M. G. & CUMBERLAND, W. G. (1997). A stochastic model for the analysis of bivariate longitudinal AIDS data. *Biometrics* 53 542–555.

TANNER, M. A. & WONG, W. H. (1987). The calculation of posterior distributions by data augmentation (C/R: P541-550). *Journal of the American Statistical Association* 82 528–540.

TAYLOR, J. M. G., CUMBERLAND, W. G. & SY, J. P. (1994). A stochastic model for analysis of longitudinal AIDS data. *Journal of the American Statistical Association* 89 727–736.

TIERNEY, L. (1994). Markov chains for exploring posterior distributions (Disc: P1728–1762). *The Annals of Statistics* 22 1701–1728.

TIERNEY, L. & KADANE, J. B. (1986). Accurate approximations for posterior moments and marginal densities. *Journal of the American Statistical Association* 81 82–86.

TREVISANI, M. & GELFAND, A. E. (2003). Inequalities between expected marginal log-likelihoods, with implications for likelihood-based model complexity and comparison measures. *Canadian Journal of Statistics* 31 239–250.

TROXEL, A. B., LIPSITZ, S. R. & HARRINGTON, D. P. (1998). Marginal models for the analysis of longitudinal measurements with nonignorable non-monotone missing data. *Biometrika* 85 661–672.

TROXEL, A. B., MA, G. & HEITJAN, D. F. (2004). An index of local sensitivity to nonignorability. *Statistica Sinica* 14 1221–1237.

TSIATIS, A. A. (2006). *Semiparametric Theory and Missing Data.* Springer.

TSIATIS, A. A. & DAVIDIAN, M. (2004). Joint modeling of longitudinal and time-to-even data: an overview. *Statistica Sinica* 14 809–834.

VAN DER LAAN, M. J. & ROBINS, J. M. (2003). *Unified Methods for Censored Longitudinal Data and Causality.* Springer.

VAN DYK, D. A. & MENG, X.-L. (2001). The art of data augmentation (Pkg: P1–111). *Journal of Computational and Graphical Statistics* 10 1–50.

VANSTEELANDT, S., GOETGHEBEUR, E., KENWARD, M. G. & MOLENBERGHS, G. (2006). Ignorance and uncertainty regions as inferential tools in a sensitivity analysis. *Statistica Sinica* 16 953–979.

VERBEKE, G. & LESAFFRE, E. (1996). A linear mixed-effects model with heterogeneity in the random-effects population. *Journal of the American Statistical Association* 91 217–221.

VERBEKE, G. & MOLENBERGHS, G. (2000). *Linear Mixed Models for Longitudinal Data.* Springer-Verlag Inc.

VERBEKE, G., MOLENBERGHS, G., THIJS, H., LESAFFRE, E. & KENWARD, M. G. (2001). Sensitivity analysis for nonrandom dropout: a local influence approach. *Biometrics* 57 7–14.

WAKEFIELD, J. (1996). The Bayesian analysis of population pharmacokinetic models. *Journal of the American Statistical Association* 91 62–75.

WANG, N., CARROLL, R. J. & LIN, X. (2005). Efficient semiparametric marginal estimation for longitudinal/clustered data. *Journal of the American Statistical Association* 100 147–157.

WANG, Y. & TAYLOR, J. M. G. (2001). Jointly modeling longitudinal and event time data with application to acquired immunodeficiency syndrome. *Journal of the American Statistical Association* 96 895–905.

WARE, J. H. (1985). Linear models for the analysis of longitudinal studies. *The American Statistician* 39 95–101.

WEISS, R. E. (2005). *Modeling Longitudinal Data.* Springer.

WILKINS, K. J. & FITZMAURICE, G. M. (2006). A hybrid model for nonignorable dropout in longitudinal binary responses. *Biometrics* 62 168–176.

WONG, F., CARTER, C. K. & KOHN, R. (2003). Efficient estimation of covariance selection models. *Biometrika* 90 809–830.

WOOLDRIDGE, J. M. (2001). *Econometric Analysis of Cross Section and Panel Data.* The MIT Press.

WU, H. & WU, L. (2002). Identification of significant host factors for HIV dynamics modelled by non-linear mixed-effects models. *Statistics in Medicine* 21 753–771.

WU, M. C. & BAILEY, K. R. (1988). Analysing changes in the presence of informative right censoring caused by death and withdrawal. *Statistics in Medicine* 7 337–346.

WU, M. C. & BAILEY, K. R. (1989). Estimation and comparison of changes in the presence of informative right censoring: conditional linear model (Corr: V46 p889). *Biometrics* 45 939–955.

WU, M. C. & CARROLL, R. J. (1988). Estimation and comparison of changes in the presence of informative right censoring by modeling the censoring process (Corr: V45 p1347; V47 p357). *Biometrics* 44 175–188.

WULFSOHN, M. S. & TSIATIS, A. A. (1997). A joint model for survival and longitudinal data measured with error. *Biometrics* 53 330–339.

XU, J. & ZEGER, S. L. (2001). Joint analysis of longitudinal data comprising repeated measures and times to events. *Journal of the Royal Statistical Society, Series C: Applied Statistics* 50 375–387.

YANG, R. & BERGER, J. O. (1994). Estimation of a covariance matrix using the reference prior. *The Annals of Statistics* 22 1195–1211.

YAO, F., MÜLLER, H.-G. & WANG, J.-L. (2005a). Functional data analysis for sparse longitudinal data. *Journal of the American Statistical Association* 100 577–590.

YAO, F., MÜLLER, H.-G. & WANG, J.-L. (2005b). Functional linear regression analysis for longitudinal data. *The Annals of Statistics* 33 2873–2903.

YAO, Q., WEI, L. J. & HOGAN, J. W. (1998). Analysis of incomplete repeated measurements with dependent censoring times. *Biometrika* 85 139–149.

ZEGER, S. L. & LIANG, K.-Y. (1992). An overview of methods for the analysis of longitudinal data. *Statistics in Medicine* 11 1825–1839.

ZHANG, D. (2004). Generalized linear mixed models with varying coefficients for longitudinal data. *Biometrics* 60 8–15.

ZHANG, D. & DAVIDIAN, M. (2001). Linear mixed models with flexible distributions of random effects for longitudinal data. *Biometrics* 57 795–802.

ZHANG, D., LIN, X., RAZ, J. & SOWERS, M. (1998). Semiparametric stochastic mixed models for longitudinal data. *Journal of the American Statistical Association* 93 710–719.

ZHANG, J. & HEITJAN, D. F. (2006). A simple local sensitivity analysis tool for nonignorable coarsening: application to dependent censoring. *Biometrics* 62 1260–1268.

ZHANG, X., BOSCARDIN, W. J. & BELIN, T. R. (2006). Sampling correlation matrices in Bayesian models with correlated latent variables. *Journal of Computational and Graphical Statistics* 15 880–896.

ZHAO, X., MARRON, J. S. & WELLS, M. T. (2004). The functional data analysis view of longitudinal data. *Statistica Sinica* 14 789–808.

ZHENG, Y. & HEAGERTY, P. J. (2005). Partly conditional survival models for longitudinal data. *Biometrics* 61 379–391.

Author Index

Index